air pollution:
the emissions,
the regulations,
& the controls

ENVIRONMENTAL SCIENCE SERIES

Asit K. Biswas, *Editor*
Department of Environment, Ottawa, Canada

Published

Dynamics of Fluids in Porous Media
Jacob Bear

Scientific Allocation of Water Resources
Nathan Buras

Environmental Policy and Administration
Daniel H. Henning

Air Pollution: The Emissions,
The Regulations, and the Controls
James P. Tomany

In Preparation

Systems Approach to Water Management
Asit K. Biswas, *Editor*

Computer Simulation in Hydrology
George Fleming

air pollution:
the emissions,
the regulations,
& the controls

James P. Tomany
Environmental Consultant

american elsevier
publishing company, inc.
NEW YORK LONDON AMSTERDAM

AMERICAN ELSEVIER PUBLISHING COMPANY, INC.
52 Vanderbilt Avenue, New York, N.Y. 10017

ELSEVIER PUBLISHING COMPANY
335 Jan Van Galenstraat, P.O. Box 211
Amsterdam, The Netherlands

International Standard Book Number 0-444-00138-7

Library of Congress Card Number 73-7686

Library of Congress Cataloging in Publication Data

Tomany, James P
  Air pollution: the emissions, the regulations & the
controls.

  (Environmental science series)
  Includes bibliographical references.
  1. Air—Pollution. 2. Air quality management.
I. Title. II. Series: Environmental science series
(New York, 1972-   )
TD883.T55   363.6   73-7686
ISBN 0-444-00138-7

Manufactured in the United States of America

This one is for Frances P. Bell.

James P. Tomany

The author, a chemical engineering graduate of The Cooper Union, N.Y.C., has been professionally concerned with air pollution and its abatement for the past fifteen years. He has authored over twenty publications relating to the assessment of specific air pollution problems and the design of economically viable emission control systems. More recently he has been involved in the evaluation of pollutant emissions as they relate to the degradation of air quality and violation of the federal, state and local standards.

As an environmental consultant, Mr. Tomany is responsible for evaluating the impact on the environment of such industrial pollution sources as fossil fuel-fired power plants, incinerators, fertilizer operations, glass manufacturing, aluminum reduction, shale oil conversion, coal mining, coal gasification processes and pulp and paper production. His assignments have involved the definition of particulate and gaseous emissions as they influence air quality, the interpretation of prevailing regulations, and the application of process modifications or economically feasible control equipment for pollution abatement. An authority in wet scrubber technology, Mr. Tomany has himself devised original and effective control systems, for which he has been awarded a number of patents.

Mr. Tomany has served as an instructor in the engineering sciences at University of Connecticut, University of Bridgeport and at The Cooper Union. He has lectured extensively on air pollution control before such groups as the Air Pollution Control Association, The American Society of Mechanical Engineers, The American Society of Chemical Engineers, The Industrial Gas Cleaning Institute, The Fertilizer Industry Round Table and the Instrument Society of America. He has also contributed to seminars at the University of New Hampshire, Rensselaer Polytechnic Institute, University of Southern California and the University of Cincinnati.

CONTENTS

Contents

CHAPTER 4

Emissions Testing

CHAPTER 5

Particulate Pollutants Control Equipment

CHAPTER 6

Gaseous Pollutants Control Equipment

Contents

CHAPTER 7

Air Pollution Control for Specific Industries

CHAPTER 8

Automotive Pollutant Emissions

CHAPTER 9

Social and Economic Aspects of Air Pollution

APPENDIX A

APPENDIX B

APPENDIX C

Contents

PREFACE

Air pollution can be defined as a condition of the earth's atmosphere which is dependent on the magnitude of pollutant *Emissions,* the *Regulations* limiting such emissions, and the *Controls* available for their abatement.

In this book these three basic elements, which determine an air pollution situation, are explained and related to various and actual industrial experiences. Thus, Chapter 1 contains an overview of the total subject, Chapters 2 and 4 describe the various pollutants and explain techniques for their identification and the determination of emission rates, and in Chapter 3 there are defined the various federal, state, and local regulations. In Chapters 5 and 6 the available control technology is described and illustrated with sample problems, while in Chapter 7 the emission problems of two major industries are defined, the pertinent regulations are cited, and the applied control systems are described and evaluated. Up to and including Chapter 7, only stationary air pollution emission sources are treated, so that Chapter 8 is exclusively devoted to automotive pollutant emissions, the prevailing regulations, and the current attempts at devising satisfactory controls. Finally, the social and economic impacts of air pollution are discussed in Chapter 9, and the vast factual shortcomings in this most important area are underscored.

In the definitive organization of this book it was the intent to provide realistic continuity in the presentation of the subject matter so as to avoid its identification as a handbook or a collection of unassociated works by individual authors. The book's foundations lie in actual industrial experiences where the emission of pollutants can be interpreted only as an air pollution problem when the regulations so prescribe it and where a viable solution to the problem becomes possible only when an adequate level of control technology is available. The various problems described in the text have been solved in accordance with present commercial practice, with due regard for economic feasibility.

In the preparation of the text, the author has attempted to bridge the gap between a highly theoretical treatise and a cursory survey of air pollution problems and currently available control technology. Many of the problems have been solved by the use of empirically determined design data because this was the only approach available to obtain a viable solution. In those areas where a knowledge of the basic principles is pertinent to an understanding of the gas-cleaning mechanism and to the problem solution, treatment of the controlling theory was included. Hopefully, this union of practice and theory will be of value to both the student and the plant engineer.

Preface

Because the scope of the text has brought the author into some uncharted areas, it was found necessary to rely on many friends, associates, and industrial personnel, all of whom I wish to acknowledge and warmly thank for their support and contributions. D. E. Reedy of Air Correction Division very kindly reviewed and edited the section on thermal incineration and catalytic combustion, while W. D. Faulkner of Pittsburgh Activated Carbon Division contributed considerably to the subject of adsorption. R. R. Koppang of TRW Systems, who has been heavily involved in automobile pollution evaluation and advanced control technology development, has reviewed and revised various elements of Chapter 8 pertaining to motor vehicle pollutant emissions. To the University of Southern California I owe a double debt for the use of its Seaver Science Library, complete with the services of its capable staff and to Professor Frank Lockhart for his review of some elements of the text.

Finally, and most importantly, my sincere thanks and appreciation are extended to the many industrial friends and acquaintances with whom I collaborated in their struggle to comply with the Regulations. Without doubt, their conscientious efforts to combat air pollution are responsible for this book.

James P. Tomany
Environmental Consultant

Los Angeles, California

Chapter 1

GENERAL INTRODUCTION

1.1. SCOPE AND OBJECTIVES

The familiar word of the day is "ecology," defined as that branch of science concerned with the interrelationships of organisms and their environment. The layman desires an improvement in that area of his environment which will allow him to breathe uncontaminated air; the naturalist is concerned with the spoilation of the forests and its effect on animal and plant life; and the sportsman decries the polluted conditions of his lakes and streams which deprive him of his fishing rights. These are problems of ecology, each dealing with a different facet of the environment.

This book is involved with the air pollution aspects of our environment; not the effects of air pollution but its causes and the efforts being exerted for its control. A description of the gas-borne contaminant emissions, the regulations enacted for their confinement, and the current industrial equipment available for their control are the subjects of this text. The major air pollution emission sources are defined, both stationary and mobile, and the pollutants identified and classified as they relate to emissions control. Federal and local regulations are reviewed to determine their influence on the attempts made by the various industries to reduce these emissions. The status of the currently available control technology for both particulate and gaseous emissions are evaluated, with a complete description of the equipment types, design parameters, advantages, limitations, and costs. Emphasis on the need for a more economical approach to air pollution control is stressed. Pertinent references are provided for more intensive study in selected areas.

This work is not a theoretical treatise. In this book the actual realistic problems of air pollution control being faced daily by industry are defined. Sample calculations are provided to illustrate methods for the optimal selection and sizing of control equipment. The major objectives of the text are to acquaint the engineering student and concerned technical personnel with these industrial air pollution problems and lead them to an economically reasonable solution. This is the area where the immediate need lies in the current struggle against air pollution.

1.2. THE TOTAL ENVIRONMENTAL PROBLEM

The control of man's environment requires the solution of those problems concerned with air pollution, water quality control, waste management, and noise abatement. In

1

many instances these problems are interrelated. Man's early efforts to dispose of his own excrement was to discharge it into the nearest stream; unfortunately, in some areas, this disposal method is still used. The health hazards of this practice were soon realized and consequently the modern sanitary sewage system was developed. In these plants, the bacteriological decomposition of human wastes causes the emission of noxious gases into the atmosphere, thereby requiring the application of an air pollution control system.

Similarly, in arresting the discharge of noxious liquid and solid wastes into streams and settling ponds by the chemical process industries, such alternative disposal processes as combustion or adsorption must be considered. However, potential air pollution problems are inherent in both processes.

In Los Angeles, California, where the authorities are particularly sensitive to air pollution problems, land-fill disposal of solid wastes, rather than incineration, is encouraged. However, this approach must be traded off against available land area and the problems of ground water contamination. A specific example involved the discharge of petroleum-based residual crudes, with a high sulfur content, into certain dump areas. Expansion of these areas involved potential ground-water contamination problems, so that other more distant sites had to be considered. Transportation costs soared, so that investigation of combustion methods at the plant site was undertaken. However, the sulfur and sand content of the crudes posed some difficult processing problems. The final approach was to reevaluate the petroleum process responsible for this residue, in an attempt to introduce modifications that would effect an economic balance between the product quality and the pollutant disposal.

This problem, besides illustrating the interrelationship between air, water, and solid waste disposal emphasizes the systems approach to environmental control problems. The economic criteria for chemical process plant design in the past were based on the costs of labor, raw materials, and utilities. Today the plant designer must consider the impact of his process on the environment. More specifically, if he cannot convince the authorities that the various plant-pollutant discharge rates lie within the values specified by the existing regulations, he is denied permission for the construction of his plant.

1.3. DEFINITION AND EFFECTS OF AIR POLLUTION

Air pollution can be defined as the emission into the atmosphere of a waste gas stream containing one or more contaminants such as dust, gases, mists, or fumes in concentrations sufficient to be injurious to human, animal, or plant health, or to affect property values adversely.

The health effects of air pollution can be determined by a number of approaches: statistical studies of past illness and death, as related to notable air pollution incidents; correlation of respiratory epidemics as a function of air pollution concentrations; and laboratory studies of the responses of animals (and, in some cases, human beings) to exposure to various pollutants. For example, it has been demonstrated that asthmatic attacks among susceptible patients correlate with air pollution caused by incomplete refuse combustion. The incidence of employee absenteeism due to respiratory illness has closely followed sulfate pollution levels. Laboratory studies with animals have proved the development of lung cancer caused by their infection with influenza virus, followed by exposure to an artificial smog comprised of ozonized gasoline. These studies, among others, have conclusively demonstrated the association of air pollution with such respiratory diseases as lung cancer, emphysema, chronic bronchitis, and asthma.

Sulfur dioxide is considered to be the principal pollutant causing injury to plant life. In special cases, other gases such as hydrogen chloride, hydrogen fluoride, the halogens, and ammonia have adversely affected vegetation, but because of the high incidence rate and the concentrations involved, sulfur dioxide is the major culprit. The ambient air concentration of this gas and the period of plant exposure are the major factors. Laboratory developed concentration-time equations have been determined for a variety of plants. For example, alfalfa exhibits incipient leaf damage at a sulfur dioxide concentration of 1.25 parts per million (ppm) for a 1 hr period. High humidity increases the plant's susceptibility to gas injury because of the increased leaf absorption rate of this noxious contaminant. Although injury to most fruit and seed plants are confined to the leaves, the product yield is reduced proportionately.

Property damage is usually associated with metals and coatings damage caused by gaseous emissions. Sulfur-bearing fuels release sulfur dioxide during combustion, which is converted to sulfuric acid in the atmosphere as a result of its reaction with oxygen and moisture. This airborne acid is responsible for considerable corrosion damage to steel buildings, bridges, and machinery in metropolitan areas. Both hydrogen sulfide and caustic soda attack paint, the former causing the surface of white lead-based paints to darken, and the latter dissolving paints. One very serious type of property damage caused by air pollution is that of rubber degradation. Because of the presence of oxides in the atmosphere, notably ozone, rubber materials will become crazed and weakened. Tires and rubber insulation are both subject to the damaging influence of these airborne pollutants.

The most apparent and objectionable effects of air pollution are the limited visibility and eye irritation characteristic of smog. *Smog,* an acronym formed from the words "smoke" and "fog," defines an abnormal weather condition in which airborne pollutants mix with haze or fog to produce a dense, murky atmosphere. The Los Angeles experience, although once considered unique, is no longer an isolated case. San Francisco, New York, Salt Lake City, Denver, and Tokyo, among others, are all plagued by this smog problem. Its major cause is automobile exhaust wherein the discharged hydrocarbons react photochemically in the presence of sunlight with the nitrogen oxides. The resulting compounds are ozone and peroxyacetyl nitrate (PAN), the latter being responsible for eye irritation.

Meteorological conditions such as rainfall, prevailing winds, geography, and inversion incidence have considerable influence on air pollution levels. Thermal inversions, encouraged by mountainous terrain, tend to trap pollutants by imposing a stable high-temperature air level on the ambient air reservoir. This entrapment causes a continuously increasing pollutant concentration within the "bowl."

1.4. AIR POLLUTION: THE EMISSIONS

As the book title implies, the total air pollution experience can be translated into a discussion of the emissions, the regulations, and the controls. There follows a summary of these three factors to illustrate how they constitute the total problem.

Air pollution emissions can be classified into two major categories: particulates and gases. Approximately 85% of the total pollutant emissions in the United States are gaseous and the remaining 15% are particulates.

The air pollution situation is growing progressively worse. The extent of the various effects of uncontrolled atmospheric emissions from stationary and mobile sources has still not been completely defined. However, it is generally agreed that at this moment in the history of our industrial development, the acceleration rate of pollutant emissions exceeds man's attempt to control them.

A breakdown of estimated particulate and gaseous emissions from various sources in the United States for the year 1969 is presented in Table 1-1.

Motor vehicles account for over 50% of the total air pollutants emitted in the United States. The relationship of particulates to gaseous emissions reflected in Table 1-1 is the basis for the increasing concern over the latter. The chemically active nature of the various gaseous pollutants and the current lack of effective control technology makes their reduction in the atmosphere one of the prime targets of technological activity.

Electric power generation that uses sulfur-bearing coals as a fuel source accounts for about 60% of the total sulfur oxide emissions. In addition to the five major pollutants listed in Table 1-1, various industrial operations emit numerous minor pollutants that present severe problems in certain areas. Examples are fluorides from steel, aluminum and fertilizer manufacture, sulfides from kraft pulp mills, and chlorides from waste liquor incineration processes.

This is the overall emissions problem for which regulations must be legislated and control technology must bring under confinement.

1.5. AIR POLLUTION: THE REGULATIONS

Air pollution regulations have developed very slowly in the United States. Prior to the 1960s, federal- and state-sponsored legislation was practically nonexistent. The federal government had enacted the Rivers and Harbors Act in 1899, prohibiting the discharge of solid refuse into navigable waters, and in 1912 the Public Health Service Act authorized the investigation of water pollution control where it affected public health. However, the first meaningful water pollution control act was not passed until 1956, and the first comprehensive air pollution act was enacted in 1963.

The Department of Health, Education, and Welfare (HEW) was first authorized to conduct an air pollution program in 1955. In that year the legislation called on state and local governments to assume the basic responsibility for preventing and controlling air pollution, and authorized the Department to conduct research and provide technical support. In 1963 the Clean Air Act was adopted to authorize HEW to expand its

Table 1.1 Total Air Pollutant Emissions in the United States
(Values for the year 1969 expressed in 10^6 tons/yr)

Source	Particulates	Hydrocarbons	SO_x	NO_x	CO
		Pollutants			
Transportation	0.8	19.8	1.1	11.2	111.5
Fuel combustion, stationary sources	7.2	0.9	24.4	10.0	1.8
Industrial processes	14.4	5.5	7.5	0.2	12.0
Solid waste disposal	1.4	2.0	0.2	0.4	7.9
Miscellaneous	11.4	9.2	0.2	2.0	18.2
Total	35.2	37.4	33.4	23.8	151.4

activities in support of state and local efforts. The Clean Air Act was amended in 1965 and again in 1967. The 1967 Air Quality Act provides for the establishment of ambient air quality standards by the federal government through the publication of a list of pollutants that adversely affect public health. These standards are then submitted to each state for their approval, implementation, and enforcement.

One important feature of the Clean Air Act was the establishment of 57 air quality regions throughout the United States. These regions were designated on the basis that two or more communities, either in the same or different states, share a common air pollution problem. Therefore, each region was accorded the responsibility for developing and implementing its own air quality standards. These regions must establish standards that are acceptable to the state or states in which they are located. The states, in turn, must satisfy the federal government criteria.

This multiple-faceted regulatory system has been the cause of some confusion among the various industries being cited as violators. In addition, the regulations are in a constant state of flux, with the regions enacting more stringent regulations than those demanded by the state, and the state in turn usually attempting to outdo the federal government. Regulations enacted two years ago have been upgraded to a degree that demands the application of an entirely different control technology by the offending industries. In this uncertain regulatory climate, many of the industries are hesitating to invest in control equipment.

Ambient air quality values established by the federal government are predicated on scientific evidence of the detrimental effects of air pollution on health and property. For example, in the air quality criteria for sulfur oxides established by the government in 1969, various ambient air concentrations of SO_2 in the range of 300 to 1500 $\mu g/m^3$ (on a 24 hr basis) were tabulated with the corresponding health effects, as shown in Table 1-2.

In reviewing the air quality data issued by the federal government, the states must consider existing levels of pollutants in their regions, meteorology, number and type of sources, control technology, and air pollution growth trends. Implementation and enforcement plans should then be developed accordingly.

Ambient air quality values define the concentration of particulates and gases in

5

Table 1-2. Sulfur Oxides Ambient Air Content and Health Effects

SO_x Concentration ($\mu g/m^3$ for 24hr)	Particulate Concentration	Possible Health Effects
300–500	Low level	Increased hospital admission of older persons with respiratory diseases
500	Low level	Increased mortality rate
600	300	Accentuation of symptoms experienced by chronic lung diseases
715	750	Increased daily death rate
1500	Some present	Increased mortality

the surrounding atmosphere. These must be related to actual stack emissions for a particular industry and so defined. Although a region may establish an air quality value of 300 $\mu g/m^3$ of ambient air for sulfur oxides, equivalent to a concentration of about 0.1 ppm, the allowable stack emissions may be on the order of 100 to 200 ppm. The relationship between ambient air and stack emission concentrations are determined by the laws of gas diffusion. The meteorology of the area, stack gas temperature, and diffusion coefficients are some of the parameters affecting this correlation. The majority of control regions usually express both ambient air quality values and allowable stack emissions quantities.

An emissions regulation essentially defines the air pollution problem. Plant emissions for a particular industry are not in violation unless they exceed the allowable values established by the regulations. Therefore, once the plant emissions are known and the regulations established, the optimum control equipment must be designed and applied to accomplish a solution to the problem.

1.6. AIR POLLUTION: THE CONTROLS

Industry has been engaged in the manufacture of air pollution control equipment for about the past 50 years. During this period, the primary concern has been dust collection and the reduction of smoke plumes. Although some performance improvements have been accomplished over the years, equipment designs have remained essentially unchanged. Novel technology has been largely unexplored because there has been little economic incentive for the development of advanced fundamental concepts.

With the advent of serious air pollution control regulations, the current technology is inadequate in many areas, particularly for gaseous emissions confinement. Specifically, proven industrial designs for the control of sulfur oxides and nitrogen oxides are presently unavailable.

6

The total technology available for the control of both particulate and gaseous pollutants is limited to seven basic items of equipment. Particulate emissions are reduced by such mechanical devices as the cyclonic separator, electrostatic precipitator, fabric filter, and wet scrubber. The cyclonic separator depends on inertial forces imposed on the dust particle through centrifugal action to separate it from the gas stream. In the electrostatic precipitator, the dust particles are electrically charged and collected on oppositely charged plates. The fabric filter is considered a positive collector for particulate matter, with the dust-laden gases passing through fabric bags on which the particulates are retained. The wet scrubber depends on intimate contact between the dust-contaminated gas stream and a suitable scrubbing liquor so as to transfer the dust from the gaseous to the liquid phase.

Gaseous pollutant control is accomplished by thermal incineration, catalytic combustion, adsorption, and wet scrubbing. In thermal incineration, gases and vapors can be either oxidized or reduced by their reaction during combustion. For example, the removal of hydrocarbon vapors can be effected by oxidizing them to carbon dioxide and water vapor. For the treatment of nitrogen oxides, their reduction to nitrogen under rich combustion conditions is utilized. To minimize fuel requirements, catalytic combustion can be applied at reduced temperatures, the choice of catalyst being dependent primarily on whether oxidation or combustion is involved. Adsorption relies on the ability of such materials as activated carbon, silica gel, and molecular sieves to attract certain gases and retain them on their surfaces by intermolecular forces. The adsorbent can then be discarded or regenerated. In wet scrubbing, advantage is taken of the absorption characteristics of certain pollutant gases in various scrubbing liquors. As in the case of particulate removal, wet scrubbing removes the pollutant from the contaminated gas stream and converts it to a liquor effluent with the attending problem of its ultimate disposal. The dual role of the wet scrubber for the collection of both particulates and gases involves it, for some industrial applications, in the simultaneous confinement of both pollutants.

In addition to these air pollution corrective devices, there is a very important possible alternative solution—that of process modification. In some areas of air pollution the offending industry is caught somewhere between the regulatory demands and the inadequacy of current control technology. For example, to control paint-spray booth emissions economically, catalytic combustion equipment is required. For some time these devices continued to fail before it was realized that the phosphorous content of the paint formulation was poisoning the catalyst. After considerable experimentation, the paint formula was modified to eliminate phosphorus and air pollution control was made possible.

Novel technology is still another area that must be expanded, particularly in the gaseous and submicron particulate control area. Equipment sales forecasts released by the U.S. Department of Health, Education, and Welfare have indicated that there will be an increase in demand for gaseous control equipment of 433% for the period 1967 to 1977, as compared to 84% for particulate removal devices. These values reflect the fact that air pollution regulations, which first recognized gaseous atmospheric contaminants, are less than 10 yr old. Although the control equipment industries have been involved in emissions control for over a half-century, they were mainly concerned with the reduction of visual smokes and dusts. The gap between regulations and adequate control technology is a serious one.

The automotive industry is making some progress in devising gaseous control systems for automobile exhaust emissions. Judging from the emissions value in Table 1-1, any advances in this direction would result in a considerable contribution to the diminution of this problem. In the fact of the declining aerospace market, many of the concerned industries are turning to the development of equipment and processes for the environmental control market. One process actively being pursued is the desulfurization of coal and oil to eliminate the need for sulfur oxides control. Among the novel approaches to the removal of submicron particulates and gases are cryogenics, foam absorption, adsorption, catalysis, and waste heat energy exchange, all of which are being investigated in laboratories and prototype plants.

With the foregoing descriptive background of the emissions, the regulations, and the controls, the interrelationship of these three elements comprising the subject of air pollution, can be summarized. The type and magnitude of pollutant emissions are a function of the type of industrial process involved. The regulations prescribe the permissible emissions discharge rate. Control equipment is selected, designed, and applied to accomplish the necessary reduction of the pollutant emissions rate at some reasonable economic cost. For example, should a process discharge particulates at a rate of 1000 lb/day and the regulations permit 4 lb/day, then control equipment capable of operating at an effectiveness level of 99.6% would be required. Would this be available for the process conditions involved? Could the process support the necessary investment? Would process modifications or a combination of modifications and controls represent the optimum solution? Could some value be credited to the collected particulates as a trade-off against operating costs? With the current and continuous upgrading of emissions regulations, would there be sufficient reserve performance in the selected equipment to serve satisfactorily for one year, two years, five years hence? These are some of the questions involved in considering the subject of air pollution emissions, regulations, and control. Hopefully, many of the answers will be provided by this text.

1.7. ECONOMICS OF AIR POLLUTION

Air pollution undoubtedly costs money in terms of health impairment and property damage, but so does its control. The more stringent the regulations, the more costly are capital and operating costs for the required control equipment.

Pollution control laws and a sense of social responsibility, or pressures exerted by the "neighbors," are the major factors influencing the decision to curtail air pollution. It lies in the province of the engineer to select and design control equipment to meet regulations as economically as possible. A systems approach is most important to determine the total economic impact for a particular plant control system. Fans, ductwork, pumps, piping, instrumentation, collected effluent disposal facilities, expanded utility services, and maintenance charges must all be considered in the total price of the package. For certain corrosive services, suitable fans can easily exceed the cost of the primary collection equipment.

The increasing severity of air pollution control regulations poses the most difficult economic problem. Consider the utilities operator who was permitted a particulates discharge loading of about 0.5 grain/ft^3 of flue gas (7,000 gr. = 1 lb.) before the enactment of regulations. In many instances he conformed to this demand with simple cyclonic separators, at a very low capital investment. As regulations were upgraded, the acceptable discharge loading was decreased to 0.1 grain/ft^3 and at the present time a more realistic value of 0.02 grain/ft^3 must be considered. The electrostatic precipitator (at about ten times the cost of the cyclonic separator) when installed in series with the cyclonic separator can usually meet this requirement. However, now there arrives the need for gaseous emissions confinement, sulfur oxides, and nitrogen oxides currently, and carbon monoxide in the future, with a requirement for gaseous control equipment. The exact type, performance levels, and costs of this equipment have still not been developed. This situation represents economic strain. Again this condition reflects the need for advanced, proven control technology.

One approach to air pollution control, as mentioned previously, is process modification. Some segments of the utilities industry are purchasing coal and oil with low sulfur content to meet the sulfur oxides emissions codes. Neither fuel is in plentiful supply, so that economic relief in this direction does not appear too promising. Another example of process modification is that being introduced to the aluminum "degassing" process. This step is necessary for the removal of solid and gaseous impurities. The use of gaseous chlorine was universally accepted for this duty. However, in agitating the aluminum melt with this gas, submicron aluminum oxide and aluminum chloride particulates together with gaseous chlorine and hydrogen chloride were vented to the atmosphere. Control equipment was expensive and somewhat uncertain in its performance. Under regulatory pressures, some of the major aluminum producers are substituting nitrogen as the degassing agent, thereby eliminating a serious source of air pollution.

In the application of air pollution equipment the possibility of product recovery should be a consideration. The aluminum industry converts considerable amounts of fluorine and alumina pollutants discharged from the reduction cells to cryolite, which is recharged as makeup to the cells. This chemical conversion is accomplished in wet scrubbers, utilizing an alkaline scrubbing liquor. Wet-type processes such as those for fertilizer, sugar, and pulp and paper can easily recycle recovered products from their air pollution control systems. In most recovery schemes, the value of the recovered material is not expected to underwrite the investment required for the control equipment. Its value could, however, have sufficient economic weight to favor the selection of a particular air pollution control system, based on annual fixed and operating costs.

In March 1972, the U.S. Environmental Protection Agency (EPA) published its economic evaluation of air pollution controls. In the EPA report to Congress, it was estimated that industry must invest about $42 billion over the next five years. An investment of this order would force substantial price increases in such major industries as automobiles, electric power, steel, cement, and sulfuric acid. Consumer prices would be increased on the order of about 1% by these predicted air pollution control activities. These estimates showed a considerable increase over similar values published in 1971, due to "the higher expected cost of emission controls to meet the more stringent standards" as defined by the 1970 Clean Air Act.

REFERENCES

1. L.L. Terry, "Let's Clear The Air." Public Health Service Publication No. 1238, December 1962.
2. W.L. Faith, *Air Pollution.* New York: Wiley, 1957.
3. "Nationwide Inventory of Air Pollutant Emissions—1968," U.S. Dept. of Health, Education, and Welfare, Public Health Service, Environmental Health Service, National Air Pollution Control Administration, Raleigh, N.C., August 1970.
4. "Air Quality Criteria for Sulfur Oxides," U.S. Dept. of Health, Education, and Welfare, Public Health Service, Consumer Protection and Environmental Health Service, National Air Pollution Control Administration, Washington, D.C., January 1969.
5. "Industrial Markets for Air Pollution Equipment," Report No. 54, Predicasts, Inc., Cleveland, Ohio, October 1969.
6. "The Economics of Clean Air," U.S. Environmental Protection Agency, March 1972.

Chapter 2

ATMOSPHERIC POLLUTANTS

2.1. INTRODUCTION

Atmospheric pollutants can be most simply classified as solid, liquid, or gaseous materials discharged from various processes in sufficient amounts so as to constitute a health hazard or a nuisance source. Air pollution can be defined as that condition where the atmosphere is overloaded with harmful or unpleasant substances. To what degree such materials being discharged to the atmosphere can be considered injurious to man and his properties, or which represent an inconvenience to the public's welfare, is defined by the various regulatory bodies responsible for atmospheric quality.

Other classifications of the various air contaminants are based on chemical composition and physical state of the pollutant. For example, in Los Angeles County, California, the major pollutant categories are organic gases, inorganic gases, and aerosols. Each of these classes includes a number of compounds that may be emitted from several industrial sources in various concentrations.

The existence of atmospheric pollution is governed by both emission source and geographical factors associated with the concerned area. Thus, both physical landscape and meteorological conditions in one area which might promote the buildup of atmospheric pollution could make it necessary to impose severe restrictions on a specific pollutant emission rate. This same contaminant might be considered relatively harmless in a more generous physical environment. In addition, atmospheric reactions might make it necessary to restrict the discharge of a particular substance, which, although harmless in itself, can react under certain atmospheric conditions to create an unacceptable air pollution condition. For example, emission restrictions are more stringent in certain areas for those hydrocarbon materials that are known to participate in photochemical reactions with nitrogen dioxide in the atmosphere to yield peroxyacetyl nitrate (PAN) and ozone. The former is the major ingredient of smog, which is responsible for eye irritation and reduction in visibility in most of the major cities of the world.

The definition of an atmospheric pollutant, therefore, is a function of the affected geographical region as influenced by the various emission sources and the physical features of the area. Thus, central Florida's pollution regulations reflect the concern of that region with fluoride emissions from its fertilizer industries, whereas (because of smog problems) Los Angeles County emphasizes hydrocarbons and nitrogen oxides control. Because of the use of coal and oil fuels that contain sulfur, in the eastern United States, flyash particulates and sulfur oxides are the major pollutants subjected to regulatory legislation. In the West, where natural gas is the major combustion fuel, particulates are not a cause for concern,* but nitrogen oxides are. As might be expected as a result of this geographical predilection for the various types of atmospheric pollutants, the majority of particulate control equipment installations applied to combustion processes can be found in the eastern United States, where solid and liquid fossil fuels are utilized.

2.2. POLLUTION DEFINITION AND SOURCE

Pollutant Classification

Classification and definitions of the more common terms used to describe atmospheric pollutants are discussed below.

Particulate matter is any material, except uncombined water, that exists as a solid or liquid in the atmosphere or in a gas stream.

An *aerosol* is a dispersion in a gaseous medium of particles, either in the solid or liquid physical state. The particles are usually sufficiently small in size as to remain suspended in the atmosphere. Some of the common aerosols and their size range, expressed as the diameter in microns, or μ (1 micron = 3.94×10^{-5} in.) are shown in Fig. 2-1.

Dust is a general term used to describe solid particles that are in the micron size range. Flyash, coarse dirt, and mechanically produced particles fall into this category.

Fumes are solid, submicron size particles. Such particles are formed by condensation, sublimation, or chemical reaction.

Mist is a gaseous dispersion of liquid particles, usually less than 50 μ diameter in size. When the concentration of suspended particles is sufficient to reduce visibility, the mist becomes a fog.

Smoke is a combination of carbon particles and liquid droplets and is usually a product of incomplete combustion.

Soot is very finely divided carbon in an agglomerated state.

Gases are molecular in size about 0.002 μ in diameter. Compounds such as ammonia, sulfur dioxide, hydrogen chloride, and methane are classified as gaseous pollutants.

Vapors are gas-phase hydrocarbons, generally derived from solvents processing.

These definitions cannot be interpreted too strictly because there exists some confusion with regard to pollutant nomenclature. For the application of control equipment to contain the various pollutant types, there is only the need to know whether particulate or gaseous emissions are involved. The contaminant will then be further defined by specifying those pertinent physical properties that influence the design of the equipment.

Pollutant Sources

The sources of particulate and gaseous atmospheric pollutants are infinite. Every major industry is beset by atmospheric emission problems. The list of pollutants is steadily

---

*This situation is rapidly changing. The shortage of natural gas is forcing western utilities to use fuel oils, with the attendant problems of SO_2 and flyash emissions control.

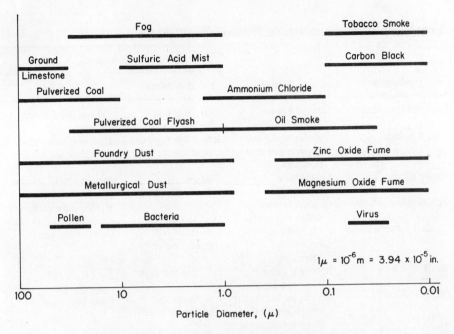

Figure 2–1. Typical aerosols and their size range.

growing as the deleterious effects of an ever-increasing number of airborne materials are being realized. At one time the criterion for atmospheric pollution control was to reduce the visible stack emissions. Therefore, up through 1950, the major emphasis was placed on particulates removal. Then the need for gaseous control became evident, with the sulfur oxides produced by the combustion of sulfur-bearing fossil fuels being the first target of the criteria established by the U.S. Department of Health, Education, and Welfare (HEW) in 1969.

The formation of the oxides of nitrogen is presently causing some concern in most of the larger cities. Both nitric oxide and nitrogen dioxide are produced during the high-temperature combustion of coal, oil, gas, or gasoline in power plants and internal combustion engines. Nitric oxide is colorless and not so nearly toxic as nitrogen dioxide. However, the photochemical conversion of the oxide to the dioxide form takes place at a fairly rapid rate in the presence of sunlight and organic materials. Furthermore, the combination of hydrocarbon and nitric oxides emissions from automotive sources reacts to produce PAN, the major constituent of smog, which is the most serious consequence of these atmospheric contaminants in heavily populated communities.

Some of the major industrial processes with the types of particulate and gaseous emissions associated with their operation are shown in Table 2-1. The combustion of coal, oil, and automotive fuels in 1969 resulted in annual emissions in the United States of about 8 million tons of particulate matter. Approximately 90% of this value was flyash from fossil-fuel fired, stationary combustion sources. Industrial processes contributed 14.4 million tons per year, while such miscellaneous sources as home

Table 2-1. Common Atmospheric Pollutants and Their Sources

Industry	Source	Emissions	
		Particulates	*Others*
Iron and steel mills	Blast and open hearth furnaces	Iron oxide and carbon fumes	CO and combustion gases
Iron foundries	Cupolas, shake-out systems and core machines	Iron oxide fumes, sand, smoke, oil	Hydrocarbons and combustion gases
Power plants	Combustion furnaces	Coal and oil flyash	SO_x and NO_x
Petrol. refineries	Catalyst furnace and sludge incinerator	Catalyst fines and flyash	H_2SO_4 mist, oil fumes and combustion gases
Cement manufacturing	Kilns, coolers, dryers and transfer equipment	Limestone and cement dusts	Combustion gases
Kraft pulp mills	Recovery furnaces, kilns, and digestors	Salt cake and lime fumes	SO_x, H_2S, and mercaptans
Asphalt plants	Dryers, coolers, batch operations, and transfer equipment	Rock dusts and oil fumes	Combustion gases
Phosphoric acid plants	Grinders, reactors and filtration equipment	Phosphate rock dust and fumes	H_3PO_4 mist, HF and SiF_4 fumes
Coke processing	Storage, transfer, oven charging and quenching	Coal and coke dusts, tar fumes	Phenols, H_2S and combustion gases
Glass manufacturing	Melt furnaces and materials handling	Raw material dusts, sulfate fumes	NO_x and combustion gases
Aluminum processing	Melt furnaces and machining operations	Al_2O_3 and $AlCl_3$ fumes, Al dusts	HCl and Cl_2 gases
Coffee processing	Roasters, spray dryers, and coolers	Coffee, chaff, and ash dusts	Oil fumes
Coal cleaning	Dryers, coolers, and transfer equipment	Coal dusts	Combustion gases
Automotive vehicles	Gasoline storage and exhaust system	Carbon fumes	NO_x, CO, and HC

Source: U.S. Dept. of Health, Education, and Welfare, Ref. 1.

heating, open refuse burning, and forest fires—together with solid-waste disposal processes—were responsible for a total annual particulate discharge of 12.8 million tons. Thus, the total annual particulate emissions in the United States amounted to 35.2 million tons in 1969 (see Table 1-1, Chapter 1).

Gaseous emissions can be most conveniently classified as organic and inorganic gases. The organic gases include all types of hydrocarbons and their derivatives. Petroleum refining and those processes involving the use of such hydrocarbons are the major emission sources. The most significant source of hydrocarbon emissions is from the combustion of gasoline by motor vehicles. Thus, the total automotive emissions in the United States for 1969 accounted for approximately 20 million tons, or about 53% of the total hydrocarbons discharged to the atmosphere. The inorganic gases included ammonia, oxides of sulfur, oxides of nitrogen, carbon monoxide, hydrogen sulfide, hydrogen fluoride, hydrogen chloride, and chlorine. A tabulation of both organic and inorganic gases and their principal sources is shown in Table 2-2. The major inorganic gas emissions, comprising sulfur oxides, nitrogen oxides, and carbon monoxide, were approximately 210 million tons for the year 1969 in the United States.

2.3. PARTICULATE POLLUTANTS

Particle Size and Stokes Law

As defined previously, particulates will be considered as solid or liquid particles with sizes extending over the entire micron (μ) range. Fumes are solid particulates in the $<$ μ (less than one micron) size range.

The most important characteristic of particulates is their size, which is generally defined as the individual particle diameter in microns. The particle diameter is related to its settling velocity by Stokes law, as follows:

$$v = \frac{gd^2\,(\rho_1 - \rho_2)}{18\mu} \tag{2-1}$$

where

v = settling or terminal velocity, cm/sec
g = acceleration due to gravity, cm/sec^2

Table 2-2. Organic and Inorganic Gaseous Pollutants

Pollutant Classification	*Sources*
Organic Gases	
Hydrocarbon paraffins	Petroleum processing, automotive vehicle operations
Olefins	Processing and transportation of liquid fuels
Aromatics	Metals processing, coatings operations
Oxygenated hydrocarbons	Solvents processing and operations, automotive vehicles operations
Halogenated hydrocarbons	Solvents cleaning
Inorganic Gases	
Oxides of sulfur	Combustion of coal and oil fuels, chemical processing metallurgical operations
Oxides of nitrogen	Combustion of coal, oil and gas fuels; automotive vehicle operation; nitric acid manufacture
HCl and Cl$_2$	Aluminum, chemical, and paper processing
NH$_3$	Fertilizer manufacture
HF, SiF$_4$	Aluminum and fertilizer processing
H$_2$S	Paper and plastic film manufacture

Source: Danielson, Ref. 2.

d = particle diameter, cm
ρ_1 = particle density, g/cm^3
ρ_2 = air density, g/cm^3
μ = viscosity of air, poise

Equation 2-1 pertains only to spherical bodies. Actually, most particles encountered in practice are irregular in shape. For practical purposes, a value for the diameter d has been adopted for a specific settling velocity and media densities, to satisfy Eq. 2-1. Therefore, the word "diameter" as applied to airborne particles is actually the Stokes diameter. If the particle density is unknown, then a "reduced sedimentation diameter" is obtained by the substitution of a unity value, which is the diameter of a spherical particle of unit density having the same settling velocity in air as the particles being evaluated. Some approximate settling velocity values in still air at 0° C and 760 mm Hg pressure for particles having a density of 1 g/cm^3 are

Micron	Cm/sec
0.1	8×10^{-5}
1.	4×10^{-3}
10	0.3
100	25
1000	390

The importance of particle size is that it is a measure of the difficulty in removing a specific particulate from a gas stream. As discussed in detail in Chapter 5, the smaller the particle diameter, the more economically burdensome is its collection, as measured by both capital and operating costs. When a dust dispersoid is acted upon by external forces in a specific type of pollution control device, such as a gravity collector or an electrostatic field, the particles will be accelerated in the direction of this force. The relative motion between the particles and gaseous medium will set up a frictional resistance that will limit the velocity of the particles in the direction of this collecting force. In order to remove the particles from the gas stream, this velocity must be sufficient to carry them to a collection surface before the gas stream carries them out of the apparatus. This is essentially the basic mechanism employed for the collection of particulate matter. For gravity type collectors the relationship of particle size with settling velocities was defined by Stokes law in Eq. 2-1. The subsequent tabulation, which shows the influence of particle size on these velocities, is an indication of the relative difficulty of collecting small particles by this mechanism.

Particle Size Distribution

Particulate matter occurs in a range of particle sizes. Therefore, to define a particular dust or fume, it is necessary to determine the relative amounts of each particle size. Procedures for determining a particle size analysis for any dust sample will be discussed more fully in Chapter 4. The various and most common methods available are sieve, microscopic, elutriation, and sedimentation. The technique to be used is a function of the estimated particle size range of the sample and the type of control equipment being considered. Except for sieve analysis, the other methods do not directly measure the particle size but rather determine some other property of the particle, which is then converted to an effective particle size by an empirical relationship. A sieve analysis is performed with dusts containing particles in the range of 5000 to 40 μ. This method is rarely applicable to current air pollution control problems, where the particulates usually measure smaller than 20 μ. A tabulation of screen mesh sizes for both U.S. Sieve and Tyler systems, with equivalent micron values, is shown in Table 2-3. In both systems the screen mesh refers to the number of screen openings per unit of length.

The occurrence of particle sizes generally follows a logarithmic distribution pattern, as shown in Fig. 2-2. The values tabulated and plotted on this curve represent pulverized coal-flyash size analysis data discussed in Chapter 5; see Table 5-8. In

Table 2-3. U.S. Sieve and Tyler Screen Sizes

	U.S. Screen Scale				*Tyler Screen Scale*		
Mesh No.	*Wire Diam, in.*	*Screen Aperature, in.*	*Particle Size, μ*	*Mesh No.*	*Wire Diam, in.*	*Screen Aperature, in.*	*Particle Size, μ*
400	0.0010	0.0015	37	400	0.0010	0.0015	37
325	0.0014	0.0017	44	325	0.0014	0.0017	43
270	0.0016	0.0021	53	270	0.0016	0.0021	53
230	0.0018	0.0024	62	250	0.0016	0.0024	61
200	0.0021	0.0029	74	200	0.0021	0.0029	74
170	0.0025	0.0035	88	170	0.0024	0.0035	88
140	0.0029	0.0041	105	150	0.0026	0.0041	104
120	0.0034	0.0049	125	115	0.0038	0.0049	124
100	0.0040	0.0059	149	100	0.0042	0.0058	147
80	0.0047	0.0070	177	80	0.0056	0.0069	175
70	0.0055	0.0083	210	65	0.0072	0.0082	208
60	0.0064	0.0098	250	60	0.0070	0.0097	246
50	0.0074	0.0117	297	48	0.0092	0.0116	295
45	0.0087	0.0138	350	42	0.0100	0.0138	351
40	0.0098	0.0165	420	35	0.0122	0.0164	417
35	0.0114	0.0197	500	32	0.0118	0.0195	495
30	0.0130	0.0232	590	28	0.0125	0.0232	589
25	0.0146	0.0280	710	24	0.0141	0.0276	701
20	0.0165	0.0331	840	20	0.0172	0.0328	833
18	0.0189	0.0394	1,000	16	0.0235	0.0390	991
16	0.0213	0.0469	1,190	14	0.025	0.046	1.168
14	0.0240	0.0555	1,410	12	0.028	0.055	1.397
12	0.0272	0.0661	1,680	10	0.035	0.065	1.651
10	0.0299	0.0787	2,000	9	0.033	0.078	1.981
8	0.0331	0.0937	2,380	8	0.032	0.093	2.362
7	0.036	0.111	2,830	7	0.0328	0.110	2.794
6	0.040	0.132	3,360	6	0.036	0.131	3.327
5	0.044	0.157	4,000	5	0.044	0.156	3.962
4	0.050	0.187	4,760	4	0.065	0.185	4.699

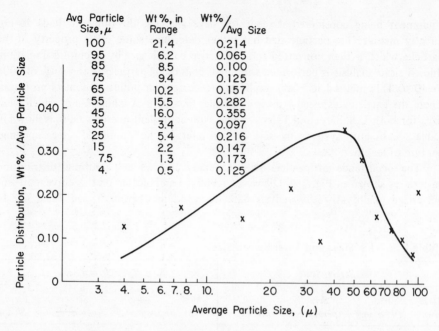

Figure 2–2. Logarithmic particle size distribution.

Fig. 2-2 the ordinate is the weight of a particle fraction per unit size interval. In actual practice a cumulative distribution curve is plotted on a logarithmic probability scale so that a straight line or one slightly curved is obtained. This type of curve, for the same pulverized coal-flyash analysis, is shown in Fig. 2-3. The agreement of the actual size analysis (obtained by elutriation) with the logarithmic distribution and logarithmic probability curves, is considered fair in this particular example. In commerical practice, usually three or four actual size-distribution values are plotted on a logarithmic probability curve, and the complete size analysis is determined by extrapolation. Investigations of test results for crushing and grinding have shown that this plot does yield a straight line. The particle size distribution method influences the shape of the curve. It is usual practice to obtain the actual distribution data used in equipment design calculations from the straight-line "smoothed" data.

Particle size analyses in Figs. 2-2 and 2-3 are based on weight values. Distribution data can also be expressed in terms of the number of particles. Thus, in microscopic analysis, the number of particles are actually counted on a measured sizing grid and their distribution is converted to a weight basis. A typical cumulative size analysis for airborne dust, based on a number count and weight values, is shown in Table 2-4. In practically all industrial applications the removal of particulates is a mass-rate function so that size analysis data are based on weight values.

Particulate Emissions and Air Quality

Particulate emissions and/or concentrations are expressed in a variety of terms. Some of the more common of these are:

18

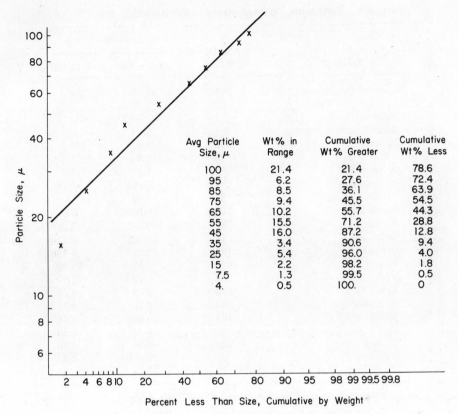

Figure 2–3. Logarithmic probability particle size distribution. (Adapted from Drinker and Hatch, Ref. 3.)

Table 2-4. Size Distribution of Airborne Dust

| | Cumulative: % less than size | |
Particle Size, μ	Number	Weight
0.04	0.1	
0.08	3.2	0.0039
0.14	20.15	0.98
0.30	72.0	2.95
0.60	96.6	11.2
1.0	99.0	15.0
2.0	99.8	33.0
4.0	99.979	60.2
8.0	99.9986	85.0
14.0	99.99991	95.6
20.0	99.999992	98.8
30.0	100.0	99.93

Source: U.S. Dept. of Health, Education, and Welfare, Ref. 4.

Table 2-5. Particulate Concentrations in the United States

Pollutant	No. of Sampling Stations	Concentrations, $\mu g/m^3$	
		Arith. Avg	Maximum
Suspended Particulates	291	105.0	1254
Fractions:			
Benzene-soluble organics	218	6.8	–
Nitrates	96	2.6	39.7
Sulfates	96	10.6	101.2
Ammonium	56	1.3	75.5
Antimony	35	0.001	0.160
Arsenic	133	0.02	–
Beryllium	100	< 0.0005	0.010
Bismuth	35	< 0.005	0.064
Cadmium	35	0.002	0.420
Chromium	103	0.015	0.330
Copper	103	0.09	10.00
Iron	104	1.58	22.00
Lead	104	0.79	8.60
Manganese	103	0.10	9.98
Nickel	103	0.034	0.460
Tin	85	0.02	0.50
Titanium	104	0.04	1.10
Vanadium	99	0.050	2.200
Zinc	99	0.67	58.00

Source: U.S. Dept. of Health, Education, and Welfare, Ref. 5.

g/Nm^3 = grams of particulates per standard cubic meter of effluent gas at $0°$ C, 760 mm Hg

grains/acf = grains of particulates per actual cubic foot of effluent gas at prevailing temperature and pressure; 7000 grains = 1 pound

grains/scf = grains of particulates per standard cubic foot of effluent gas at *$60°$F, 1 atm pressure

lb/hr = pounds per hour of particulates being emitted

$\mu g/m^3$ = micrograms of particulates per actual cubic meter of air at ambient conditions

In Table 2-5 the concentrations of certain specific contaminants are defined as they are found in airborne suspended particulate matter. Some of these values can be related to commercial processes. Thus, the vanadium emissions can be correlated with the type and combustion rate of residual oil fuels, and the lead fraction is found to be a function of gasoline consumption and its tetraethyl lead content. Another emissions

---

*Throughout the text, both $60°$F and $70°$F have been used in calculations as the "standard" temperature. More recently, the $60°$F value seems to be favored by industry and the local regulatory agencies. EPA uses $70°$F.

value, defined as the *dustfall,* is usually expressed in tons per square mile per month, and is an index of the fallout of particulates in the +10μ range. Such values are determined by the extrapolation of sampling data from jars a few inches in diameter. Typical values for cities are 10 to 100 short-tons/mile^2-month. For particularly heavy emissions areas, dustfall rates as high as 2000 short-tons/mile^2-month are not uncommon.

Harmful Effects of Particulate Pollutants

The major objective of air pollution control regulations is to ensure an acceptable level of contaminant emissions to the atmosphere which will not interfere too seriously with human activities. The air quality criteria for particulate matter published by the U.S. Department of Health, Education, and Welfare [5] are defined as an "expression of the scientific knowledge of the relationship between various concentrations of pollutants in the air and their adverse effects on man and his environment." These criteria lead to the establishment of acceptable air quality particulate concentrations, which in turn are converted to allowable discharge emissions from a specific source. Air quality criteria are concerned with effects on health, odor levels, visibility, and property damage factors.

The effect of particulates on human health is determined not only by their chemical composition but also by their size. The particle size determines the method of entry into the human body. Thus small particles will penetrate deeply into the respiratory system, whereas coarse ones have a tendency to remain in the upper respiratory tract or on the skin. During periods of heavy pollution, the average daily particulate intake per person is about 1 milligram (mg). The larger particles above 10 μ lodge in the nostrils; those above 5 μ are probably collected by the mucous lining of the upper respiratory passages. Particles less than 2 to 3 μ usually reach the deep structures of the lung. Such inert particles as soot or soil are known to aggravate the symptoms of those individuals suffering from chronic bronchitis. Soot also acts as a carrier of other liquid and gaseous pollutants into the deeper passages of the lungs. There is also a class of harmful metal particulates such as lead, beryllium, and selenium. The level of atmospheric lead associated with automobile exhaust in Los Angeles, California, is high, with an arithmetic average value of 5 μg per cubic meter of air. Of this contaminant concentration, 60% comprises particles below about 0.5 μ. With the oral ingestion of < 1μ lead particles, it is estimated that 50% is retained by the body, with the remainder being carried out by the exhaled air. Although there is evidence that long-term exposure to lead can result in chronic kidney disorders and affect the reproduction of hemoglobin, the association between high ambient air concentration and the physiological effects of lead has still not been established. Beryllium and selenium have similar adverse effects on humans when drawn into the lungs. Beryllium is particularly toxic, and in recognition of its hazard to health the American Conference of Governmental Industrial Hygienists has issued a single ambient air standard, which states: "The average daily concentration of beryllium in the vicinity of a plant producing and processing beryllium should be limited to 0.001 microgram per cubic meter."

Odor-producing particulates generally do not cause organic diseases, although

Table 2-6. Common Odors and Their Sources

Source of Odor	Number Reported
Animal processing	
Meat packing and rendering plants	12
Fish oil odors from mfg. plants	5
Poultry ranches and processing	4
Combustion processes	
Gasoline and diesel engine exhaust	10
Coke oven and coal-gas odors from steel mills	8
Maladjusted heating systems	3
Food processes	
Coffee roasting	8
Restaurants and bakeries	5
Paint and related industries	
Paint, lacquer and varnish mfg.	8
Paint spraying and drying	4
Solvents handling	3
General chemical processes	
Hydrogen sulfide	7
Sulfur dioxide	4
Ammonia	3
General industrial processes	
Burning rubber from smelting and debonding	5
Dry cleaning shops	5
Fertilizer plants	4
Asphalt odors: roofing and street paving	4
Plastic manufacturing	3
Foundries	
Core oven odors	4
Heat treating, oil quenching, and pickling	3
Smelting	2
Combustible wastes	
Home incinerators and trash fires	4
Garbage-burning incinerators	3
Open dump fires	2
Refineries	
Sulfur and mercaptans	3
Crude oil and gasolines	3
Sewage disposal	
Sewage treatment plants	2
Industrial waste sewers	3

Source: U.S. Dept. of Heath, Education, and Welfare, Ref. 5.

extreme levels of discomfort may cause some temporary ill effects. Some particles can stimulate the human olfactory sense per se because of their volatility or because they are desorbing a volatile odorant. For example, in the case of diesel exhaust, the particulate emissions do yield diesel odor, which is believed due to the release of the adsorbed acroleins and formaldehydes. Many volatile particulates, such as camphor, are vaporized on entering the human nasal cavity, thereby producing sufficient gaseous material to be detected by smell. However, by far the major sources of odor-producing emissions are the liquid and gaseous pollutants. Table 2-6 lists common industrial odor sources with an index of the reporting frequency. These data were compiled from a survey of state and local air pollution control personnel in the United States. Of the total listings, about 25% are believed to be related to particulate emissions.

Poor visibility is most often related to the presence of airborne particulates. Visibility is reduced by two optical effects that air molecules and suspended particulates impose on visible radiation. One is the attenuation of the light passing from the object to the observer via air molecules and aerosol particulates. The other effect, which reduces the contrast between the object and background, is the illumination of the intervening air when the sunlight is scattered by the molecules and particulates in the line of sight. The word "visibility" means the distance at which it is just possible to perceive an object against the horizon, with the sky as a background. Suspended particles found in the atmosphere cover a broad size range. Visibility is mainly affected by a relatively narrow segment of this size distribution, usually from about 0.2 to 2 μ. Once particulate matter exists as a dispersoid in the atmosphere for some periods of time, the size distribution of the particles assumes a definite pattern. Thus, the value of visibility can be related to the particulate concentration by the following expression:

$$L_v = \frac{A \times 10^3}{G'}$$ (2-2)

where

L_v = visibility, miles
A = constant, 0.74 average with a range of 0.38 to 1.5
G' = particle concentration

Because water vapor in the air influences visibility, deviations from Eq. 2-2 might be expected when the relative humidity values exceed 70%. Generally, this equation can be used to estimate the expected visibility for different levels of particulate concentrations in the atmosphere. Thus, with a typical rural particulates concentration of 30 μg/m^3, the visibility is about 25 miles. For an urban concentration of 100 μg/m^3, the expected visibility would be 7.5 miles. Aside from esthetic considerations, the effect of reduced visibility on commercial aircraft operations in metropolitan areas is a serious problem. Federal Air Regulations prescribe increasingly severe limitations on aircraft operation as visibility decreases below 5 miles.

The effects of particulate matter on plant life is a relatively unexplored subject. Gaseous pollutants are more readily recognized as being harmful to a variety of plants, and a considerable number of toxicology studies concerned with this relationship has been undertaken. Certain airborne particulates have been applied to plant leaves under laboratory conditions and the deleterious effects recorded. Thus, when bean plant leaves were dusted with cement kiln particulates at an equivalent dustfall rate of 400 tons/mile^2-month, some moderate damage of the leaves was observed and plant growth was inhibited. However, the mechanisms producing these injuries are not fully understood. Other studies attempted to relate the effects of cement kiln dusts on plant growth as a function of the soiling precipitation level, but some inconclusive results were obtained. Fluoride and soot particulates were also investigated. Fluorides deposited on vegetation caused little damage, but were responsible for health injuries to animals feeding from these plants. Soot may be injurious to plant life only if it carries with it a soluble acidic pollutant.

Airborne particles may damage surfaces merely by settling on them and making frequent cleaning necessary. The extent of chemical deterioration depends on the chemical activity of the particles that might act as condensation nuclei for the retention of adsorbed gases and acids, or which might be corrosive in themselves. For example, carbonlike particles in the atmosphere, at relatively high humidities, cause increased corrosion rates in the presence of SO_2 traces due to concentration of the acid values by adsorption. Laboratory investigations [7] in this area were based on particulate concentrations equivalent to 0.3 tons/mile^2 and SO_2 concentrations of 100 ppm. Hygroscopic particulate matter such as sulfide and chloride salts serve as nuclei for corrosive substances, and in the presence of SO_2 at humidities above 70% will produce considerable corrosion. The rate of corrosion of various metals is accelerated in industrial and urban areas because of the greater concentrations of particulates and sulfur compounds in the atmosphere. A plot illustrating the relationship of dustfall in a number of different geographic areas with the corrosion rate of mild steel is shown in Fig. 2-4.

Soiling of painted surfaces and textiles is another form of property damage caused by atmospheric particulate pollutants. Water-soluble chlorides and sulfates are commonly found in particulate samples taken in urban areas. These particles are potential sources of paint blistering. In various laboratory studies, the presence of 0.1 ppm iron sulfate in the water produced yellow staining of paint panels. Curtains are particularly susceptible to soiling and deterioration from airborne particulates. The curtain material acts essentially as a filter for acid-laden dust and soot, which causes it to be weakened and ultimately to disintegrate. Airborne particulates also cause damage to electronic and switchgear equipment. The contact surfaces in electrical switches are subject to chemical or mechanical deterioration by particulates soiling. Particulate types and loadings characteristic of urban and industrial areas are responsible for the failure of electrical connectors and circuits.

Plume Visibility and the Ringelmann Numbers

Aside from the damaging aspects of discharging particulate pollutants into the atmosphere, the visibility of the plume associated with aerosol emissions is rapidly assuming an exaggerated position in the most recent round of air quality regulations. When reason and technical competence prevailed, particulate emissions were quantitatively limited, either by concentration values or mass rate flows. The offending source was

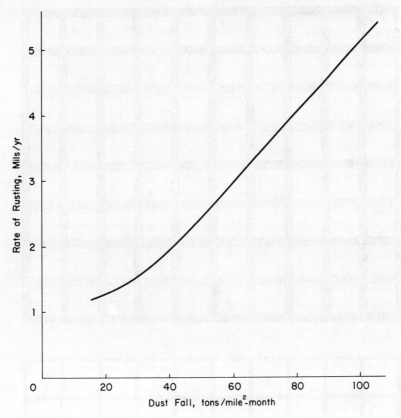

Figure 2-4. Corrosion rate versus dust fall. (J.D. Hudson, Ref. 8.)

subjected to a series of source tests which defined the type and quantities of the discharged contaminants. A comparison of these values with the prevailing regulatory limits would then define the required degree of emissions control to bring the contaminants discharge rate into compliance. Within the past year a series of regulations has been passed which makes it necessary for the corrected emissions source to meet certain visual requirements based on the use of Ringelmann numbers. The Ringelmann chart, shown in Fig. 2-5, is the basis for these regulations. The offender is required to evaluate the plume as it leaves the stack by comparing it with various Ringelmann numbers viewed against the background of the sky. The values of 1,2,3, and 4 correspond to 20,40,60, and 80% obscurity of the background light. Ringelmann numbers of 0 and 5 correspond to full white or full black. Unfortunately, there is just no connection between this visual evaluation and quantitative emissions data. However, the most currently demanded Ringelmann numbers of 1 and 2 do give an esthetically appealing stack discharge and provide the ecology-minded public with maximum satisfaction. Estimated emission levels equivalent to Ringelmann numbers are tabulated in Table E-3 of Appendix E. Values for the various emission sources lie in the range of 0.01 to 0.02 grain/acf for a Ringelmann reading of 0 to 2.

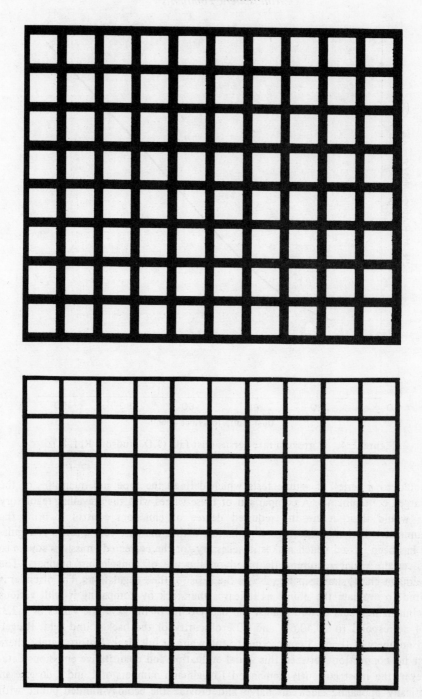

Rdg. No. 2 at a light obscurity of 40%

Rdg. No. 1 at a light obscurity of 20%

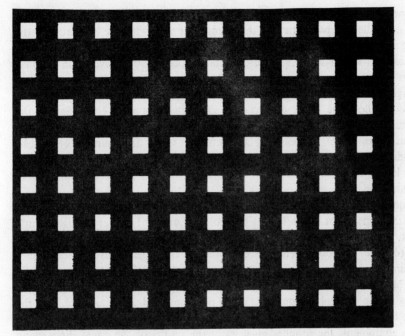

Rdg. No. 4 at a light obscurity of 80%

Rdg. No. 3 at a light obscurity of 60%

Figure 2–5. The Ringelmann Chart.

The Ringelmann reading is subjected to a number of uncontrolled variables. Visual effects are not inherent properties of the plume but vary with the background, the illuminating source, and viewing conditions. These variations are much greater with white plumes than with black. Light obscuration as specified by Ringelmann numbers is a function of the diameter of a stack rather than its area. Thus, for two sources—say, one discharging four times the gas flow rate of the other—the use of Ringelmann numbers would allow the larger one to emit about twice as much particulate as the smaller source; but there would also be a parallel quantitative demand that its flue gas shall be four times better in measured concentration terms. In a number of "smoke studies" undertaken by the British, it was found that the loss of light in the plume over a path length equal to the diameter of the stack will vary between 60 and 70% at a Ringelmann value of 2.

The major problem posed by the use of Ringelmann numbers is that it provides no valid quantitative basis for the design of control equipment. To meet a Ringelmann 1 requirement, the supplier must make a chance estimate of the equivalent mass loadings on which to base his equipment design. Usually, the quantitative concentration emission being considered as equivalent to Ringelmann 1 is arbitrarily chosen as 0.02 grain/acf. This value imposes a severe performance requirement. For example, if the inlet loading to the control equipment is in the neighborhood of 5 grains/acf, then the required collection efficiency would be about 99.6%. Having thus designed and installed his equipment, the supplier must await the results of the visual Ringelmann test before the equipment is accepted by his client. Ringelmann values are particularly difficult to interpret when applied to the saturated steam plume discharged from a wet scrubber. One obvious but slightly prejudiced solution to this difficulty is to discourage the use of wet scrubbers where such visual emission evaluations are practiced. In some areas of the country, this step has already been undertaken

Ringelmann judgments can be used as an aid in controlling air pollution. The Ringelmann chart should be used as an index of noncompliance to quantitative regulations, thus furnishing a quick and uncomplicated prejudgment technique. However, quantitative emissions must be the decisive parameter in the measurement of pollutant emissions. Even as a pretesting tool, extensive study is required to relate Ringelmann values to quantitative data.

Dispersion of Particulates and Diffusion Equations

Air pollution control regulations are concerned with the ground-level concentrations of pollutants. These levels can be reduced by emitting gas-borne contaminants from increased heights so as to improve their degree of dispersion into the atmosphere. At one time the use of very tall chimneys, up to 1000 ft in height, to dilute pollutants to an acceptable air quality level was considered a viable solution to the problem. This practice is rapidly being discouraged. However, there is still the need to consider the relationship between the pollutants discharged from a stack source and the corresponding values at ground level. Many regulations incorporate the stack height and location of plant property lines as major parameters in the determination of the desired air quality in the neighborhood of the emissions source. Thus, the allowable stack emissions are increased proportionally with the stack height and plant property dimensions.

The relationship between stack emissions and ground-level concentrations of pollutants is mainly controlled by the prevalent meteorological conditions. Essentially, the required reduction of air contaminants discharged from a stack is a function of the ability of the atmosphere to disperse and dilute them. For a given stack height and a knowledge of local weather conditions, a number of relationships has been developed for the prediction of diffusion effects. The most basic equations are those developed by Bosanquet and Pearson [9] and Sutton. [10] The Bosanquet and Pearson equation is

$$C_0 = \left[\frac{M(10)^6}{(2\pi)^{1/2} pqux^2} \right]^{-H/px - y^2/2qx^2} \tag{2-3}$$

where

C_0 = ground-level concentration, mg/ft^3
M = particulates emission rate, kg/sec
p and g = diffusion coefficients related to wind force (see Table 2-7)
u = mean wind speed, fps
x = downwind distance from emission source, ft
y = perpendicular distance to x, ft
H = effective stack height, ft

In this equation, H refers to the effective stack height. When stack gases are at an elevated temperature or have an appreciable exit velocity from the stack, the effective stack height is greater than its physical height, due to buoyancy and jetting action. The rise of the discharged plume can be computed by the method of Bosanquet, Carey, and Halton [12] [13] as follows:

(a) For the stack exit velocity rise H_v,

$$H_v = H_v(\text{max}) \frac{(1-0.8H_v(\text{max}))}{x} \quad \text{when } x > 2H_v(\text{max}) \tag{2-4}$$

Table 2-7. Bosanquet and Pearson Diffusion Coefficients

	Diffusion Coefficients		
Climatic Conditions	*p*	*q*	*p/q*
Low turbulence	0.02	0.04	0.50
Average turbulence	0.05	0.08	0.63
High turbulence	0.10	0.16	0.63

Source: Manufacturing Chemists Assoc., Ref. 11.

and

$$Hv_{(max)} = \frac{4.77}{1 + 0.43u/v} \times \frac{(Qv)^{1/2}}{u} \qquad (2\text{-}5)$$

where

$$
\begin{aligned}
H_{v(max)} &= \text{maximum momentum rise of plume, ft} \\
x &= \text{downwind distance from emission source, ft} \\
u &= \text{mean wind speed, fps} \\
v &= \text{stack exit velocity, fps} \\
Q &= \text{gas flow rate from stack (ft}^3\text{/sec) measured at} \\
&\quad \text{temperature at which stack gas density would be equal} \\
&\quad \text{to that of ambient atmosphere}
\end{aligned}
$$

(b) For the buoyancy rise H_T,

$$H_T = \frac{6.37g_cQ}{u^3\,T_1}\left(\ln J^2 + \frac{2}{J} - 2\right) \qquad (2\text{-}6)$$

where

$$
\begin{aligned}
g_c &= \text{gravity acceleration constant, 32.2 ft/sec}^2 \\
T_1 &= \text{absolute temperature (}^\circ\text{K) at which density of stack gases} \\
&\quad \text{equals density of atmosphere}
\end{aligned}
$$

and

$$J = \frac{\mu^2}{(Qv)^{1/2}} \quad 0.43\left[\left(\frac{(T_1)}{g_c G}\right)^{1/2} - 0.28\frac{v}{g_c} \times \frac{T_1}{(T_s - T_1)}\right] + 1 \qquad (2\text{-}7)$$

where

$$
\begin{aligned}
G &= \text{potential temperature gradient of the atmosphere, }^\circ\text{C/ft} \\
T_s &= \text{stack exit gas temperature, }^\circ\text{K}
\end{aligned}
$$

Thus, to obtain the plume rise above the stack due to the gas discharge velocity, Eq. 2-4 would be used. Equation 2-6, expressing the influence of temperature on plume rise, indicates the strong influence of the wind speed on the plume temperature rise. Thus, doubling the wind speed reduces the rise by a factor of 8. The effect of temperature on plume rise for smelter operations has been estimated at 2.5 ft of extra stack height per degree Fahrenheit of smoke temperature above ambient. Equations 2-6 and 2-7 can be simplified by representing ($\ln J^2 + 2/J - 2$) by Z and determining Z as a function of X in Fig. 2-6, where

$$X = \frac{ux}{3.57\,(Qv)^{1/2}} \qquad (2\text{-}8)$$

30

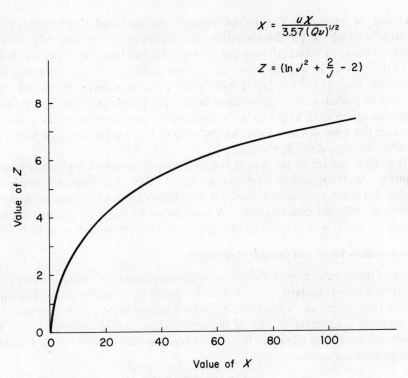

$$X = \frac{u X}{3.57 (Qv)^{1/2}}$$

$$Z = (\ln J^2 + \frac{2}{J} - 2)$$

Figure 2–6. Correlation for smoke plume buoyancy rise, for use with Eq. 2–6. (Adapted from O.H. Bosanquet et al, Ref. 12)

In summary, the problem of estimating ground-level concentration must be considered in two steps. First, the plume path and maximum height reached, known as the effective stack height H, must be determined and the diffusion of particulates (or gases) considered from this point. Thus, Eqs. 2-4 or 2-5 can be utilized to calculate the rise of the plume caused by the velocity of the gases issuing from the stack. Then the buoyancy rise can be calculated from Eqs. 2-6 and 2-7. As a good approximation, the effective height can be taken as the sum of these two values:

$$H = H_v + H_T \tag{2-9}$$

Actually, the velocity and buoyancy rises are not strictly additive. Factors such as air entrainment in the plume, spreading of the plume, and loss of momentum and heat by lateral diffusion cause the combined rise to be somewhat less than the sum indicated by Eq. 2-9. However, as a first approximation, the value of H can then be substituted in Eq. 2-3 as step 2, to estimate ground-level concentration values.

In the application of this equation to the dispersion of aerosols, the terminal velocity of the individual particles must be negligible as compared with the eddy velocities in the atmosphere. When the free-falling speed of the particulates cannot be neglected (because of the appreciable size of the particles), then a local particulate fallout can be expected. This is the situation for coal-fired flyash emissions. A theoretical method for the determination of the average rate of dust fallout under these conditions, in terms of tons per square mile, has been developed by Bosanquet [13]. It would be expected that the location of maximum deposition would be closer to the stack as the free-falling speed of the particles increases. Bosanquet found it possible to predict such results within a factor of 2. His investigations also indicated that increasing the stack height causes a considerable improvement in fallout close to the source; but some distance away, on the order of 1 to 2 miles, the stack height had little effect on particulates deposition.

The major concern of the federal and state control agencies is the maintenance of air quality standards, usually expressed as $\mu g/m^3$. Thus, a considerable number of advanced dispersion computation methods have been developed to relate emissions to atmospheric pollutant concentrations. A compilation of the most recent dispersion equations, together with relevant sample problems, has been published by HEW [19].

Source Emission Terms and Control Regulations

Because of the complexities in relating ground concentrations to source emissions, the majority of control regulations either directly specify an allowable stack emission in mass-rate terms or present a simplified diffusion equation based on the most pertinent plant-operating parameters. In one of the air pollution control regions [14] in the Midwest, the maximum allowable flyash discharge rate was defined by the following simplified diffusion equation:

$$C_{max} = \frac{70 Pt_f Q_m^{0.75} n^{0.25}}{ah_s} \qquad (2\text{-}10)$$

where

C_{max} = maximum ground level concentration with respect to distance from the point source at "critical" wind speed for level terrain. This value shall not exceed 50 $\mu g/m^3$

Pt_f = pounds of particulate matter emitted per million Btu heat input

Q_m = total plant-operating capacity rating in million Btu heat input per hour

n = number of stacks in fuel-burning operation

a = plume rise factor. The value of 0.67 shall be used for fuel-burning equipment ratings of less than 1000 million Btu/hr. No value greater than 0.8 for larger equipment ratings shall be used

h_s = stack height in feet. If a number of stacks of different heights exist, the average stack height to represent n stacks shall be calculated by weighing each stack height with its particular matter emission rate

A further restriction cites that the particulate matter discharged from all existing boilers shall in no case be greater than 0.8 lb/million Btu of heat input. Thus, for an industrial coal-fired boiler rated at 200 million Btu/hr with a single stack 80 ft in height, the allowable flyash emissions would be computed as follows:

$$Pt_f = \frac{50 \times 0.67 \times 80}{70 \times 200^{0.75} \times 1^{0.25}} = 0.72 \text{ lb}/10^6 \text{ Btu}$$

$$\text{Allowable emissions} = \frac{0.72}{10^6} \times 200 \times 10^6 \text{ Btu/hr} = 144 \text{ lb/hr}$$

The expected flyash emissions for a plant of this capacity would be very approximately 900 lb/hr, so that a collection efficiency of $100(900 - 144)/900$, or 84%, would be required. This is not a particularly difficult performance level to achieve, the allowable emissions being equivalent to approximately 0.19 grain/acf. The type of control equipment required for this service and the pertinent process parameters necessary for its design are explained in Chapter 5.

Should a Ringelmann 2 rating be the controlling regulation for this particular emissions problem, then an approximate value for the revised discharge loadings would be somewhere in the range of 0.05 to 0.15 grain/acf, so that it would be necessary to upgrade the control equipment performance level.

2.4. LIQUID POLLUTANTS

Occurrence

The most common source of liquid pollutant types is that segment of the manufacturing industries concerned with wet processing. Some typical examples are acid mists and carryover from plating operations and spray droplets discharged from a wet process phosphoric acid reactor. Combustion processes and drying operations also emit noxious gases and particulates, which combine with water vapor in the air to form toxic and/or odorous aerosols. In the combustion of residual fuel oils, the sulfur oxides are absorbed in atmospheric moisture to produce sulfurous and sulfuric acids. Nitric, hydrochloric, hydrofluoric, and fluosilicic acids are similarly formed from the corresponding inorganic gases. Some of the organic acids discharged into the atmosphere by the chemical process industries are acetic, propionic, and butyric. To qualify as a pollutant, these various liquid emissions must be of a harmful nature. A cooling tower may discharge sufficient mist under very humid atmospheric conditions, causing a fallout problem in the plant vicinity and thus being considered as a nuisance source. On the other hand, a highly concentrated NO_2 gas stream emitted under similar atmospheric conditions would cause precipitation of acid droplets in the area and therefore be definitely classified as an air pollution source.

One important potential source of liquid pollutants is the use of wet scrubbers for the reduction of particulate or gaseous emissions. Water-scrubbing systems removing pollutants from combustion processes usually generate a supersaturated water vapor. In the presence of residual particulates, which act as nuclei, the size and mass of water vapor droplets may be sufficient to cause their precipitation in the area. In scrubbing SO_x contaminants the effluent saturated gas stream at relatively low temperatures has a reduced buoyancy so that effective dispersion of the residual gaseous pollutants is diminished. This problem is usually solved by preheating the saturated gas above its dewpoint before releasing it to the atmosphere. An alternate solution involves cooling the scrubbing liquor so that the saturation temperature and humidity of the discharged gases can be reduced. This method, when combined with preheating, provides the most effective solution. In all wet scrubber designs a deentrainment section is provided to remove liquid droplets formed in the contacting zone. These spray droplets have a size range of 10 to 1000 μ, with the majority of them being greater than 50 μ in diameter. Such droplets are nearly saturated with the pollutant being scrubbed so that their discharge to the atmosphere would present a serious air pollution problem. However, these particles are comparatively easy to remove from the entraining gas streams by centrifugal or impingement type separators.

Liquid Particle Size

Liquid droplets 50 μ and greater in size do not present a serious pollution problem because of the ease with which they can be separated from a gas stream. Such droplets follow Stokes law, and their size distribution can be determined accordingly.

The mists produced by the condensation of acid particles are usually very fine in size, and their removal from a gas stream is difficult. Unlike solid particles, aqueous mists do not remain stable but are influenced by changes in temperature and humidity of the carrier gas stream. As stated previously, the size distribution of aerosols is a measure of the difficulty involved in their removal from gas streams. Acid mists from almost identical processes are variable in their particle size analysis. Table 2-8 gives a particle size distribution comparison for sulfuric acid mist being discharged from similar processes. It is believed that some agglomeration of the discrete particles is partly responsible for the larger size distribution. Sampling and the determination of mist size analysis is an exacting technique involving microscopic particle counts. Typical size analysis of various acid mists are shown in Table 2-9. The fineness of acid mists can also be affected by the presence of particulates. Thus, in indirect-fired

Table 2-8. Sulfuric Acid Mist Particle Size Comparison

Particle Size, μ	Weight % Less than Size	
	Process 1	Process 2
26	–	100
13	100	80
6.6	87	24
3.3	58	1
1.6	24	–
0.8	6	–

Source: G. Nonhebel, ed., Ref. 15.

Table 2-9. Typical Size Analysis for Various Acid Mists

	Weight % Less than size							
Particle Size, μ	*H_2SO_4 in Concentrator Tail Gas*	*H_2SO_4 in Acid Stripper Tail Gas*	*H_2SO_4 in Calciner Tail Gas*	*H_2SO_4 in Anhydrite Plant Tail Gas*	*Shock-cooled HCl Plant A*	*Shock-cooled HCl Plant B*	*Humidified P_2O_5 Fumes*	*Formic Acid in Vacuum Still Tail Gas*
19	–	–	–	–	100	–	–	–
13	–	–	–	–	79	100	–	–
9.5	–	–	–	–	47	91	–	–
6.5	100	–	100	–	24	74	100	100
4.7	90	–	78	–	13	49	98	77
3.3	69	100	25	100	6	23	57	38
2.3	60	72	5	70	2	7	25	18
1.6	32	30	–	29	1	2	8	9
1.1	18	6	–	6	–	–	2	4
0.8	9	1	–	1	–	–	–	1

Source: G. Nonhebel, ed., Ref. 15.

evaporator operation the phosphoric acid mist is more coarse than for the direct-fired vessels. In the latter, solid aerosols from the products of combustion provide nuclei that encourage the formation of very fine stable particles. This experience has also been noted for sulfuric acid concentration processes.

Large aerosol particles are easier to collect than smaller ones. Therefore, it is customary to encourage mist particle growth before subjecting the polluted gas stream to control equipment. Mechanical devices such as fans or ejectors have been used to promote collision between HCl mist particles. In one application the mist particles in an effluent gas passing through an air ejector were increased from an average size of 7 to 50 μ. Other acid mists are more resistant to particle agglomeration by mechanical methods unless large power inputs are provided.

Acid Mist Hazards and Plume Formation

Sulfuric acid mist formed by the reaction of sulfur dioxide with oxygen and water vapor in the atmosphere is the only liquid aerosol that has been investigated for its toxicological effects on man [16]. Although both SO_2 and acid mist seldom attain concentrations in ambient air above a few parts per million, both pollutants under extraordinary source and meteorological conditions are considered the prime causes for a number of air pollution disasters. Sulfuric acid mist can penetrate deeply into the lungs and attack sensitive tissues. In addition, droplets of sulfuric acid can carry absorbed sulfur dioxide farther into the respiratory system than the free gas could penetrate by itself, thereby increasing the effective toxicity level. However, in view of particle sizes of the sulfuric acid mist, previously discussed, its settling velocity in accordance with Stoke's law would not allow it to persist in ambient air for more than a few days.

Although there is a need for more quantitative data on the effects of the oxides of sulfur on materials and property damage, preliminary results indicate that most of the damage is due to the more highly reactive sulfuric acid produced in the atmosphere from the gaseous sulfur oxides. Simultaneous measurements of sulfuric acid mist and sulfur dioxide in the United States have indicated average ambient air concentrations of 43 and 1220 $\mu g/m^3$, respectively, equivalent to 0.015 and 0.43 ppm. [16] In laboratory tests it was found that steel test panels corroded at an increased rate under the synergistic influence of an atmosphere containing SO_2, high humidity, and particulates. Field studies indicated that the corrosion rates of such metals as iron, steel, and zinc were accelerated in industrial areas where air pollution levels, including sulfur dioxide, were higher. In assessing the damage to plants and vegetation, it was found that the mechanism causing serious injury involved the plants' ability to convert atmospheric sulfur dioxide to sulfuric acid and then to sulfates, which cause damage to the cells. From various studies it has been tentatively concluded that damage will not be inflicted if the maximum sulfur oxides concentration for the year is maintained at less than 0.3 ppm.

Sulfur dioxide emissions and the attendant acid mist formation do create visible and persistent plumes. The visibility of a plume is a function of its light-scattering effect caused by reflection from the projected area of the suspended particles in the atmosphere. As the particle sizes fall below 20 μ, the surface area is increased greatly. For example, at equal atmospheric concentrations, an acid mist comprised of 1 μ particles would have roughly ten times the projected area of one containing 10 μ particles, so that the finer mist would produce the more objectionable plume. Therefore, the obscuring power of a fine mist is out of proportion to its concentration in the atmosphere. Wet scrubbers for the abatement of acid mists preferentially remove the larger particles. In an application where 98% of the mist is thus removed, it is conceivable that the scrubber effluent, containing 2% of the mist with all the particles in the submicron size range, could produce only a very slightly improved visual plume. This situation is also relevant to the control of particulates where high removal efficiencies are no assurance of an improvement in the visual appearance of the plume. A Ringelmann assessment of emissions under these conditions can be misleading.

2.5. GASEOUS POLLUTANTS

Classification and Toxicity

Gaseous pollutants were categorized earlier as organic and inorganic. The organic gases can be further broken down into true gases and vapors. Thus, ethane, having a boiling point of $-89°$ C, is classified as a true gas, whereas toluene with a boiling point of $111°$ C and therefore existing as a liquid at ambient conditions, is considered as a vapor. All the inorganic gases encountered in air pollution activities are true gases.

The sources and types of the most common gaseous pollutants are shown in Table 2-2. Toxicity effects and odor problems are mainly caused by gaseous contaminants. A summary of the results of odor surveys for the various industries is given in Table 2-6. Some of the specific odor-producing gases with corresponding odor thresholds are shown in Table 2-10. The assessment of odors involves a very subjective technique, relying on the human "nose" for the determination of threshold values. This level of an odor-producing substance is defined as that concentration of the pollutant in ambient air which is just detectable by the olfactory senses. One of the most common sources of unpleasant odors is that of automotive diesel engines. Incomplete combustion of diesel fuels yields the irritating and sickening aldehydes, with formaldehyde as the major ingredient. In general the higher states of oxidation for odorous organic compounds are the most offensive. Thus, as butanol is oxidized progressively to butyraldehyde and butyric acid, the corresponding odors are characterized as mild, bad, and very bad. Except for ammonia and hydrogen sulfide, the inorganic gases do not contribute too seriously to the odor problem. Hydrogen sulfide

Table 2-10. Odor Thresholds of Various Gases and Vapors

Compound	Odor Threshold, Vol ppm	Odor Description
Acetaldehyde	0.21	green sweet
Acetone	100.0	chemical sweet, pungent
Acrolein	0.21	burnt sweet, pungent
Amine, Dimethyl	0.047	fishy
Amine, Trimethyl	0.00021	fishy, pungent
Ammonia	46.8	pungent
Aniline	1.0	pungent
Benzene	4.68	solvent
Benzyl Sulfide	0.0021	sulfidy
Bromine	0.047	bleach, pungent
Butyric Acid	0.001	sour
Carbon Disulfide	0.21	vegetable sulfidy
Chlorine	0.314	bleach, pungent
Dimethyl Sulfide	0.001	vegetable sulfidy
Diphenyl Sulfide	0.0047	burnt rubber
Ethanol	10.0	sweet
Ethyl Mercaptan	0.001	earthy, sulfidy
Formaldehyde	1.0	hay/straw-like, pungent
Hydrogen Sulfide	0.00047	eggy sulfide
Methanol	100.0	sweet
Methyl Ethyl Ketone	10.0	sweet
Methyl Mercaptan	0.0021	sulfidy, pungent
Nitrobenzene	0.0047	shoe polish, pungent
Paraxylene	0.47	sweet
Perchloroethylene	4.68	chlorinated, solventy
Phenol	0.047	medicinal
Phosgene	1.0	haylike
Pyridine	0.021	burnt, pungent, diamine
Styrene	0.1	solventy, rubbery
Sulfur Dioxide	0.47	pungent
Toluene	4.68	floral, pungent, solventy
Trichloroethylene	21.4	solventy

Source: Adopted from G. Leonardos et al, Ref. 6.

and the mercaptans are objectionable because of their pungent and unpleasant odor even at the very low concentrations of 0.10 to 0.15 ppm. Some mercaptans are detectable at 0.05 parts per billion. In addition to the odor nuisance, hydrogen sulfide is responsible for tarnishing silver and copper and causing lead-based paints to blister.

The toxicity effect and property damage characteristics of sulfur oxides were discussed in Section 2.4, where they were introduced because of the atmospheric conversion of this gas to sulfuric acid mist, which is the active agent responsible for most environmental damage. The dangerous combination of the nitrogen oxides and hydrocarbon gases and vapors to produce peroxyacetyl nitrate, the major constituent of smog, has been previously explained. A by-product of this reaction is ozone, which is considered responsible for tobacco and grape crop damage. High levels of ozone in ambient air, in excess of 1 ppm, cause deterioration of rubber products and adversely affect fabric colors. An expression of the permissible upper limits for three of the most common gaseous pollutants is contained in the alert system for toxic air pollutants established by the Los Angeles County Air Pollution Control District [17]. For carbon monoxide, the first, second, and third alert values are 100, 200, and 300 (expressed as ppm of the pollutant in ambient air); for nitrogen oxides, 3, 5, and 10; for sulfur oxides, 3, 5, and 10; and for ozone, 0.5, 1.0, and 1.5. The significance of the three alert stages is as follows:

First Alert: Close approach to maximum allowable concentration for the population at large. Still safe, but approaching a point where preventive action is required.
Second Alert: Air contamination level at which a health menace exists in a preliminary state.
Third Alert: Air contamination level at which a dangerous health menance exists.

Recommended preventive actions corresponding to these three situations are also described in the regulations. Because three of the four gaseous pollutants specified mostly represent automotive emissions, actions to be taken under the first alert include the rights "to request the public to stop all unessential use of vehicles in the L.A. Basin and to operate all privately owned vehicles on a pool basis." At the third alert, actions recommended in the first two stages are to be enforced and "the Air Pollution Control Board shall request the Governor to declare the existence of a state of emergency." One typical set of data for the urban center of Los Angeles, representing an average of the four highest daily maxima for the summer months of 1967 and 1968, were 28 ppm for carbon monoxide, 0.78 ppm for nitrogen oxides, 0.11 ppm for sulfur oxides, and 0.41 ppm for ozone.

Airborne gaseous fluorides damage certain plants and are harmful to domestic animals that have eaten contaminated crops. The concentration of gaseous fluorides in the ambient air is usually too low to cause health damage to humans. Both hydrogen fluoride (HF) and silicon tetrafluoride (SiF_4) will affect plant life over a wide range of concentration. For example, gladiolus will suffer leaf tissue damage at concentrations as low as 0.0001 ppm with continued exposure for five weeks. More resistant plant species are unaffected at concentrations up to 500 ppm. Some plants have the ability to concentrate fluorides, which makes them very susceptible to damaging effects even at low concentrations.

Organic vapors and gases per se do not pose a serious air pollution problem. Some petroleum off-gases and solvent cleaning operations do create an odor nuisance problem. In certain geographical areas, gaseous organic pollutants contribute to the smog problem through their photochemical reaction with the nitrogen oxides. Some laboratory investigations have tentatively identified airborne aromatic hydrocarbons adsorbed on soot as carcinogenic agents.

Dispersion and the Regulations

In some instances gas dispersion in the atmosphere is more of a problem than an assist to the maintenance of air quality. The emission of hydrogen sulfide and dimethyl sulfide from kraft paper mill operations is detectable at distances up to 10 miles from the mill, depending on the climatological conditions. The ease with which these malodorous gaseous compounds can be airborne, plus the deficiency of practical control technology, makes their abatement one of the prime targets in the pursuit of environmental control. However, this particular example represents the evils associated with gaseous dispersion and pertains only to highly odoriferous compounds.

Because most of the public's concern has been directed toward the air pollution potential of the sulfur oxides, considerable strides have been undertaken to determine the dispersion characteristics of these compounds. Numerous investigations have been made of plume rise and dispersion from power plants as measured by sulfur dioxide ambient air concentrations. In one such study [18], air monitoring data were used at one of the TVA plants to evaluate the applicability of the Bosanquet and Pearson dispersion formulas (Eq. 2-3) in predicting ground-level concentrations of sulfur dioxide. It was found that the Bosanquet and Pearson model predicted measured conditions 90% of the time, whereas a comparative dispersion model developed by the U.S.

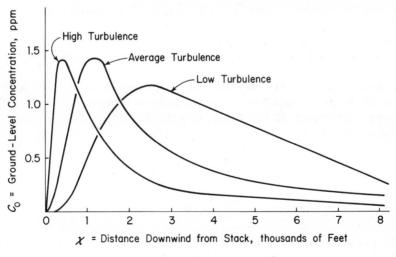

Figure 2–7. Typical gaseous dispersion pattern at varying climatic controls. (Adapted from *Air Pollution Abatement Manual,* Ref. 11.)

Public Health Service (USPHS) and TVA personnel yielded accurate predictions 94% of the time.

A curve representing typical gaseous dispersion patterns, calculated for different climatic conditions by the Bosanquet and Pearson equations, is shown in Fig. 2-7. By substituting Q for M in cubic feet per second in Eq. 2-3, the value of C_0 can be calculated as parts per million.

The use of the various diffusion equations to relate gaseous ground-level concentrations to stack emissions is complicated because of the complexity of the various meteorological parameters involved in their solution. As we have seen in the case of particulates dispersion, many of the control regulations presented a simplified diffusion expression based on plant-operating parameters. A similar approach has been adopted for gaseous emissions; in the same Midwest air quality region [14], the following gaseous diffusion equation was developed:

$$C_{max} = \frac{90 \times S_f Q_m^{0.75} n^{0.25}}{ah_s} \tag{2-11}$$

where

C_{max} = maximum ground-level concentration with respect to distance and at "critical" wind speed for level terrain, in micrograms per cubic meter, resulting from the point source. This value shall not exceed 200. Lower values may be selected where terrain and other conditions dictate

S_f = pounds of sulfur dioxide emitted per million Btu of heat input value of the fuel

Q_m = total equipment capacity rating, fuel heat input in millions of Btu per hour

n = number of stacks or chimneys in fuel-burning or process operations

a = plume rise factors; 0.67

h_s = stack height in feet. Same conditions as in Eq. 2-10.

A further restriction limits maximum SO_2 emissions to 6.0 lb/10^6 Btu. Assuming the same coal-burning boiler considered for particulates emissions, rated at 200 million Btu/hr with a single stack 80 ft in height, then the allowable sulfur dioxide emissions would be

$$S_f = \frac{200 \times 0.67 \times 80}{90 \times 200^{0.75} \times 1^{0.25}} = 2.25 \text{ lb } SO_2/10^6 \text{ Btu}$$

$$\text{Allowable emissions} = \frac{2.25}{10^6} \times 200 \times 10^6 \text{ Btu/hr} = 450 \text{ lb/hr}$$

The flue gas flow from a plant of this rated capacity would be approximately 60,000 scfm, so that the permissible SO_2 concentration at the stack discharge would be 750 ppm. This is a relatively generous allowance; most of the codes require emissions of about 200 to 300 ppm SO_2. If this plant were burning coal with a sulfur content of 3% then the expected uncontrolled SO_2 concentration would be about 2100 ppm. Therefore, a proposed wet scrubber system would have to provide an absorption efficiency of 64%. Details of the selection and design of this equipment are explained in Chapter 6.

REFERENCES

1. "Control Techniques for Particulate Air Pollutants." U.S. Dept. of Health, Education, and Welfare, Public Health Service, Washington, D.C., January 1969.
2. J. A. Danielson, ed., *Air Pollution Engineering Manual*, 2d ed., Air Pollution Control District, County of Los Angeles. Environmental Protection Agency, Research Triangle Park, N.C. May 1973.
3. P. Drinker and T. Hatch, *Industrial Dust*, 2d ed. McGraw-Hill Book Co., Inc., New York, 1954.
4. "Source Sampling and Analysis—Air Pollution Training," U.S. Dept. of Health, Education, and Welfare, Public Health Service, Cincinnati, Ohio, April 1958.
5. "Air Quality Criteria for Particulate Matter," U.S. Dept. of Health, Education, and Welfare, National Air Pollution Control Administration, Washington, D.C., January 1969.
6. G. Leonardos, D. Kendall, N. Barnard, "Odor Threshold Determinations of 53 Odorant Chemicals," *J-APCA*, Vol. 19 No 7, 91—95, February 1969.
7. W. H. J. Vernon, "A Laboratory Study of the Atmospheric Corrosion of Metals," *Trans-Faraday Soc.*, Vol. 31 (1935), pp. 1668—1700.
8. J. D. Hudson, "Present Position of the Corrosion Committees Field Tests on Atmospheric Corrosion," *J. Iron Steel Inst.*, Vol. 148 (1943), pp. 161—215.
9. C. H. Bosanquet and J. L. Pearson, "The Spread of Smoke and Gases from Chimneys," *Trans-Faraday Soc.*, Vol. 32 (1936), p. 1249.
10. O. G. Sutton, "The Theoretical Distribution of Airborne Pollution from Factory Chimneys," *Quart. J. Roy. Meteor. Soc.*, Vol. 73 (1947), pp. 426—436.
11. *Air Pollution Abatement Manual*, Manufacturing Chemists Assoc. Inc., Washington, D.C., chap. 8, 1953, pp. 9—21.
12. O. H. Bosanquet, W. F. Carey, and E. M. Halton, "Dust Deposition from Chimney Stacks," *Inst. Mech. Engrs.*, London, Vol. 162 (1950), p. 355.
13. C. H. Bosanquet, *J. Inst. Fuel*, Vol. 30 (1957), p. 322.
14. State of Indiana Air Pollution Control Board Regulation APC 4-R, Section 1, June 1966; Regulation APC-13, Section 1, July 1972.
15. G. Nonhebel, ed., *Gas Purification Processes for Air Pollution Control*, 2d ed. Butterworth and Co. Ltd., London, 1972, chap. 15.
16. "Air Quality Criteria for Sulfur Oxides," U.S. Dept. of Health, Education, and Welfare, National Air Pollution Control Administration, Washington, D.C., January 1969.
17. "Rules and Regulations," Los Angeles County Air Pollution Control District, December 1969.
18. T.L. Montgomery and M. Corn, "Adherence of Sulfur Dioxide Concentrations in the Vicinity of a Steam Plant to Plume Dispersion Models," *J-APCA*, Vol. 17 (1967), p. 512.
19. D.B. Turner "Workshop of Atmospheric Dispersion Estimates," U.S. Dept. of Health, Education, and Welfare, Public Health Service Rev., 1969.

Chapter 3

AIR POLLUTION CONTROL REGULATIONS

3.1. INTRODUCTION

History and the Air Pollution Problem

The recognition of air pollution and efforts to control it are not restricted to contemporary times. The use of bituminous coals in England in the fourteenth century prompted some severe smoke control regulations. One suggested solution for the abatement of smoke emissions, made in the seventeenth century, was to relocate all industries downwind of the city. Although this early proposal recognized the role of meteorology in air quality maintenance and strove to take advantage of the prevailing climatic situation, it unfortunately ignored the rights of the neighboring downwind town to its share of clean air.

It is an established fact that the condition of our atmosphere is deteriorating. The progressive worsening of the quality of the air we breathe has been a direct function of the industrial development of every country in the world. Technological growth in the United States assures us of a continuously aggravated atmospheric pollution level unless effective control programs are instituted. Figure 3-1 shows a history of carbon monoxide emissions in tons per day in Los Angeles County for the period 1940 through 1970. The potential fourfold increase in emission values for this single pollutant during the period 1945 to 1970 illustrates the need for regulations. The problem is how to institute adequate regulations to protect human health and property within some reasonable technological and economic framework.

Unfortunately, the desecration of man's environment has had a considerable head start, being seriously initiated with the advent of the Industrial Revolution in the mid-eighteenth century. Real and serious nationwide efforts toward air and water quality improvement have been in effect only since early 1960. Because the purchase of emissions control equipment brings almost a zero rate of return on the investment requirement, and since very modest pressures were exerted on industries to control their emissions prior to 1960, it follows that very limited efforts were expended by control equipment manufacturers to improve the economies and performance level of their wares. In addition, the current activity in identifying an almost infinite compendium of air pollutants, particularly of the gaseous type, has strongly emphasized the impoverished state of control technology. For example, although guides [2] for the control of sulfur oxides were issued by the federal government in January 1969, as yet a commercially acceptable control system has not been developed for the abatement of this pollutant from power plant sources.

42

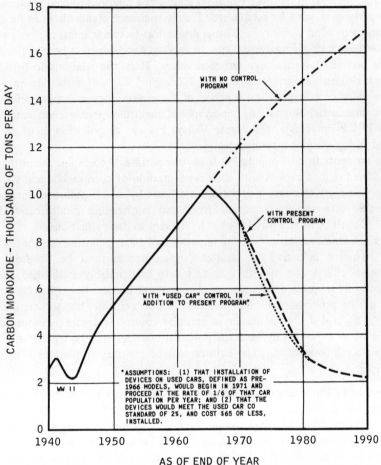

CARBON MONOXIDE
Emissions from Motor Vehicles

Figure 3–1. History of carbon monoxide emissions in Los Angeles County. (Courtesy of Los Angeles County Air Pollution District, Ref. 1.)

Bases for Air Pollution Control Regulations

Three of the major elements to be considered in the development of an effective air pollution control program are discussed below.

   1. Definition of an Air Pollution Condition.   Because the protection of human health is the major objective of air pollution control, attempts must be made to define the type and permissible quantity of a specific pollutant in ambient air as they relate to the comfort and well-being of a total population. This is a complex undertaking, involving the establishment of sufficient physiological experiences to justify a statement of the permissible pollutant levels. Information regarding the health effects of air pollution is presently very meager.

Similar definitions of those emissions considered injurious to plants, animals, and real property must also be established. Finally, nuisance standards must be defined with respect to odor, visibility, and other discomfort factors inimical to man's welfare.

2. Influence of Geography. Once air quality standards have been defined, they must be converted into source emission values. Thus, the relationship between an allowable ambient air concentration of 200 $\mu g/m^3$ for sulfur dioxide to a stack emission value of 600 ppm is a function of the meteorology and topography of the area. Unusual meteorological and topographical conditions greatly contributed to the Donora [3], Pennsylvania, and Meuse Valley, France, air pollution disasters, which occurred within the past three decades.

The concentration of population is another pertinent geographic factor that must be considered. Urban centers with their concentration of both mobile and stationary emission sources and dense populations are beset by an air pollution environment considerably different in magnitude from that confronting rural areas. Effective regulations must be sufficiently broad to be relevant to these situations.

3. Economic Feasibility. In the imposition of an air pollution control regulation on an offending industry, the economic consequences must be weighed. If the toxicological effects of a specific pollutant have been reliably established, then the source must be suppressed, regardless of the economic burden. If, on the other hand, a firm basis for prohibition of the pollutant is not available, but the demise of an industry is assured if the installation of emission control equipment is enforced, then efforts must be expended by both the potential offender and the regulation body to determine a middle position. The enforcement of arbitrary regulations must not be permitted.

There is a definite and crying need for environmental improvement. However, the damage to our atmosphere, which has been committed over the past 200 years as the price for an expanding economy, cannot be undone tomorrow. Control regulations are necessary, but they must be based on rational grounds and not be dictated by emotionalism. Although air pollution is a problem, it is not presently a serious or critical menace to public health. Industry must certainly cooperate in the overall effort to clean the air. It can no longer rely on its economic worth to the community as a rationale for polluting the environment.

3.2. AIR POLLUTION CONTROL LEGISLATION

In the consideration of legislation requirements to control air pollution, the responsibility for enforcement lies with governmental agencies whereas proof and responsibility for compliance is the obligation of the individual or company involved. Laws passed should combat air pollution in an efficient and reasonable manner so that they do not unnecessarily restrict any single segment of industry. Restriction in this sense refers to economic pressure. Engineering can provide solutions to many industrial emission problems, but the cost in many instances can be prohibitive to the companies involved and ultimately to the community that relies on the services being provided.

It can be concluded from prior discussions that a specific air pollution condition is very much a localized problem. Therefore, the concentration limits established for each air contaminant at its source cannot be applied uniformly to all areas. Before any community adopts an ideal set of control ordinances, each section must be analyzed to determine its validity for local conditions. Thus, fuel specifications in the Mid-west United States would hardly be applicable to Canadian power plants which generate mostly hydroelectric power. In the majority of European countries, clean air legislation is applicable to the entire country whereas in the United States, although the prescribed air quality standards are essentially identical across the nation, the emission controls vary regionally.

The technologically advanced countries of the world have established air pollution control regulations. What follows is a description of the overall air pollution situation in some of these countries.

Great Britain. [4] Control of atmosphere pollutions in England are based on two sets of legislation. The first demands reduction of "noxious or offensive gases" from 59 different types of industrial operations. These must be registered under the Alkali etc. Works Regulations Act of 1906. Revised in 1966 and again in 1971, these regulations require:

1. Scheduled Processes involving the production or utilization of sulfuric acid, chlorine, arsenic, zinc, fluorine, aluminum, copper, electricity, etc. must be registered annually.
2. Prior to initial registration the Process must be equipped with "best practicable means" for preventing the escape of noxious or offensive gases.
3. The "best practicable means" must be continuously operated and maintained in good and efficient working order.
4. In the case of certain processes, upper emission limits are specified for the concentration of total acidity in effluent gases, i.e., HCl gas, 0.2 grain/ft^3, SO$_3$ gas from Chamber Process, 4.0 grains/ft.^{3}; SO$_3$ gas from Contact Process, 1.5 grains/ft^3.

Emissions from processes or plants not covered by the Alkali Acts and those from domestic and commercial furnaces are controlled by the second set of legislation, the Clean Air Acts of 1956 and 1968. These acts essentially are:

1. Set smoke emission limits for both domestic and industrial fuel usage.
2. Limit grit and dust emissions for all non–domestic sized furnaces.
3. Specify chimney heights at two and one half times that of nearest buildings or a minimum of 120 ft.
4. Arrange to set up Smoke Control areas subject to approval of Housing Ministry.

The application of the provisions of the Clean Air Acts is largely the responsibility of local authorities. If, in their opinion, a problem does not exist, then there is no need to take action. As of the end of December 1970, there were approximately one million acres of land, encompassing five million dwellings, under the jurisdiction of smoke control orders. As a result of the marked improvement in the ambient air quality during the past ten years, the citizens have cooperated whole-heartedly in the control plan. In some of the recently established smoke areas over fifty percent of the populations had achieved voluntary compliance before a survey was made. As a further incentive for citizen cooperation, local home-owners may obtain a grant for conversion of their heating facilities to a reduced-emissions system. The cost of conversion is divided into 30% by the owner, 30% by the local control authority and 40% by the Exchequer.

As a result of the regulations and the social consciousness of industry and the populace, the particulate emissions have been reduced substantially. For example power stations have experienced a 50% increase in the use of coal but a 500% decrease in the discharge of dust. Cement, iron and steel works, potteries and railroads have effected similar reductions in particulate emissions. The total sulfur dioxide emissions trends have not enjoyed a similar decline and it is predicted that a slight increase might be suffered.

The London Smog of 1952 has prompted the local authorities to make daily observations of smoke and sulfur dioxide as well as monthly measurements of dust fall. As of March 1970 there were about 1200 ambient air monitoring stations throughout Great Britain. Data from these stations are collated, interpreted and published as the National Survey of Air Pollution.

No effective legislation exists in England for the control of motor vehicle exhaust pollutants. The Motor Vehicles Regulations of 1966 require that every vehicle eliminate the discharge of avoidable or visible vapor. More recent regulations, effective in 1974, would reduce carbon monoxide emissions up to 30% and hydrocarbons up to 10% by carburetor and timing adjustment.

France. [4] France has a ministry of Environment which was organized in 1970. It performs similarly to EPA, having divisions for air, water, noise and solid wastes. Although the national environmental problems have been assessed, specific laws or implementations plans are still non-existent. A review, on a local level, of any proposed manufacturing process is mandatory where potential toxic or dangerous emissions are identified. In this review the proposed emissions control equipment is evaluated and either approved or disapproved by the review board.

Because of the continuous prevailing winds, the French view air pollution as a multiple specific source problem rather than an overall air quality situation. There are 318 air monitoring stations in Paris alone to detect and evaluate point source emissions. The basic measurements are for carbon monoxide and sulfur dioxide. Considerable efforts are being expended to develop a continuous monitoring instrument for sulfur dioxide as a replacement for the present detection devices which are based on solution acidity measurement. Presently about 50% of the electricity demand in France is being met by fossil fuel plants and their sulfur oxide emissions are very closely monitored.

The majority of air sampling stations in Paris are located near schools and other public installations. Efforts to locate and identify individual industries responsible for a local air pollution problem are supported by a mobile air sampling van operated by the Prefecture de Police Laboratory. Any necessary enforcement actions are considered a normal police action under the control of the Prefecture.

46

Motor vehicle emissions and safety are continuously evaluated by the Technical Association of Automobile, Motorcycle and Cycle Industries. Certification tests and motor vehicle regulations for French and European cars are in effect to maintain air quality despite the rapidly increasing automobile population in the major French cities.

Ringelmann, inventor of the smoke opacity index system that bears his name, was a French professor of agricultural engineering. The French have endorsed the use of his chart, limiting particulate emissions to smoke densities of Ringelmann 1 or less, except for about five percent of the source operating period.

Japan. [5]   The Japanese first proposed an ambient air quality standard for sulfur oxides in 1969 in order to protect human health. The new standard, equivalent to an annual mean hourly average of 40 $\mu g/m^3$ was to be attained within five years.

Emission standards were also set forth in the Enforcement Ordinances of the Air Pollution Control Law of 1972. Although the regulations are somewhat similar to those enacted in the United States, there are no penalty provisions. The meteorological/topographical factors in Japan favor rapid dispersion of air-borne pollutants so that the problem is local rather than national.

In view of the accelerating motor vehicle population in Japan, continuous monitoring systems have been set up in Tokyo, Osaka and other large cities for the

Table 3-1. Summary of Comparative Ambient Air Quality Standards for Various Countries

Country	Standards Enacted	Averaging Period	Pollutant Concentrations					
			Particulates $\mu g/m^3$	SO_x ppm	CO ppm	O_x ppm	HC ppm	NO_x ppm
United States	April, 1971	Annual	60					
		24 hr		0.10				0.05
		8 hr			9.0			
		3 hr					0.24	
		1 hr				0.08		
Canada	Jan, 1973	Annual	60					
		24 hr		0.07			—	—
		8 hr			5.0			
		1 hr				0.05		
Japan	May, 1973	24 hr	100	0.04			—	0.02
		8 hr			20.			
		1 hr				0.06		
U.S.S.R.	1969	24 hr	150	0.02	0.9	—	—	0.04
Sweden	Oct, 1970	24 hr	—	0.10	—	—	—	—

Source: S. Yanagisawa, *J-APCA;* Ref. 5.

detection of photochemical pollutants. The air quality standard for nitrogen oxides is the most stringent in the world, demanding an annual mean hourly value of 20 $\mu g/m^3$, which is to be attained during the period 1974-1977. This standard value represents half the present concentration of nitrogen dioxide in Tokyo. Since May, 1971, the oxidant concentration in Tokyo has easily exceeded the warning level of 300 $\mu g/m^3$ on one out of every three summer days. Both oxides of nitrogen and photochemical smog are considered to be Japan's most serious air pollution problem.

There are some advantages in establishing worldwide air quality standards. International standards would permit valid comparisons of air pollution levels among all countries and would provide a basis for assessment of the earth's pollutant dispersion mechanism. Table 3-1 summarizes comparable ambient air quality standards for a number of countries. This compilation would indicate that, despite differences in overall objectives and definitions, the near-agreement of the concentration values would confirm a common recognition of the tolerable health levels for the various pollutants.

3.3. UNITED STATES FEDERAL AIR POLLUTION CONTROL PROGRAM

History and Intent

Air pollution control legislation was initially enacted by the United States Congress in July 1955. Its major objective was to authorize a federal program of research in air pollution and provide technical assistance to state and local governments. In December 1963, the Clean Air Act [6] was legislated. Federal participation in environmental activities was mainly based on the needs of the nation's "rapidly expanding metropolitan and other urban areas, which generally cross the boundary lines of local jurisdictions and often extend into two or more states." Although the Act recognized that the prevention and control of air pollution at its source was the primary responsibility of state and local governments, it underscored the need for federal assistance toward the development of cooperative federal, state, regional, and local air pollution control programs. Specifically, the purposes of the Act were stated as follows:

1. To protect the nation's air resources so as to promote the public health, welfare, and productive capacity of the population.
2. To initiate and accelerate a national research and development program to achieve the prevention and control of air pollution.
3. To provide technical and financial assistance to state and local governments in connection with the development and execution of their air pollution prevention and control programs.
4. To encourage and assist the development and operation of regional air pollution control programs.

With the marked increase of motor vehicle pollution, the Act was amended in October 1965 to enable the federal government to establish automobile emission standards. Another amendment enacted at this time empowered the Secretary of Health, Education, and Welfare (HEW) to investigate and seek to prevent potential air pollution problems. Additional amendments in October 1966 authorized federal grants to the state and local agencies to assist them in maintaining effective air pollution programs. The procedures relating to federal government activities in conjunction with state and municipal agencies are illustrated in Fig. 3-2. The federal government's role in the various possible control actions emphasizes the major responsibilities of the states in developing an effective air pollution control program. In only one area is action initiated by HEW and that involves interstate pollution problems.

Air Quality Act

In November 1967, Congress enacted a further amendment to the Clean Air Act, to define specific implementation procedures. The Air Quality Act of 1967 served notice "that no one has the right to use the atmosphere as a garbage dump and that there will be no haven for polluters anywhere in this country" [7]. The new provisions of this Act included:

1. Establishment of atmospheric areas throughout the United States on the basis of climate, meteorology, and topography.
2. Designation of air quality control regions, based on jurisdictional boundaries and the concentration of industrial activities.
3. Development of air quality criteria that would serve as a guide for the definition of acceptable ambient air pollutant concentrations.

The federal budgeting plan for air pollution abatement activities maintained pace with the government's increasing involvement. From 1956 to 1963, annual appropriations for the federal program were gradually increased from less than $2 million to about $11 million. Monies made available under the first Clean Air Act in 1963 reached a peak of slightly more than $35 million. The 1966 amendments to the Act authorized expenditures of $46 million for the 1967 fiscal year, which ended June 30, 1967. Additional sums of $66 million and $74 million were promised at this time for the next two fiscal years.

The three major objectives of the Air Quality Act of 1967 were: to designate air quality control regions, issue air quality criteria, and issue reports on control techniques. From the air quality data, the state governments were then expected to establish air quality standards for the designated air quality control regions. These standards and implementation plans had to be submitted to HEW for review and approval. A flow diagram indicating the timing of the various HEW actions to control air pollution on a regional basis, under the Air Quality Act, is shown in Fig. 3-3.

Unlike the passive role illustrated in Fig. 3-2, based on 1965 legislation, the federal government has now become the initiator of air pollution control actions by defining the air quality criteria and requesting the states to respond. Specifically, the state governments are allowed a total period of 450 days to define their standards, as guided by the HEW criteria, and establish a detailed plan for implementing these standards.

Air Quality Criteria

Basically, there are two critical factors in the overall plan: air quality criteria and availability of control techniques. Air quality criteria for a number of particulate and

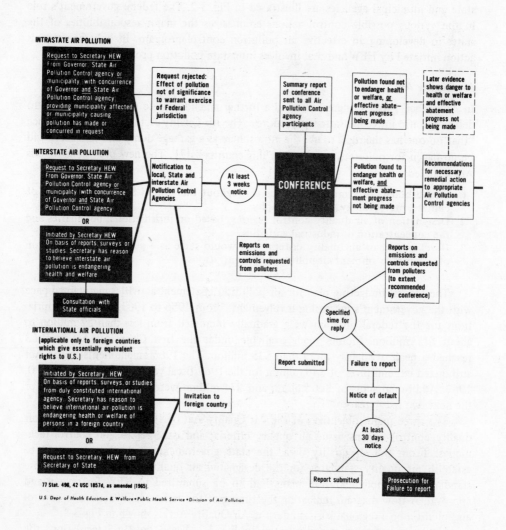

INTRASTATE AIR POLLUTION

Request to Secretary HEW From Governor, State Air Pollution Control agency or municipality (with concurrence of Governor and State Air Pollution Control agency providing municipality affected or municipality causing pollution has made or concurred in request

INTERSTATE AIR POLLUTION

Request to Secretary HEW From Governor, State Air Pollution Control agency or municipality (with concurrence of Governor and State Air Pollution Control agency)

OR

Initiated by Secretary HEW On basis of reports, surveys or studies, Secretary has reason to believe interstate air pollution is endangering health and welfare

Consultation with State officials

INTERNATIONAL AIR POLLUTION

(applicable only to foreign countries which give essentially equivalent rights to U.S.)

Initiated by Secretary, HEW On basis of reports, surveys, or studies from duly constituted international agency, Secretary has reason to believe international air pollution is endangering health or welfare of persons in a foreign country

OR

Request to Secretary, HEW from Secretary of State

77 Stat. 496, 42 USC 1857d, as amended (1965).

U.S. Dept. of Health Education & Welfare • Public Health Service • Division of Air Pollution

Request rejected: Effect of pollution not of significance to warrant exercise of Federal jurisdiction

Notification to local, State and interstate Air Pollution Control Agencies

Invitation to foreign country

At least 3 weeks notice

Summary report of conference sent to all Air Pollution Control agency participants

CONFERENCE

Pollution found not to endanger health or welfare, or effective abatement progress being made

Pollution found to endanger health or welfare, and effective abatement progress not being made

Later evidence shows danger to health or welfare and effective abatement progress not being made

Recommendations for necessary remedial action to appropriate Air Pollution Control agencies

Reports on emissions and controls requested from polluters

Reports on emissions and controls requested from polluters (to extent recommended by conference)

Specified time for reply

Report submitted

Failure to report

Notice of default

At least 30 days notice

Report submitted

Prosecution for Failure to report

Figure 3–2. Air pollution control procedures of the U.S. government. (U.S. Dept. of Health, Education, and Welfare.)

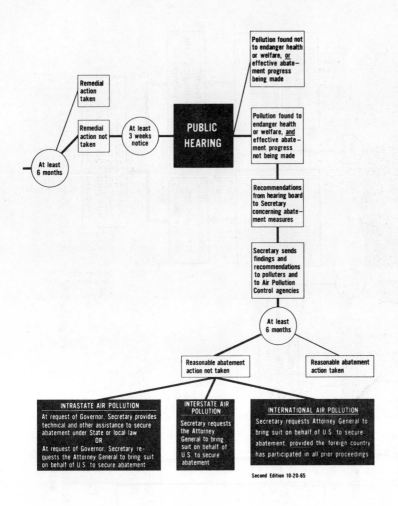

Second Edition 10-20-65

51

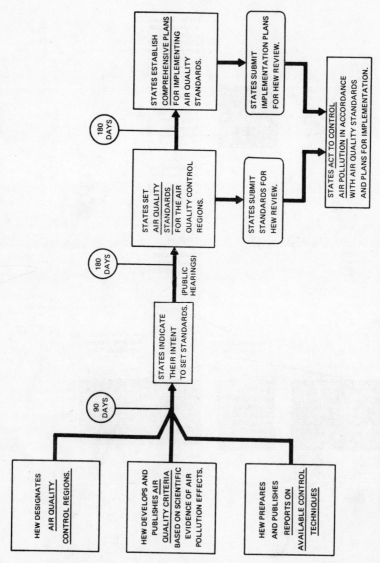

Figure 3–3. Actions for the control of air pollution by the U.S. Dept. of Health, Education, and Welfare.

gaseous types of emissions were determined and issued by HEW early in 1969. Included in these publications were the following criteria:

Air Quality Criteria for Particulate Matter, AP-49 [8], January 1969.
Air Quality Criteria for Sulfur Oxides, AP-50 [2], January 1969.
Air Quality Criteria for Photochemical Oxidants, AP-63, March 1970.
Air Quality Criteria for Hydrocarbons, AP-64, March 1970.
Air Quality Criteria for Nitrogen Oxides, AP-84, January 1971.

In each of these HEW reports an attempt has been made to measure the deleterious effects of air pollution on man and his environment. Thus, air quality criteria are established to determine to what point the pollution levels must be reduced to protect the public health and welfare. The effect of each pollutant type on visibility, vegetation, and materials is assessed. Most importantly the toxicological effects on man and animals are determined and the result of these findings are converted to allowable ambient air concentrations. This is an extremely nebulous area of investigation, complicated by consideration of pollutant mixtures such as particulates and sulfur oxides and by responses from special human groups that might be suffering from respiratory diseases. The air quality standards defined by these criteria are essentially uniform across the nation.

In defining air quality criteria for particulate matter, HEW concluded that there were insufficient laboratory data from which to develop suitable quantitative ambient air pollutant levels. However, in *Air Quality Criteria for Particulates Matter* [8] it is noted that:

1. Daily averages of smoke in the range of 300 to 400 $\mu g/m^3$ have been associated with acute worsening of chronic bronchitis patients in England.
2. Increased absences due to illness (type undefined) occurred among British workmen when smoke level exceeded 200 $\mu g/m^3$.
3. Two recent British studies showed increases in selected respiratory illness in children to be associated with annual mean smoke levels above 120 $\mu g/m^3$.
4. In studies made in Buffalo, New York, it was clearly shown that increased death rates from selected causes in males and females 50–69 years old occurred at annual geometric average ambient air particulate levels of 100 $\mu g/m^3$ and greater.

It would appear from these data that the maximum particulates level permitted in ambient air to protect human health should be set at about 100 $\mu g/m^3$. Based on further evaluation of the effects of particulates on visibility, odor, materials, and public awareness, it was concluded in the HEW criteria report "that ambient concentrations of particulates of 80 $\mu g/m^3$ (annual average), or more in the atmosphere may produce adverse health effects in particular segments of the population."

A similar "criteria" report [2] was issued by HEW for sulfur oxides. The conclusions comprised evaluations of multiple ambient concentrations for the various environmental factors:

1. Adverse health effects were noted when 24 hr average levels of sulfur dioxide exceeded 300 $\mu g/m^3$ (0.11 ppm) for 3 to 4 days.

53

2. Adverse health effects were also noted when the annual mean level of sulfur dioxide exceeded 115 $\mu g/m^3$ (0.04 ppm).
3. Visibility reduction to about 5 miles was observed at 285 $\mu g/m^3$ (0.10 ppm).
4. Adverse effects on materials were observed at an annual mean concentration of 345 $\mu g/m^3$ (0.12 ppm).
5. Adverse effects on vegetation were observed at an annual mean value of 85 $\mu g/m^3$ (0.03 ppm).

HEW then "reasonably" concluded that, when promulgating ambient air quality standards, consideration should be given to requirements for margins of safety that take into account long-term effects on health, vegetation, and materials occurring below the specified air quality levels. As a comment to these requirements, it should be noted that the allowable ambient air sulfur oxide concentration specified at 150 $\mu g/m^3$ for a 24 hr period is relatively severe. In the urban center of Los Angeles, the sulfur dioxide concentration in ambient air (representing an average of the four highest daily maximum values for the summer months of 1967 and 1968) was determined at 0.11 ppm, equivalent to 300 $\mu g/m^3$.

Air Pollutant Control Techniques

In keeping with the objectives of the Air Quality Act, air quality criteria for the various pollutants were followed by reports describing control techniques. These reports, issued by HEW, provided information on the availability and applicability of techniques for the prevention and control of air pollutants at their source and the cost and effectiveness of such techniques. Some of the more important control technique reports are as follows:

Control Techniques for Particulate Air Pollutants, AP-51 [9], January 1969.
Control Techniques for Sulfur Oxide Air Pollutants, AP-52 [10], January 1969.
Control Techniques for Nitrogen Oxide Emissions from Stationary Sources, AP-67 [11], March 1970.
Control Techniques for Carbon Monoxide Emissions from Stationary Sources, AP-65, March 1970.
Control Techniques for Hydrocarbon and Organic Solvent Emissions from Stationary Sources, AP-68, March 1970.

The purpose of these publications is to provide technical data to air pollution control agencies and process operators to permit them to make an informed and intelligent choice among the various techniques available for the abatement of air pollution from any given source. Unfortunately, as previously noted, proven control techniques for many of the more serious pollutant emission problems are commercially unavailable.

In the discussion on HEW's control techniques for particulates [9], the major emission sources are listed and the various control systems are described. Such process innovations as energy substitution, good operating practice, and dispersion are discussed. However, the major theme is a description of the commercial equipment available for particulates collection, including performance characteristics, operating limitations, physical description, capital costs, and operating costs. An expanded version of this aspect of pollution control abatement is treated in Chapter 5.

HEW's "Control Techniques for Sulfur Oxide Air Pollutants" [10] describes the sources of this pollutant and the available control techniques. Sulfur oxides, primarily SO_2, are produced from the combustion of any sulfur-bearing fuel and by many industrial processes that utilize sulfur-bearing raw materials. In 1969 about 33 million tons of SO_2 were discharged to the atmosphere in the United States. Of this amount, roughly 24 million tons were emitted by fossil-fuel stationary combustion sources, the majority of these being electrical utility operations. Sulfur removal from coal and oil is presently being explored, and the substitution of natural gas or naturally occurring low-sulfur coal and fuel oil is possible and has been considered. However, both solutions are not practical at the present time. Processes for sulfur removal from fossil fuels are barely commercially available and the substitutions of low-sulfur natural fuels on any reasonable scale is economically prohibitive.

The only near-commercial means for the abatement of sulfur oxides from power-plant operations is flue gas cleaning. At least 20 to 30 processes have been seriously investigated over the past 15 years for the control of power-plant sulfur oxide emissions. Of these, the most promising from the viewpoint of capital costs, operating charges, and performance has been a wet scrubbing system utilizing various alkaline liquors. To maintain a reasonable chemical makeup cost level and prevent contamination of ground waters from the discharged liquor, the use of calcium alkaline slurries has been favored. However, the carbonate precipitates in process pose a severe plugging problem in the scrubber and associated piping. This operating problem, plus the difficulties of liquor effluent disposal and plume dispersion, has decidedly classified this process as not being quite ready for troublefree commercial operation. At the present time there have been installed about 20 full-scale systems, almost all of which are demonstration units to be further evaluated. A number of the most promising scrubbing systems is being evaluated at a Tennessee Valley Authority (TVA) power-plant facility, with funds allocated by HEW.

The control techniques for nitrogen oxide emissions from stationary sources as discussed in the HEW publication of March 1970 [11] are mostly concerned with combustion control variations rather than with flue gas treatment. The unavailability of effective control systems is not too surprising, since the role of the nitrogen oxides in the formation of smog and the initial regulations for their control have been a consideration in the maintenance of air quality only since the early 1960s. The total nitrogen oxide emissions in the United States in 1969 was about 24 million tons, of which nearly 40% was contributed by motor vehicles. Stationary combustion sources, mostly comprising electric generation plants, were responsible for a little over 40% of the total. The recommended nitrogen oxide control techniques for stationary combustion sources in the HEW publication are fuel substitution, burner design modifications, stoichiometric variation, wet scrubbing, and catalytic reduction. Fuel substitution is based on the recognized high NO_x emissions with coal, intermediate values with oil, and low values with gas. Thus, the use of gas is favored over coal and oil, and—naturally—nuclear fuel is favored over fossil fuels.

Various types of burner designs favoring "cool" combustion conditions have been developed. The use of two-stage combustion in oil- and gas-fired boilers is actually the most acceptable control technique, reducing NO_x emissions by 30 to 50%. Rearrangement of burner nozzle patterns has also resulted in considerably reduced NO_x

emissions. The treatment of the flue gases by either wet scrubbing or catalytic reduction methods has thus far not been seriously considered for large-scale fossil-fuel combustion boilers, and operating data are nonexistent. The most promising technique is a combination of two-stage combustion with low excess air, resulting in reported nitrogen oxide reductions up to 90%. [12] However, these methods of control, even when applied to gas- and oil-fired boilers, seem to give varied and unpredictable results. Large pulverized-coal fired boilers cannot be controlled by combustion variations because of the difficulties in controlling CO and hydrocarbon emissions, and because of the attendant operational hazards.

At the present time, control methods for the more recently defined pollutants such as sulfur oxides and nitrogen oxides are not too adequately defined because of a deficiency in technology development. Today, equipment vendors are not readily guaranteeing control equipment performance to the large utilities for SO_x abatement systems. Nitrogen oxide control regulations are still in the formulation stage in many parts of the country, and stringent enforcement is not being exercised—probably because of the performance uncertainties associated with currently available control systems. Despite HEW's guidance in providing the process operators with control technique information, it is still the responsibility of those who pollute to select the control equipment and make it perform in accordance with the prevailing air pollution regulations.

Air Quality Control Regions

One of the first actions of HEW was to designate air quality control regions throughout the United States on the basis of meteorological and topographical factors, urban-industrial concentrations, jurisdictional boundaries, and other factors relevant to effective implementation of air quality standards. Thus, it was recognized that air pollution would not respect geographical boundaries but would be carried by air currents in the atmosphere as moved by natural forces. By July 1967, a number of air quality control regions had been established in several large metropolitan areas. A list of 57 major cities where additional regions would soon be defined was also published [7] at that time. By November 1971, Section 107 of the amended Clean Air Act had expanded to include 100 Interstate Air Quality Control Regions. A definition of the region comprising New Jersey, New York, and Connecticut, as stated in this amendment, is listed below.

INTERSTATE AIR QUALITY REGION FOR NEW YORK, NEW JERSEY, AND CONNECTICUT

The New York—New Jersey—Connecticut Interstate Air Quality Control Region has been revised by an Act of the Department of Health, Education, and Welfare to consist of the territorial areas encompassed by the boundaries of the following jurisdictions, including the territorial area of all municipalities as defined in Section 302(f) of the Clean Air Act [42 U.S.C. 1857h(f)] and geographically located within the outermost boundaries of the area so delineated.

In the State of New York, the counties of

Bronx	New York	Rockland
Kings	Queens	Suffolk
Nassau	Richmond	Westchester

In the state of New Jersey, the counties of

Bergen	Middlesex	Passaic
Essex	Monmouth	Somerset
Hudson	Morris	Union

In the State of Connecticut, the townships of

Bethel	Monroe	Stratford
Bridgeport	New Canaan	Trumbull
Brookfield	New Fairfield	Weston
Danbury	Newtown	Westport
Darien	Norwalk	Wilton
Easton	Redding	
Fairfield	Ridgefield	
Greenwich	Stamford	

Before each of the designated regions is established or modified, HEW arranges a consultation with state and local officials. Designation of a region is officially announced by HEW through the publication of a notice in the Federal Register. Once a region has thus been designated and approved by state and local governments, and when the air quality criteria for specific pollutants have been issued and control techniques reviewed and published, state governments must respond to HEW by indicating their intent to set air quality standards and establish plans to implement these standards; see Fig. 3-3.

Environmental Protection Agency

On Jan. 1, 1970, the President signed into law the National Environmental Policy Act (NEPA). This Act represented a further extension of the Air Quality Act, with the major objectives being to establish in the Executive Office of the President a Council on Environmental Quality (CEQ) charged with the responsibility of studying the nation's total environment, developing new environmental programs and policies, and coordinating the numerous federal activities in this field.

One of the most encompassing features of the new Act requires that each federal agency prepare a statement of environmental impact in advance of each proposed action, recommendation, or report on legislation that may affect the quality of the environment. Such actions may include new highway construction, harbor modifications, housing construction, urban development, etc. The primary purpose of this impact statement is to disclose the environmental consequences and the effects on ecological systems that might result from the proposed action. The Environmental Protection Agency (EPA) is to review these impact statements if they touch on any aspects of its responsibilities regarding air and water pollution, drinking water supplies, solid wastes, pesticides, radiation, and noise. If EPA determines that any proposed activity is deleterious to environmental quality, its decision is published in the Federal Register and the matter referred to CEQ.

Impact statements are made available to the public by their announcement in the Federal Register, thus involving the public in federal decisions that may affect the environment. As a result of this exposure, more than 200 legal actions were filed, up to September 1972, against federal agencies. Some of these cases have underscored the broad range of considerations that must be weighed in evaluating the environmental aspects of federal projects. In one case involving the granting of off-shore oil drilling rights by the Department of the Interior, the Court ruled that Interior should have given greater consideration to potential alternatives to the lease sale, even including increased nuclear energy development. Stimulated by the NEPA process, the Department of Housing and Urban Development is reevaluating its design standards for family dwellings to include improved insulation that will reduce energy consumption, thereby assisting in the abatement of air pollutants from fossil-fuel combustion. Better control of erosion and sedimentation at individual homesites, to reduce water run-off and improve water quality, has also been included in the specifications.

This facet of EPA activity is the most far-reaching and extensive effort yet developed to make the public aware of environmental consequences. Every federal action must be accompanied by the assurance that the environmental as well as the technical and economic factors are taken into account. A number of states have passed statutes requiring environmental impact statements for both private and state activities, similar to the statements required from the federal agencies by EPA. This nobly motivated legislation is currently responsible for some bitter confrontations and innumerable unresolved crises between ecologically minded concerned citizens and project planners.

In keeping with its role as guardian of the environment, the federal government, through EPA, makes awards to organizations and industrial companies to participate in air pollution control activities. During the fiscal year 1972, $179,300 [13] was awarded to local citizens' groups to assist them in informing the public about state plans to carry out the Clean Air Act. In this same year, a total of $1,828,521 [13] was awarded to 15 universities, research organizations, and industrial concerns to investigate various technological aspects of air pollution control.

3.4. STATES AIR QUALITY PROGRAM

Response to Federal Actions

As shown in Fig. 3-3, the states must respond to federal government activities in specific time increments in the establishment of air quality standards and the development of comprehensive plans for their implementation. Specifically, the governor of the state must notify HEW of his intent to adopt air quality standards applicable to the specific pollutant in the designated air quality control region.

Air Quality Standards

In setting an air quality standard for a given type of pollutant, a state is defining an air quality goal in terms of the desired limit on levels of that pollutant in the air. Such standards must be consistent with the intent of the Federal Air Quality Act. Air quality standards should be expressed on a weight per unit-volume basis, such as micrograms per cubic meter, as previously expressed in the Federal Air Quality Criteria. Appropriate expressions of air quality for particulate matter, for example, would include an average annual concentration as well as the highest 24 hr concentration in the year. For a gaseous pollutant such as sulfur dioxide, a standard would specify an annual average concentration, say, the highest 24 hr concentration in the year and the highest 1 hr concentration in the year. The methods by which pollutant levels are measured must also be specified. Thus, a high-volume sampler is preferred for particulate matter and the modified West-Gaeke method for sulfur dioxide. The test equipment and procedures for each pollutant are indicated in the air quality criteria publications [6–10].

An example of ambient air quality standards as amended by the State of New York on Nov. 3, 1968, under Public Health Laws 1271 and 1276, is presented in Appendix F. In Section 501.1 the objectives of the standards are described. In Section 501.2 the bases for the standards describe the industrial activities, varying from recreation, farming, etc., to extensive areas of heavy industry similar to that at Niagara Falls, which are associated with the five arbitrarily assumed permissible contamination levels. These levels are specified in Table 1 of Section 501.4 for the various pollutants. The "multiple levels" approach is not applicable for extremely toxic pollutants such as beryllium particulates and gaseous and particulate fluorides. Table 2 generally acknowledges the need for controls to eliminate odorous and other toxic and deleterious substances not specifically listed in Table 1. In Sections 501.3 and 501.4, methods for establishment of the standards and their application are discussed.

In establishing ambient air standards, the affected region usually includes sections of two or more states. New York must consider the standards defined by the states of New Jersey and Connecticut. Usually, state governments assume the responsibility for achieving the necessary coordination with each other. Because interstate activities are involved, the Air Quality Act includes provisions for financial support of the air pollution activities of the concerned state agencies or commissions.

Source Emissions

Ambient air quality standards are established to protect people, vegetation, animals, and property from the adverse effects of air pollution. The various states, in addition to formulating air quality standards, generally enact source emission regulations as well. Permissible particulate and gaseous emissions are so defined as to be consistent with the ambient air quality requirements. Thus, as we have seen in Chapter 2, the State of Indiana Air Pollution Regulations set maximum ambient air concentration levels for particulates and sulfur dioxide at 50 and 200 $\mu g/m^3$. Incorporated in these regulations were simplified diffusion formulas which permitted the process operator to predict the allowable source emission concentrations. Thus, to maintain the air quality at the levels prescribed by the Indiana State APC Regulations, the allowable stack emissions for the combustion process described in Chapter 2 were calculated to be 144 and 450 lb/hr for particulates and sulfur dioxide, in that order. The interpretation of air quality concentrations and their conversion to source emission values is a very complex technical area and the subject of considerable debate and disagreement among the various authorities.

Another example of a source emissions law is the Contaminant Emissions Ruling, Part 187, for the state of New York, shown in Appendix G. This law became effective on Feb. 6, 1968, and defines source emission limitations consistent with the desired air quality.

Reference is made to Public Health Laws 1271 and 1272 (in Appendix F), concerned with ambient air quality, for the statutory authority to enact these source emission regulations. The degree of compliance required by this law is based on an environmental rating that considers the properties and quantities of pollutants; their effect on human, plant, and animal life; meteorological parameters; stack heights; community characteristics; and ambient air quality classification of the designated area. These ratings are categorized and labeled by the letters A, B, C and D as shown in Table 1 of Appendix F. For each designation, gaseous, liquid, and solid particulate emission limits are described in Table 2. Allowable solid particulate emissions are further specified in Table 3, in pounds per hour, as a function of the production capacity or process weight of the facility.

The intent of this law is to have the various concerned industries prepare and submit an Environmental Analysis Report to the State of New York Department of Health for its evaluation of the degree of compliance. Because of the relative complexities of the law, addenda entitled "Guidelines for the Preparation of the Environmental Analysis Report for Compliance with 10 NYCRR 187" are included with the legislative issue. See Appendix H; Graphs 1 and 2 and sample calculations contained in this issue assist the process operator in determining the permissible emissions for his plant.

Implementation Plans

Once the state has set air quality standards, it must respond to the federal government (180 days later) with the submittal of implementation plans. Such a plan is the blueprint of the steps that will be taken to ensure the attainment of an air quality standard. It includes all the steps to be taken to abate and control pollutant emissions from existing sources in an air quality control region and to prevent urban and economic growth from aggravating the region's air pollution problems.

Some of the elements comprising an effective implementation plan are as follows [14] :

1. Knowledge of the existing air quality measured in the region and an estimate of its rate of deterioration based on future industrial growth.
2. Determination of existing and projected future emission levels for both stationary and mobile sources.

3. Establishment of reasonable and effective ambient air quality standards for specific pollutants.
4. Calculation of the degree of emission abatement required for specific sources to achieve and maintain the air quality standards.
5. Evaluation of the available control technology as it pertains to the practicable attainment of the proposed abatement limits for each pollutant.
6. Establishment of legally enforceable compliance schedules for each pollutant for the various industrial stationary and mobile sources.
7. Formulation and publication of rules and regulations that limit specific pollutant emissions to the atmosphere.
8. Organization of an effective staff with capabilities for field surveillance and enforcement of the regulations and provision for engineering and technical services.

Each of the foregoing elements of the implementation plan must be described in detail when the plan is submitted for review by the HEW. This description must cover all projected requirements for the prevention and control of air pollution, the time-table for accomplishing these requirements, provision for surveillance, statement of the rules and regulations, etc. Funding and manpower resources must also be defined. In the development of this material, the state officials are made to understand that its submittal to HEW is for the purpose of this agency's assessing its ability to attain the specified air quality standard.

The Environmental Protection Agency is aware of the difficulties involved in the preparation of implementation plans. On April 6, 1971, the EPA proposed regulations [15] to be followed by the states in developing and submitting their plans. These guides requested the states to consider the installation of emission control devices on motor vehicles now in use and the conversion of motor fleets to low-pollution fuels such as natural gas, petroleum gas, or electricity. The use of "attainable" emission limits for controlling stationary sources was recommended—probably as a caution against the adoption of severe and impractical ambient air and emission standards. A description of the information required in the implementation plans, procedures for its development, and compliance timetables were included in these proposed regulations. EPA promised to lend assistance to the states in the preparation of their plans, and in an indirect but unmistakable manner urged them to accelerate their efforts.

3.5. LOCAL AIR POLLUTION CONTROL AGENCY

Responsibility and Activities

The scope or jurisdiction of the local control agency is usually governed by the number and types of air pollution sources in its area. Municipalities and counties having large population and industry concentrations generally have strong comprehensive air pollution programs. In keeping with the provisions of the Clean Air Act, state agencies are assuming increasing responsibility and authority for air pollution control and the support and coordination of local agencies.

Direct enforcement of the regulations is most ideally achieved at the local level. In general, the local agency implements relevant portions of the state control plan. It often adopts standards consistent with or more stringent than those specified by the state. One definite partitioning of functions for the state and local authorities is that of motor vehicle emissions. In keeping with its jurisdictional control over the automobile, the state has total authority in all matters concerning automotive pollutant emissions.

The Regulations

The most important function of an effective air pollution control program is that of enforcement. The interpretation of the regulations, surveillance of the emission source, evaluation of emission data, and notification of the operator as to his compliance or noncompliance to the regulations are the various actions comprising enforcement. The rules and regulations are the core of this procedure and generally contain the following sections:

1. Emission limitations that prohibit the discharge of pollutants in excess of specified standards.
2. Equipment design standards that specify permissible process equipment design features. For example, regulations concerning minimum stack heights would be included under this item.
3. Equipment operation prohibition denying the process operator the right to conduct certain operations which even if provided with control equipment could not meet the prescribed emission levels. Thus, the beehive incinerator has been legislated out of existence throughout the United States.
4. Raw material regulations which specify the type of fuel and fuel properties. The allowable sulfur and ash content of both coal and oil fuels are thus controlled in most communities.
5. Emergency actions which enable the control agency to prepare and use contingency plans in the event of episodes of high pollutant concentrations. The joint preparation by the agency and concerned industries for emergency reduction or shut-down of production facilities is treated in this section.
6. Regulatory powers, defined as a group of supporting rules and regulations which establish the right of entry, policing actions, access to facilities for source emissions testing, etc.

One of the most comprehensive air pollution control programs in the United States is that developed by the County of Los Angeles Air Pollution Control District. A table of contents and statements of some of the rules and regulations issued by this agency in 1972 are shown in Appendix I. These regulations contain all elements considered as requisites for a strong, enforceable code. The heart of these regulations is Section IV, "The Prohibitions." It is interesting to note how the various titles under this section reflect the special industrial profile of the Los Angeles County basin.

Some comments with regard to the interpretation of some of these rules are:

Particulate matter: Defined in these regulations as any material, except combined water, which exists in fairly divided form as a liquid or solid at standard conditions.

The three most relevant rules pertaining to the emissions of particulates are Rule

50, Rule 52, and Rule 54. The maximum emission condition defined by each rule must not be exceeded, so that in effect the most stringent condition represented by any single one of these three rules will prevail. For example, taking each rule in order of its severity, consider an aggregates-treating plant, processing crushed stone in a kiln dryer at the rate of 30 ton/hr. Then, according to amended Rule 54, the process chart on sheet 5 of Appendix I would indicate a maximum allowable particulates emission of 26.6 lb/hr. For a plant of this capacity, the gas effluent rate would be about 30,000 dscfm, so that in accordance with the chart for Rule 52, on sheet 4 of Appendix I, the maximum allowable particulates concentration in the effluent gas would be 0.0544 grain/dscf calculated to a discharge rate of 14 lb/hr. Finally, a Ringelmann 1 is imposed in Rule 50. As discussed earlier, this optical index value is very difficult to convert to quantitative emissions. However—say, as a very rough guess—it might be equivalent to a value of 0.03 grain/scf for the relatively white smoke expected from this type of operation. This final value would convert to an emissions value of approximately 7.5 lb/hr and would represent the maximum allowable discharge rate from this particular plant.

Sulfur compounds (defined as gaseous sulfur oxides): Rule 67, concerned with sulfur oxides emissions, is not too detailed because SO_x emissions in the Los Angeles area are not as much a matter of concern as the oxides of nitrogen. One of the major sources of sulfur oxides is from power plants, and only during that period when it becomes necessary to substitute oil for natural gas. During these periods of operation, currently representing about 50%* of the total combustion time, low sulfur oil is used. The rule is rather stringent; for example, for a moderate size 300 MW utility boiler, the expected SO_x discharge rate, using fuel oil containing about 0.2 wt % sulfur, would be about 300 lb/hr, which is well above the maximum emission value of 200 lb/hr demanded by Rule 67 for new installations. Actually, Rule 62 limits the sulfur content of fuels to 0.5 wt % sulfur, but for a 300 MW boiler, the lower limit of 200 lb/hr would prevail. For this size boiler it would be necessary to fire fuel oil containing no greater than 0.13% sulfur to meet the code. In firing fuel oil, particulate and NO_x emissions must also comply with the pertinent regulations.

Timing and plant status: It is interesting to note the manner in which the dates are posted for compliance and the various amendments upgrading the compliance measures. Thus, for Rule 54, sheet 4, compliance is required by Jan. 6, 1972, for equipment being planned or designed, and therefore does not pertain to existing plants. However, all plants in the county, regardless of age, must comply by Jan. 1, 1973. In this same rule, which actually was in existence prior to 1969, the tables defining the maximum discharge rate have been upgraded as recently as Jan. 6, 1972.

The oxides of nitrogen emission limits in Rule 68 have recently been addended to define emission concentrations. Furthermore, the schedule for compliance is prorated up to Dec. 31, 1974, with the emission value to be halved over that posted for Dec. 31, 1971. Thus, emission controls grow ever tighter.

---

*This value is continuously increasing because of natural gas shortages.

REFERENCES

1. "Profiles of Air Pollution Control in Los Angeles County Air Pollution Control District," Los Angeles County, 1971.
2. "Air Quality Criteria for Sulfur Oxides," U.S. Dept. of Health, Education, and Welfare, Public Health Service, Consumer Protection and Environmental Health Service, National Air Pollution Control Administration, Washington, D.C., January 1969.
3. W. F. Ashe, "Acute Effects of Air Pollution in Donora, Pennsylvania," in L. C. McCabe, ed., *Air Pollution Proceedings of the United States Technical Conference on Air Pollution.* McGraw-Hill Book Co., New York, 1952.
4. H. C. Wohlers, R. N. Meroney, D. L. Brenchley, "Where We Are?—International Air Pollution" J-APCA Vol. 23, 7, July 1973 (573-579).
5. Yanagisawa, "Air Quality Standards—National and International" J-APCA Vol. 23, *11,* Nov. 1973 (945-948).
6. The Clean Air Act, U.S. Public Law. 88-206: approved Dec. 17, 1963.
7. "Guidelines for the Development of Air Quality Standards and Implementation Plans," U.S. Dept. of Health, Education, and Welfare, Public Health Service, Consumer Protection and Environmental Health Service, National Air Pollution Control Administration, Washington, D.C., May 1969.
8. "Air Quality Criteria for Particulate Matter," U.S. Dept. of Health, Education, and Welfare, Public Health Service, Consumer Protection and Environmental Health Service, National Air Pollution Control Administration, Washington, D.C., January 1969.
9. "Control Techniques for Particulate Air Pollutants," U.S. Dept. of Health, Education, and Welfare, Public Health Service, Consumer Protection and Environmental Health Service, National Air Pollution Control Administration, Washington, D.C., January 1969.
10. "Control Techniques for Sulfur Oxide Air Pollutants," U.S. Dept. of Health, Education, and Welfare, Public Health Service, Consumer Protection and Environmental Health Service, National Air Pollution Control Administration, Washington, D.C., January 1969.
11. "Control Techniques for Nitrogen Oxide Emissions from Stationary Sources," Public Health Service, Environmental Health Service, National Air Pollution Control Administration, Washington, D.C., March 1970.
12. J. P. Tomany, R. R. Koppang, and H. L. Burge, "A Survey of Nitrogen Oxides Control Technology and the Development of a Low NO_x Emissions Combustor," ASME Meeting, November 1970.
13. *Environmental Reporter,* The Bureau of National Affairs (Washington), Vol. 2, No. 11 (July 16, 1971).
14. *Field Operations and Enforcement Manual for Air Pollution Control,* Vol. I, Environmental Protection Agency, Office of Air Programs, Stationary Source Pollution Control Programs, Research Triangle Park, N.C., August 1972.
15. *Environmental Reporter,* The Bureau of National Affairs (Washington), Vol. 1, No. 50 (April 9, 1971).

Chapter 4

EMISSIONS TESTING

4.1. INTRODUCTION

The characteristics and emission rates of the various pollutants responsible for the contamination of the atmosphere were reviewed in Chapter 2 and the regulations enacted for their control were discussed in Chapter 3. The definition of an air pollution problem, whether in ambient air or from a specific source, requires the identification of the responsible pollutants and a measurement of their concentration. The techniques of ambient air testing are quite different from those of source emissions testing. Because of the direct impact of source test results on control equipment design, that phase of emissions testing will be emphasized in this chapter. Thus, should source emission values indicate a need for the abatement of one or more pollutants for regulation compliance, then process modifications or control equipment must be considered. If and when such abatement actions have been accomplished, additional source testing is required to confirm compliance with the prevailing regulations.

The reasons for source emissions testing can be considered under two categories. The first and most important is the identification of an air pollution condition. The second is concerned with the evaluation of process-operating data necessary for the design of suitable control equipment. Although there is some overlap in the data requirement for both functions, emissions testing for equipment design considerations is more demanding. Thus, the measurement of gas flow rate, temperature, and pollutant concentrations are common to the pollution condition and equipment design. However, such additional data as dewpoint, trace contaminants, physical properties of the pollutant, and process cyclical characteristics are some of the additional factors necessary for the design of effective control equipment. Regardless of the final uses for emissions-testing information, the immediate objective in testing an air pollution source is to obtain reliable data as to the composition of the effluent and its rate of emission to the atmosphere.

4.2. GENERAL PROCEDURES

Basic Requirements

The following objectives are relevant to any source test:

1. The gas stream being sampled from a source should represent either the total or a known portion of the source flow rate.

2. Samples of the emissions collected for analysis should be representative of the gas stream being sampled.
3. The volume of gas sample being withdrawn for analysis should be measured to permit calculation of the concentration of the analyzed constituents in the sampled gas stream.
4. The gas flow rate from the sources must be determined in order to calculate the emission rates for the various constituents.

Methods for attaining these overall objectives are discussed in this chapter. The need for reliable source test data is receiving increased attention as the number and type of potential pollutants continue to proliferate. In every stage of emissions definition, from the criteria published by HEW to the regulations enacted by pollution control agencies, the source test methods are specified.

Source Test Planning

The purpose for a source test should be clearly specified before a procedure can be formulated. As discussed previously, the test data requirements for emissions compliance is considerably different from that necessary for control equipment performance evaluation. The source type and some knowledge of the process is also required. Cyclical operation, very high temperatures and combination particulate, and gaseous pollutant measurements are some of the factors that must be considered. A preliminary visit to the test site by testing personnel to determine test station locations is also recommended. In many plants it may be necessary to prepare test holes, hooding, or access platforms to obtain adequate and representative samples. The actual testing period should be carefully coordinated with plant management personnel to ensure operation of the facilities under the conditions of test.

The selection of the optimum sampling, analytical, and flow rate measurement methods is extremely important to the accuracy of the results. Interpretation of the results will depend on whether dry or wet collection of the particulates is practiced. For example, wetting certain soluble or agglomerative particulates would negate the results of any size analysis. Test conditions at the plant must be conducted in accordance with the operating characteristics of the equipment, usually at full capacity. A drying kiln operated at half-capacity will register particulate emissions far below those obtained at its rated capacity. Efforts to conduct source testing at one set of operating conditions, with the expectation of extrapolating to other conditions, are of dubious value.

The application of statistics to sampling procedures, to minimize the number of tests, should be considered. Where data are extremely variable, and samples and analysis difficult to obtain, a statistical approach is invaluable. The use of arithmetic averages is generally applicable, but in many problems the mean value is not indicative of the range or distribution of the data. For example, dustfall data from various receptor stations are subjected to statistical analysis to determine their significance and reliability.

The frequency of sampling is usually dictated by the process characteristics. Thus, in batch operations there is usually a peak emission value which must be determined for valid results. In such an operation, with a cycle period of 2 hr, as many as eight individual samples would probably be necessary to establish peak conditions. The sample size must be sufficiently large for the subsequent analyses. For microscopic size analysis, a 1-g sample would be sufficient, whereas for elutriation size analysis up to 50 g would be required.

The Source Test Report

A source test report should be sufficiently complete and clear to be understood by the various interested personnel, from the plant operator to the air pollution engineer. The major elements comprising the report should be:

1. *Introduction:* purpose of the test, equipment tested, name and location of company and equipment, personnel involved, and the date of testing.
2. *Discussion:* description of process, operating conditions during test, comments on unusual conditions, and location of sampling stations.
3. *Summary of results:* presentation of results in tabulation or prose form correlated with time increments and process comments such as batch number, nonstandard plant practices, etc., description of samples and their identification, discussion of results, and an indication of their significance.
4. *Sampling and analytical procedures:* description of standard methods with detailed comments and reasons for modification.
5. *Appendix:* original test data and sample calculations.

To illustrate the principles of source emissions testing, as described in the following sections, a sample source test report is presented at the end of this chapter.

4.3. GAS FLOW RATE AND TEMPERATURE MEASUREMENT

Need for Measurement

In the determination of particulate or gaseous emissions, it is necessary to know the flow of gas transporting these various pollutants. The sample of the pollutant collected in the source test equipment must be correlated with both sampling gas stream and total gas flow to permit computation of the stack emissions, expressed as pounds per hour. Thus, in a typical flyash particulates source test, the total gas sampled was 148 ft^3 at standard conditions, and 22.8 g of solid were collected. The dust loading was computed as (22.8 $\times$ 7000/454 $\times$ 1/148), or 2.38 grains/scf. The total flue gas flow discharged from the stack was 122,200 scfm, so that the flyash emission rate was determined to be (2.38 $\times$ 122,200 $\times$ 60/7000), or 2750 lb/hr. Source testing is usually carried out at process temperatures, and because emission data are preferably expressed at standard conditions (defined as 70° F and 760 mm Hg pressure), the observed gas measurements must be converted from one set of conditions to the other.

For control equipment performance tests or for the determination of operating data for the design of such equipment, gas flows and temperatures are the two major parameters. The total volumetric gas flow rate determines the size of both particulate and gaseous control equipment. The temperature of the gas is important because of its effect on the volumetric flow and because of the high-temperature limitations characteristic of many types of control equipment.

The dewpoint temperature of the gas stream, as a measure of its moisture content, becomes an important consideration when corrosive gases are present. At the dewpoint, water droplets (which may be saturated with such gases as SO_2, SO_3, HCl, and HF) begin to precipitate out of the gas stream, thereby producing corrosive conditions and reducing the life of the equipment. The importance of gas flow rates and the temperature and moisture content of the gas stream to the design of the various classifications of emissions control equipment is explained in Chapters 5 and 6.

Principles of Gas Flow Measurement

A gas traveling at a specific velocity will create a definite pressure, which is known as the *velocity pressure*. The expression relating gas flow to its velocity pressure is based on the Bernoulli theorem, and can be stated as:

$$V_s = (2gh)^{1/2} \tag{4-1}$$

where

V_s = gas velocity, fps
g = gravitational acceleration constant, 32.2 ft/sec^2
h = velocity pressure, feet of flowing gas

Assuming the gas to be air, measured at standard condition (70° F, 760 mm Hg), the formula becomes

$$V_s = 2.90 \ (HT_s)^{1/2} \tag{4-2}$$

where

H = velocity pressure, in. water
T_s = absolute temperature of gas, °R

If the pressure of the gas differs from 760 mm Hg or if its density differs from that of air, then this equation can be written as

$$V_s = 2.90 \ \left(\frac{760}{P_s} \times \frac{1.00}{G_d} \times HT_s \right)^{1/2} \tag{4-3}$$

where

P_s = absolute pressure of sampled gas, mm Hg
G_d = specific gravity of gas, referred to air at 32° F, 1 atm press.

By determining the average velocity pressure of a gas in a duct, together with its temperature and pressure, its average linear velocity can be determined by use of Eq. 4-3. The volume of gas passing any cross section of the duct in unit time is the product of the linear velocity and the area of the cross section. Thus,

$$V_0 = V_s \times A \times 60 \tag{4-4}$$

68

where

V_0 = flow of gas, cfm at actual conditions (acfm)
A = effective area of the duct, ft^2

The most direct, simplest, and generally accepted method for measuring gas flows is based on these formulas. The gas velocity pressure, temperature, and static pressure are directly measured in the duct, and the gas flow is calculated from Eqs. 4-3 and 4-4.

Methods and Equipment

Gas Flow Measurement. There are a number of velocity-pressure measuring devices, such as the Pitot tube, orifice plate, or venturi tube. However, the most simple and favored instrument is the Pitot tube. Fig. 4-1 shows a Pitot tube and inclined manometer for measuring gas flows. The Pitot tube comprises an impact and static pressure tube. The differential pressure sensed by this design nulls the static pressure component so that the manometer reading is a direct measure of the velocity pressure. When using the Pitot tube, velocity measurements must be recorded at a number of well-distributed points in the duct, at least 10 diameters in length both upstream and downstream of the Pitot tube location. A sampling traverse arrangement for a circular and rectangular duct is shown in Fig. 4-2. Table 4-1 tabulates a number of velocity-pressure readings and the corresponding velocities, based on the use of a standard Pitot tube at various temperatures. These values have been calculated from Eq. 4-3 and are valid for both air and flue gas flows.

Temperature Measurement. In determining the temperature of a gas flowing in a duct, an allowance must be made for thermal radiation effects. This is particularly true for high temperatures in the range of 600 to 1800° F, where a large temperature

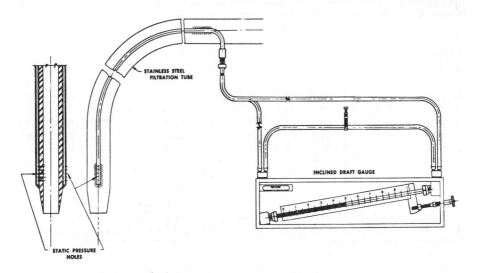

Figure 4—1. A Pitot tube and an inclined manometer.

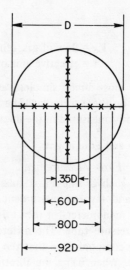

Note: Gas sampling positions located
in equal area segments

Figure 4–2. Pitot tube transverse patterns for rectangular and circular ducts.

differential exists between the gas stream and duct walls. A temperature traverse will permit a closer estimation of the true gas temperature under these conditions, but reduced surface area thermocouples or shielded sensing elements are recommended when high accuracy is required.

Table 4-1. Velocity Head Values

Velocity Head, in H_2O	Temperature, °F										
	200	250	300	350	400	450	500	550	600	650	700
0.10	23.6	24.4	25.3	26.0	26.9	27.7	28.4	29.1	29.8	30.6	31.2
0.20	33.4	34.6	35.8	37.0	38.1	39.2	40.2	41.2	42.2	43.2	44.2
0.30	40.8	42.3	43.8	45.2	46.6	47.8	49.2	50.4	51.7	52.9	54.2
0.40	47.2	48.9	50.6	52.2	53.8	55.3	56.8	58.3	59.7	61.0	62.4
0.50	52.7	54.7	56.6	58.3	60.1	61.8	63.5	65.1	66.7	68.4	69.9
0.60	57.8	59.8	61.9	64.0	65.9	67.7	69.6	71.4	73.1	74.9	76.5
0.70	62.4	64.6	66.9	69.1	71.1	73.2	75.2	77.1	79.0	80.8	82.6
0.80	66.6	69.1	71.4	73.8	76.0	78.2	80.3	82.3	84.4	86.4	88.3
0.90	70.7	73.2	75.8	78.3	80.7	82.9	85.2	87.4	89.5	91.7	93.6
1.00	74.5	77.2	79.9	82.5	85.0	87.5	89.8	92.1	94.3	96.6	98.7
1.10	78.2	81.0	83.8	86.6	89.2	91.8	94.3	96.7	99.0	101.0	103.0
1.20	81.6	84.6	87.5	90.4	93.1	95.8	98.5	101.0	103.0	106.0	108.0
1.30	84.9	88.0	91.1	94.1	96.9	99.8	102.0	104.0	107.0	110.0	112.0
1.40	88.2	91.4	94.6	97.6	100.0	103.0	106.0	109.0	111.0	114.0	116.0
1.50	91.3	94.6	97.9	101.0	104.0	107.0	110.0	112.0	115.0	118.0	121.0

Source: Velocity Tables, Ref. 1.

Note: Velocities expressed in feet per second, based on the use of the standard type Pitot tube.

Table 4-2. Temperature Ranges for Various Sensing Devices

Sensing Device	Fluid/Material	Temperature Range, °C
Thermometer	Alcohol	−80–250
	Mercury	0–350
	Platinum resistance	−200–600
Thermocouples	Copper-constantan	−200–400
	Iron-constantan	0–800
	Chromel-alumel	0–1100
	Platinum-platinum/rhodium	0–1600
	Iridium-tungsten	1000–2200
	Tungsten-tungsten/rhenium	Up to 2500

Mercury-filled thermometers, platinum resistance thermometers, and thermo-couples are the sensing elements most commonly used in the determination of process gas temperatures. The overall temperature range covered by these instruments is 20 to 4500° F. The specific range for each type of sensing device is shown in Table 4-2.

The moisture content of a process gas can be measured directly in a suitable indirect condenser by condensing the water from the sampled gas stream. From the dry-bulb temperature and the moisture content, the dewpoint can be obtained from a psychrometric chart [2]. If the gas temperature is in the range of 100 to 300° F, dry- and wet-bulb temperatures of the main gas stream can be determined as a measurement of the dewpoint. Above these temperatures, the sample gas stream may be cooled and wet/dry bulb values obtained and converted to the dewpoint temperature.

The presence of very small concentrations of sulfur oxides in a gas stream will increase the dewpoint of the gas. The active agent that affects the dewpoint value is the sulfur trioxide component, which might represent from 1 to 10% of the total sulfur oxides. Thus, a concentration of 0.010% of SO_3 in a flue gas containing about 10 vol. % water vapor will raise the dewpoint of the gas by about 190° F, from 115° F to 305° F. This relationship between the SO_3/H_2O content and dewpoint elevation is shown in Fig. 4-3.

Gas Flow Estimates

Although the direct measurement of gas flow rates is essential to source testing, there are a number of simple approximate checks that can be made to confirm the measured results. Fan nameplate data affords the most direct of these methods for estimating the gas flow through the system. If fan characteristic curves are not available, then a power-demand meter applied to the electric motor drive and a determination of the static pressure developed by the fan will allow the calculation of a reasonably approximate gas flow quantity. Another very approximate check is that of comparing the measured gas velocity through duct work with those velocities adopted by industrial plant designers. The latter generally lie in the range of 3000 to 4500 fpm. These comparative values can be used to obtain a better than an order of magnitude verification of measured gas velocities. In fact, this method of reckoning gas flows has often been applied by equipment vendors for preliminarily sizing and estimating

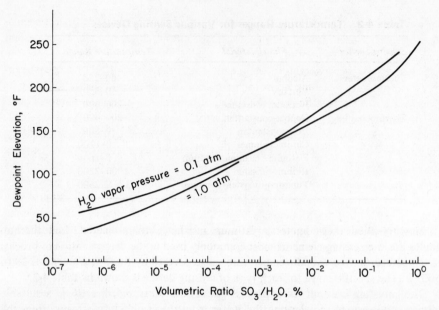

Figure 4–3. Dewpoint elevation as a function of SO$_3$ concentration. (Adapted from Muller, Ref. 3.)

control equipment. Still another estimating approach, which can be used with combustion processes, is to determine the fuel-burning rate and compute the expected flue gas flow.

These various estimating methods are often used to establish confidence in the measured gas flow values. They do serve one other useful purpose. As in all areas of human activity, there are those who measure and those who are measured, and the interests of each party are often at variance. The source emissions testing of a plant effluent to determine compliance with prevailing regulations presents a very special interest situation. The operator of an aggregates-processing kiln is most anxious to demonstrate to the APC testing personnel that his control equipment does indeed perform sufficiently well to reduce the atmospheric discharge rate of particulates to the prescribed level. In his desire to "meet the code," he may be tempted to diminish surreptitiously his processing rate so as to reduce the particulates loading to the emissions control facilities. However, such process cutbacks are reflected in the process gas flows being measured. Thus, if a fan rating indicates a potential capacity of 30,000 cfm and the measured flow is somewhere in the neighborhood of 15,000 cfm, then there is some basis for questioning the plant-processing level. As recommended earlier, source test measurements are generally conducted at plant rated capacity.

4.4. COLLECTION AND ANALYSIS OF PARTICULATES

Collection Methods and Equipment

In sampling a gas stream, either particulate matter or gases, or a combination of both pollutant types, must be identified and quantified. The collection of particulate matter from a process gas stream during the sampling period may be made for the determination of weight concentration, particle size distribution, chemical analysis, or certain chemical and physical properties.

The methods of collecting a sample of particulate matter from a gas stream involves continuously passing a known portion of the stream through a sampling or collection train for a specific period of time. In the train the particulates are separated from the carrier gas and a measurement of the total volume of sampled gas is provided. Generally, sampling rates on the order of 0.5 to 1 cfm are used to ensure efficient performance of the collection equipment in the train. The length of the sampling period is predicated on the need to obtain a representative sample of the emission and a sufficient particulate sample to permit the necessary testing procedures such as size analysis. The components of a sampling train consist of a sampling probe and nozzle, a means for separating the particulates from the sampled gas stream, a flow-metering instrument, and a vacuum pump for producing and regulating the flow of gas through the components of the train. Such a sampling train is shown in Fig. 4-4. In this arrangement, dry collection of the particulates is achieved by a paper thimble filter enclosed in a metal holder. The temperature limitation for this type of collector is 250° F. For higher temperatures, in the range of 500 to 600° F, Alundum thimbles are used. Both these collectors are of the dry filter type in which the sampled gases must be above their dewpoint so as to be free of condensed moisture droplets. An Alundum thimble and holder are shown in Fig. 4-5.

Where wet collection of particulates does not present any testing problem, the use of wet impingers is preferred. Usually, three standard Greenburg-Smith glass impingers

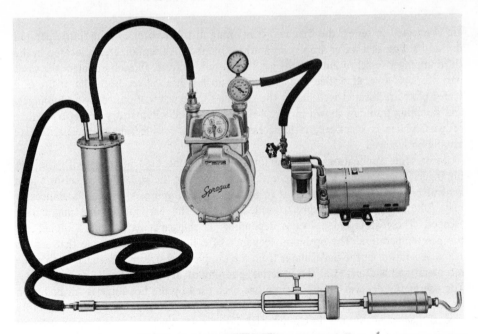

Figure 4-4. Particulates sampling train.

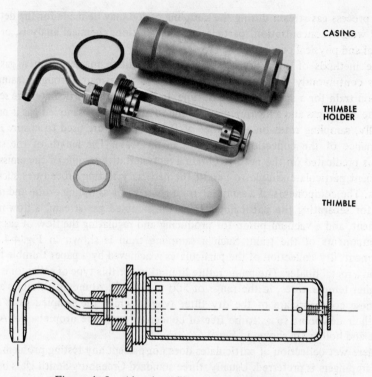

Figure 4—5. Alundum filter thimble and holder.

are connected in series, the first two containing distilled water and the third operated dry and acting as a water droplet separation chamber. A thermometer is located in the third impinger vessel to indicate the gas stream temperature. In each impinger the gases pass through a small orifice and impinge against a glass plate. Both the orifice and target plate are located well below the surface of the water. Four such wet impingers and sampling train are shown in Fig. 4-6. Regardless of whether dry or wet collection of particulates is practiced, it is necessary that 100% collection of the particulate sample be achieved.

For some applications, the dry filter may be connected in series with the wet impingers. The thimble may precede the impingers for the collection of sulfuric acid aerosol or finely divided fume that might escape the impingers. In some instances, it may follow the wet impingers to collect any of the particulates that might have escaped. These arrangements vary, depending on the particulates being sampled and the gas temperature. The optimum sequence of collection devices is one that assures 100% separation of the particulates from the sampled gas stream. There has been some recent criticisms directed at the in-series grouping of dry and wet collector types [4]. The use of the dry-wet sampling train has been used by the Los Angeles Air Pollution Control District [5] for some time. However, when this type of system was more recently adopted by the Environmental Protection Agency, opposition developed mainly because of the false definition of particulates as determined by this version of

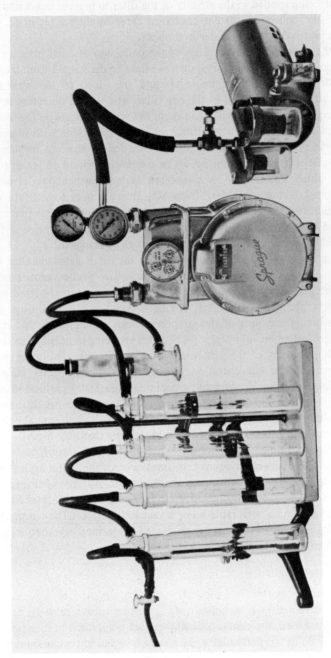

Figure 4–6. Wet collection impingers and sampling train.

the sampling train. The front half of the train comprised a sampling probe and dry filter, with heat being applied in the vicinity of the filter to prevent condensation. The second half of the collection system contained three impingers set in an ice bath together with a fourth dry vessel for deentrainment.

Actually, the controversy revolves about the definition of particulates. Critics of this system maintain the organic vapors, such as might be discharged from incineration processes, condense in the chilled impingers and are added to the weight of the discrete solid particulates collected in the dry filter. The second objection is that any SO_2 in the sampled gas stream will dissolve in the wet impingers to form sulfurous acid, which then will be oxidized to sulfuric acid and/or combine with ingredients in the dust to form metal sulfates. Even with elimination of the wet impingers, it is claimed that sufficient condensation occurs in the neighborhood of the dry filter to yield sulfates, which should not be considered as true particulates. Hemeon and Block [4] obtained comparative data from SO_2 flue gas tests to confirm that the weight of "particulates" as collected by the EPA train was two to four times as much as that collected with a dry filter when condensation was avoided. This is certainly an important aspect of emissions control, which has yet to be resolved.

As shown in Fig. 4-4, the other major items in the particulates sampling train are the sampling probe, gas meter, and vacuum pump. The probe consists of a tube inserted into the gas stream and which is provided with a properly sized nozzle to maintain isokinetic sampling conditions. Isokinetic flow conditions through the sampling nozzle, involving matching of the sampling stream velocity with that in the gas duct, is not so critical when submicron particulates are being sampled. This is because the Stokes settling velocity of the particulates is not sufficiently significant to prevent obtaining a representative concentration through the sampling probe. However, to avoid any misinterpretation of test results, sampling is generally performed isokinetically; there are formulas available [5] for determining the proper nozzle diameters to obtain this condition. Probes are fabricated of such materials as glass or stainless steel. For temperatures above 1200° F, quartz glass and stainless steels are used.

The gas meter recommended is the dry type; it is usually located on the suction side of the pump. The collection equipment must protect the meter against moisture and high temperatures. The vacuum pump should have a capacity of at least 2 cfm at vacuums up to 8 in. Hg. Generally, positive displacement types are favored, although in some areas the pump has been replaced by a vacuum aspirator. The temperature and pressure of the sampled gas stream are indicated by a mercury thermometer and manometer located in the sampling train.

Concentration Determination

When the particulate matter is collected, either by dry and/or wet methods, in sufficient amounts to satisfy subsequent tests, its concentration must be determined. If wet impingers are used, the contents of all three are evaporated to dryness to obtain the dry weight of collected particulates. This value is added to that amount contained in the filter, if it is used, and the total value is divided by the total gas sampled. The sampled gas volume is expressed at standard conditions (70° F, 1 atm) after a correction is made for any moisture that might have been condensed out, if wet

76

collection was involved. Thus, the initial distilled water used in the impingers, subtracted from the total volume of liquid at the end of the run, is recorded as the condensate volume. Many regulations express allowable emission concentrations in grains per dry standard cubic feet (grain/dscf). The use of a dry gas basis is to eliminate the dilution effect of unusually high-moisture gas streams; for example, the discharge of a saturated gas stream from a wet scrubber at elevated temperatures. For combustion applications, the regulations demand a specific CO_2 content in the flue gas carrier gas so as to prevent dilution of the discharged gas stream with air which would result in reporting misleading reduced emission concentration values. Thus, many of the codes demand that particulate concentrations in flue gas streams be expressed as grain/dscf at 12% CO_2 content. Calculation of particulate concentration values is demonstrated in the Sample Source Test Report presented at the end of this chapter.

Particle Size Determination

Prior to the advent of serious air pollution abatement efforts, screen analysis was almost the sole method for the determination of particle size distribution. This technique was limited by the smallest opening screen (sized at 400 mesh, which is equivalent to approximately 37μ). See Table 2-3. Today's emissions control technology requires particle size data down to 0.1 μ and even lower in some cases. The difficulty of removing particulate matter from an effluent gas stream increases as the particle size decreases. The recent emphasis on the control of these particles has lent impetus to the improvement of measurement techniques for this size range.

There are four basic methods of size analysis: sieving, sedimentation, classification, and microscopy.

1. Sieving. Sieving is limited to 400 mesh (37 μ) wet and 200 mesh (74 μ dry). The speed and convenience of this method is good, but the size range is inadequate.

2. Sedimentation. This operation is based on Stokes law of settling particles. Either liquid or gas sedimentation can be used. In the liquid technique, sizes of +2 μ can be determined by gravity and 20 to 0.2 μ by centrifugation. The major problem is the need for a large selection of liquids and dispersants and the difficulty of obtaining an acceptable dispersion. Its practical application seems limited to obtaining such crude cuts as a − 5 μ, a 5 to 10 μ, a 10 to 20 μ, and +20 μ fraction.

Complex and expensive commercial equipment is available for gas sedimentation. The greatest range is achieved with an air centrifuge, yielding size analysis distribution data in the 3 to 0.1 μ range. The distribution data are microscopically obtained.

3. Classification. The Roller elutriator using air as the classification fluid is the most common of the various designs available. It is restricted to relatively coarse separations such as 0 to 3 μ, 3 to 6 μ, 6 to 12 μ, 12 to 24 μ, and 24 to 28 μ. Although analysis of the 0 to 3 μ fraction is time consuming, requiring about 1 hr, complete separation of the individual fractions is achieved, which permits a more detailed examination of the various size cuts. Other types of size analysis devices merely analyze the sample without physical separation.

The Roller elutriator is shown in Fig. 4-7. In operation, the instrument directs a current of air at a specific and measured velocity toward the particulate sample, which

77

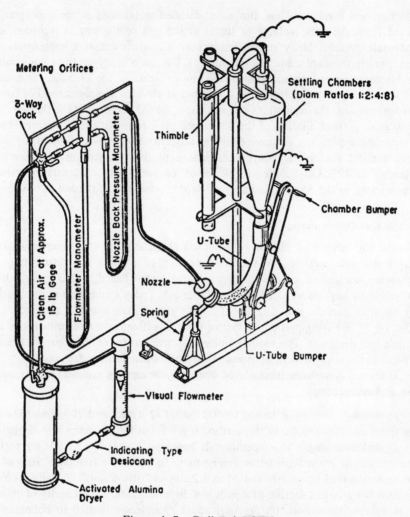

Figure 4–7. Roller elutriator.

is contained in an oscillating U-shaped sample tube above which there are located a settling chamber and particle size collection chamber. The smallest particles are separated out first, lifted on the air stream into the settling chamber, and thence transferred to a paper thimble in which they may be weighed and/or examined microscopically.

All particles larger than those in the selected micron range fall back into the sample tube, ready for subsequent reanalysis that will separate particles in progressively larger micron ranges, with air flow adjusted upward in accordance with Stokes law. A chart is provided to show the desired micron separation size and sample density as a function of air flow in liters per minute. A total of 12 nozzles are provided to cover the entire range of possible flow rates from 2.5 to 35 liters/min. A typical size analysis for a limestone sample with an elutriation instrument is given in Table 4-3.

Table 4-3. Air Elutriator Size Analysis

Micron Size	Average Size	Wt %	Cum. Wt %
0–1	0.5	6	6
1–2	1.5	15	21
2–3	2.5	13	34
3–4	3.5	10	44
4–5	4.5	9	53
5–6	5.5	5	58
6–7	6.5	6	64
7–8	7.5	3	67
8–9	8.5	3	70
9–10	9.5	3	73
10–11	10.5	2	75
11–12	11.5	2	77
+12		23	100

4. Microscopy. The ordinary optical microscope has a resolving power on the order of 0.4 to 500 μ. Rather than weight measurement of the various size fractions, a particle count and particle dimensions are measured in microscopic-size analyses. For highly accurate results with irregular particles, separate measurement of the length and width is usually made.

An electron microscope has a range of 0.01 to 2 μ. The effectiveness of either optical or electron microscopy depends largely on the method of collecting the sample to give a properly dispersed particle distribution on the slide or film. For optical microscopic measurements, samples with a few particles below 5 μ can be air-dispersed directly onto a dry microscope slide. Another technique involves the use of millipore filter elements, which have a pore size range of 0.01 to 5.0 μ. By aspirating small gas samples through this type of filter, the particles collected on its surface can be examined under a microscope. The use of calibrated reticles in the microscope eyepiece is used to count the number of particles of given size ranges for a given area. As an alternate, grids can be furnished as a part of the filter so that each square represents a certain fraction of the total volume of the gas sample passed through the filter.

Although particles can be deposited directly from an air stream onto the slide of the electron microscope, consisting of a collodion film supported on the wire grid, two sampling instruments have been used. These are the electrostatic precipitator and the oscillating thermal precipitator [6]. The latter was developed primarily for the collection of dust samples for electron microscopy. The number concentration of particles determined by electron microscopy is usually on the order of five times greater than that obtained by optical microscopy. Table 4-4 gives a comparative size analysis for the two microscopic methods. For those particles in the range of 0.4 to 1 μ, the results are in close agreement. However, below 0.4 μ, the optical microscope yielded inconsistent readings because its size limitation had been exceeded. The relationship between particle distribution by count and by weight was illustrated in Chapter 2, Table 2-4.

Table 4-4. Comparative Particle Size Distribution

(Number per cubic centimeter greater than stated size)

Size, μ	Optical Microscope	Electron Microscope
0	310	1728
0.022	–	1229
0.043	–	987
0.087	–	565
0.174	–	203
0.26	–	87
0.28	138	
0.35	–	45
0.40	39	
0.43	–	26
0.52	–	14.5
0.57	12.2	
0.70	–	4.66
0.80	4.8	
0.87	–	(0.86)
1.04	–	(0.11)
1.13	1.8	
1.22	–	(0.11)
1.60	0.2	

Source: Drinker and Hatch, Ref. 7.

Measurements in one system can be converted to the other through the use of a theoretical equation [8] involving the standard deviation by weight or by count.

There are two additional, interesting methods for particle size analyses. The first utilizes an electric gating principle. In the Coulter counter, a suspension of particles in an electrically conductive liquid flows through a small aperture between two electrodes. If a relatively nonconductive particle passes between the electrodes, a voltage decrease occurs which is proportional to the size of the particle. The apertures may be between 10 and 1000 μ in diameter and the minimum particle size measured is about 0.3 μ.

As mentioned previously, particle size distribution, particularly in the submicron range, is very important to the selection and design of abatement equipment. Techniques for the determination of particle characteristics and their application to equipment design have not been too adequately developed. A recently developed method for assessing particle size analyses involves the use of a portable submicron classifier, which operates on the venturi-scrubbing principle. The unit is essentially a miniature scrubber that cleans a portion of the effluent gas stream at a known set of operating conditions. The observed data permit the determination of particle size characteristics from a series of collection efficiency and pressure drop values determined at various gas and liquor flows to the scrubber.

Chemical Analyses

Chemical analyses [5] are performed on the same particulate matter sample that is collected for weight concentration and size analysis determinations. Inorganic constituents such as metals, metal and nonmetallic ions, and radicals are determined by standard wet chemical methods described in analytical texts. The emissions spectrograph is used for qualitative and semiquantitative metal analyses, especially at low concentrations.

The soluble portions of the particulates collected in the dry filters can be determined by standard Soxhlet procedures. The solvent-soluble constituents are extracted (using such solvents as ether, methyl chloroform, or carbon tetrachloride) from an aqueous suspension of these particulates. After extraction and removal of the solvent, the water portion can be filtered in Gooch tared crucibles to determine the insoluble solids. The combustible volatile components and the ash content of the total collected solids can be determined in a muffle furnace at an ignition temperature of 900° C. Lower temperatures and longer heating periods must be used for the Alundum thimbles. The paper thimbles should not be fired because of their considerable ash content.

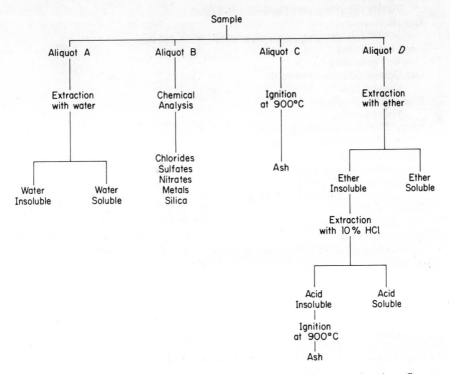

Figure 4–8. Analytical system for particulates in oil-fired combustion flue gas. (Adapted from Holmes, Ref. 5.)

An example of a system used for the analysis of particulate matter contained in oil-fired combustion flue gases is shown in Fig. 4-8. By reference to this diagram, free carbon can be estimated from the loss on ignition of the acid-insoluble portion of aliquot *D*. The total loss on ignition can be corrected for organics to estimate the volatile or partially volatile inorganic portion, such as carbonates. In many source tests, only the combustible and ash content of the particulate matter is of interest because of the type of control equipment that might be considered. Thus, the carbonate content might be of interest to the application of wet scrubbers, whereas the combustible/ash analysis would be invaluable for consideration of an afterburner control system. In the determination of combustion products, the fuel type should be identified and analyzed as a check on the flue gas analysis.

Physical Properties

Certain physical properties of particulate matter and the effluent gas stream represent important parameters in the selection and design of control equipment. Therefore, in many source tests these data are sometimes determined as "extras." There follows a listing of the more important properties and the type of control equipment operations that rely most heavily on these data.

> Gas Temperature
>   Fabric filtration, combustion, adsorption
> Gas Moisture
>   Fabric filtration, cyclonic separation, wet scrubbing
> Gas Corrosiveness
>   Wet scrubbing
> Gas Combustible Contaminants
>   Combustion
> Particulates Solubility
>   Wet scrubbing
> Particulates Conductivity
>   Electrostatic precipitation
> Particle Hardness
>   Cyclonic separation
> Particulates Flammability
>   Fabric filtration, combustion
> Particle Density
>   Cyclonic separation
> Particle Viscidity
>   Cyclonic separation, fabric filtration
>   electrostatic precipitation, wet scrubbing

4.5. COLLECTION AND ANALYSIS OF GASES

Collection Methods and Equipment

The method for the collection of gas samples utilizes the same general type of sampling train as shown in Fig. 4-4. The gases can be collected by liquid absorption or chemical reaction in the wet impingers shown in Fig. 4-6. The dry filter should be located ahead of the wet impingers to remove particulates. As an alternate method, particularly for gases with slow absorption and/or reaction rates, evacuated flasks may be used. For continuous sampling, the train consists of a simple particulates filter (such as a plug of glass wool), a control cock, and a flow meter. As many as five flasks connected in parallel, each with a capacity up to 50 liters, may be used in a single sampling run.

There are available portable gas sampling and/or analytical equipment that operate directly from the effluent gas duct, thereby eliminating the need for volumetric samples. The Orsat apparatus, various colorimetric gas analyzers, combustible gas indicators, and infrared analyzers are examples of such in-line analytical devices. The total combustibles gas indicator is used for the preliminary detection of industrial solvent vapors in the flammable and toxic ranges: 0 to 1.0 of the lower explosive limit and 0 to 1000 ppm of aromatic hydrocarbons. The detector probe is inserted directly into the gas stream, the gas continuously sampled and oxidized by a platinum filament, and the concentration values read directly from the instrument.

Analytical Methods

There are recommended analytical methods available for the various gaseous pollutants presently identified. These methods and their limitations are listed in Table 4-5 for the more prominent gaseous contaminants.

There follows a general description of the sampling and one of the analytical procedures for sulfur dioxide. The total gas volume sampled is converted to standard conditions after being corrected for condensed moisture, if necessary. The sulfur dioxide is collected by continuous sampling through an impinger absorption train, with each of two impingers containing 100 ml of a 5% sodium hydroxide solution. Protection from particulates is achieved by the use of a dry filter ahead of the impingers. Assuming the gas stream to be free of solid sulfates, the solution or aliquot from both impingers is checked for alkalinity; then bromine water is added until an excess is present. The solution is then boiled until a pale yellow color is attained, and is then neutralized by the addition of concentrated hydrochloric acid. A slight excess is added, and boiling is continued to expel the free bromine, with the volume of the solution being maintained by the addition of water. When the solution is free of bromine, as indicated by a drop of methyl red not being decolorized when added, it is filtered to remove any particulates carryover. Hot 10% solution of barium chloride is added for complete precipitation; the precipitate is filtered and washed free of chloride ion. The filter paper and precipitate is maintained in a furnace at 800° C and the weighed, dried precipitate is reported as barium sulfate. The sulfur dioxide is gravimetrically computed from the barium sulfate precipitate and expressed in grams. With this value and the total corrected gas volume, in standard cubic feet (scf), the concentration of sulfur dioxide is calculated as grains per standard cubic feet (grain/scf). The grams of sulfur dioxide can also be converted to standard cubic feet and, with the total sampled gas volume concentration, can be expressed in volumetric parts per million. The emission rate for sulfur dioxide, expressed in pounds per hour, can be computed from the total effluent gas flow rate and the SO_2 concentration. These various calculations are illustrated by the following simple hypothetical problem.

Table 4-5. Summary of Gaseous Pollutant Analytical Methods

Gas	Collection Method	Analytical Procedure	Limitations
Ammonia, NH_3	Impinger train with dilute sulfuric acid	Treatment with Nessler's reagent with color intensity measurement by colorimeter with blue filter; concentration determined from scale calibrated with std. NH_4Cl	1 μg/vol tested in colorimeter
Sulfur Oxides, SO_2 and SO_3	Impinger train with dilute caustic	Bromine water addition, neutralization with HCl; filtration and reaction with $BaCl_2$ with gravimetric determination	1 ppm SO_2 or SO_3 in 30 cf sample
	Alternate: impinger train with solution of tetrachloromercurate	West and Gaeke method: Addition of solution of p-rosaniline hydrochloride in HCl and HCHO solution. Determination made by absorption level in spectrophotometer	0.005–0.2 ppm with gas sample of 40 liters.
Nitrogen Oxides, NO and NO_2	Evacuated flask	Peroxide oxidation to HNO_3; evaporation and reaction with phenoldisulfonic acid, colorimetric determination	5–5000 ppm measured as NO_2
Carbon Monoxide CO	Evacuated flask or direct connection to source	Orsat analysis: by acid cuprous chloride absorption	0.2% vol. accuracy
	Alternate: evacuated flask	Infrared analysis of CO_2 from combustion of CO	5 ppm measured as CO_2
Fluorides, HF and SiF_4	Impinger train with dilute caustic	Perchloric acid addition to yield H_2SiF_6; distillation followed by thorium nitrate titration with alizarin red S indicator	0.02 ppm in 60 cf gas; results expressed as fluoride ion
Chlorine, Cl_2	Impinger train with dilute caustic	Neutralization with sulfuric acid and addition of o-tolidine in HCl colorimetric determination	1 μg/vol tested in colorimeter
Hydrogen Sulfide H_2S	Impinger train with alkaline suspension of cadmium hydroxide	Addition of solution of N,N-dimethyl-p-phenylenediamine in H_2SO_4 and $FeCl_2$; determination by optical density of methylene blue with spectrophotometer	0.05 μg/ml of test solution
Hydrogen Cyanide, HCN	Impinger train with potassium hydroxide solution	Formation of chelate compound by addition of potassium bis-(5-sulfoxino)-palladium, glycine and magnesium chloride; cyanide concn measured by fluorescence intensity	0.02 μg/ml of test solution
Formaldehyde, HCHO	Evacuated flask	Sodium bisulfite reaction, chromotropic/sulfuric acids addition with colorimetric measurement	1 ppm in 2 liters of gas
Hydrocarbons	Evacuated flask	Infrared spectrophotometric measurement with normal hexane as the calibration standard. Total hydrocarbon determination	10 ppm hydrocarbons as n-hexane

Source: Katz, Ref. 12.

These methods are for the "free" gas only. For gaseous mixtures, supplementary and/or alternate techniques may be required.

Data:

Molecular weight SO_2 = 64

Molecular weight $BaSO_4$ = 233.4

Molar volume at std. cond. (70° F, 1 atm) = 387 ft^3/lb-mole

Total gas volume sampled = 90 scf

Total $BaSO_4$ collected = 38.1 g

$$SO_2 \text{ weight} = 38.1 \times \frac{64}{233.4} = 10.5 \text{ g}$$

$$SO_2 \text{ conc} = \frac{10.5}{90} \times \frac{15.4 \text{ grains}}{1 \text{ g}} = 1.80 \text{ grains/scf}$$

$$SO_2 \text{ vol} = \frac{10.5}{454} \times \frac{387}{64} = 0.140 \text{ scf}$$

$$SO_2 \text{ conc} = \frac{0.140}{90} \times 10^6 = 1560 \text{ ppm}$$

The Environmental Protection Agency (EPA) cites references for sampling and analytical procedures for each gas they identify as a pollutant. Thus, in the Control Techniques for Sulfur Oxide Air Pollutants (AP-52) issued in January 1969, a number of references for detailed source test methods for the sulfur oxides are listed. Among these references is the Los Angeles County Air Pollution Control District Method [5], which was the basis for the procedure and demonstration calculations given above.

4.6. ODOR MEASUREMENT

Odor is considered a nuisance type pollutant by many of the regulations. Chemical or physical methods have not yet been devised for the quantitative definition of odors, and reliance on the sense of smell is still the accepted practice. A sampling technique developed by the American Society for Testing Materials (ASTM) is entitled "ASTM Standard Method D1391-57, Standard Method for Measurement of Odor in Atmospheres." Additional information is available from other publications [9].

The ASTM method is based on a relative odor measurement using the human olfactory senses for the determination of odor level or intensity. A sample of the odor-producing gas is diluted with air to a threshold concentration at which only 50% of a selected "smell-testing" panel can detect its presence. The ratio of the total volume of diluted sample to the volume of the original sample is a measure of the odor concentration in the sample. A grab sample of about 200 ml of effluent gas to be tested is usually sufficient. It is aspirated into a collection tube and a known portion of it is transferred to a calibrated syringe. Measured quantities of dilution air are incrementally introduced and smell-tested after each dilution until the threshold concentration is reached. The major variability in this method is the odor sensitivity and consistency of the panel members.

4.7. EMISSION FACTORS

In the identification and measurement of both particulate and gaseous emissions, some knowledge of the expected results, however approximate, would be useful. In evaluating the air pollution potential of a specific industry, some prior idea of the types and general magnitudes of the various uncontrolled pollutants could serve as a guide for the application of emissions abatement equipment. A description of pollutant emission types and concentrations for the various industries is available in Reference 10. This publication gives estimated particulate and gaseous emission values for the various industries as a function of the plant-processing rate. For example, for the manufacture of wet-process phosphoric acid from phosphate rock and sulfuric acid, a brief description of the process is followed by an identification of the major air pollutants and their expected emission rates. The two principal operations in this process which are responsible for pollutant emissions are the reaction and acid concentration areas. The gaseous fluorides emitted are hydrogen fluoride (HF) and silicon tetrafluoride (SiF_4). The emission rates for both compounds are defined as 18 lb/ton of processed phosphate rock for the reaction station and 20 lb/ton of rock from the evaporator condensers. Thus, for an average size plant processing 150 ton/day of phosphate rock, the expected emissions for the reaction and concentration operations would be 113 and 125 lb/hr of total fluorides. These approximate, uncontrolled emission values are very useful to technicians about to perform source emission tests for this industry.

4.8. AMBIENT AIR MEASUREMENTS

As stated earlier, the foregoing discussion was concerned with stationary source emission testing because of its importance to the selection and design of air pollution abatement equipment. The measurement of the ambient air pollutant level is equally important because an acceptable air quality level is the primary goal of the overall air pollution control effort.

Atmospheric sampling requires considerably different techniques than those needed for source emission testing. This fact is reflected in the units used to express particulate concentrations, such as micrograms per cubic meter, or dust fallout rates expressed as tons per mile^2-year. For particulate concentration measurements, one type of instrument used is a high-volume sampler consisting of an air blower, a filter, and flow recorder housed in a protective shelter. The blower has an air-flow capacity in the range of 20 to 60 cfm. The filter has a surface area of about 80 sq in. and is supported on a screen. Both are clamped in a frame from which they can easily be released for sample weighing.

A recent ambient air study [11] performed to determine ambient air particulate concentrations at a jet aircraft facility utilized a high-volume sampler. Multiple sampling runs were conducted on a 24 hr/day basis for a six-month period. For a single month, the particulates were measured outdoors at one location at a concentration range of about 60 to 200 $\mu g/m^3$. In parallel with these measurements, carbon monoxide concentrations were continuously monitored with a Mines Safety Appliance, Model 200, recording, nondispersive, infrared gas analyzer. Hourly average data, taken at the same location where the particulates were measured, indicated a concentration range of about 1 to 20 ppm of CO.

The reduced particulate and gas concentrations encountered in air quality measurements invariably require continuous and automatic sampling equipment so as to ensure obtaining a sufficient amount of total sample to yield valid and significant results.

4.9. SAMPLE SOURCE TEST REPORT

An actual source test report containing the results of a particulates and gas emissions test program is presented below. An industrial boiler utilizing a sulfur-bearing fuel oil was the air pollution source. The pollutants to be identified and quantified were flyash and sulfur oxides in the combustion gas effluent.

SOURCE TEST REPORT

INTRODUCTION

Performance and analytical tests were conducted on the cyclonic collector for flyash emissions control serving the oil-fired boiler at the municipal power plant in the town of Commerce, Georgia. The purpose of the test program was to determine the performance efficiency of the existing collector and to obtain data for the design of equipment to control the sulfur oxides.

Nameplate data for the boiler involved in the tests are as follows:

Boiler Number	3
Manufacturer	Erie City
Rating, lb/hr steam	175,000
Superheat temperature, °F	900
Maximum pressure, psig	1000

The boiler was located at 330 Spruce Street, Commerce, Georgia. The chief engineer responsible for its operation is M. E. Bell. The testing period extended from Dec. 9 through Dec. 13, 1969.

DISCUSSION

The subject boiler is usually operated at peak conditions so that there was no need for variable load tests. All tests were performed at near-design capacity, as tabulated below.

Rating, lb/hr steam	165,000	160,000
Fuel-oil firing rate, lb/hr	13,000	13,150
Steam temperature, °F	880	880
Steam pressure, psig	870	870
Excess air, %	10	12
Date of tests	12/11/69	12/12/69

The boiler was fired with Venezuelan residual oil, containing 2.7% sulfur. The flyash collectors were Aerotec Corp., Type 6 MP. The collected ash is not reinjected into the boiler. The oil ash content was 0.3%.

The arrangement of the boiler and collector is shown below. Locations of the test stations are indicated by the symbol *T*.

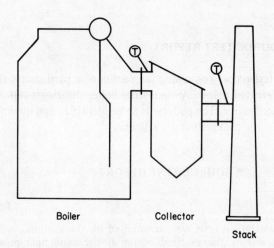

The test program comprised the following procedures at each test port:

1. Mass flow and temperature traverse.
2. Isokinetic sampling with thimble filter.
3. Sulfur oxides sampling with wet impingers preceded by thimble filter.
4. Determination of total sulfur oxides by chemical analysis, using barium chloride precipitation method.

SUMMARY OF RESULTS

A summary of test results is tabulated below:

	Test of 12/11/69		Test of 12/12/69	
Process Quantity	*Collector Inlet*	*Collector Discharge*	*Collector Inlet*	*Collector Discharge*
Flue gas rate, scfm	52,200	53,000	51,500	51,800
Flue gas temp., °F	650	650	650	650
Flyash concentration, grain/scf	0.070	0.021	0.081	0.023
Flyash mass rate, lb/hr	31.2	9.5	36.8	10.2
Collector efficiency, %		69.6		72.3
SO_x concentration, ppm		1250		1320
SO_x emissions rate, lb/hr		660		680

The process values obtained were fairly consistent for the two series of tests undertaken. The flyash loadings are within the expected range dictated by experience for this type of operation. The total uncontrolled emissions for the particulates were checked by material balances and reasonable agreement obtained. Based on the oil-firing rate of about 13,000 lb/hr and an ash content of 0.3%, the calculated total ash content of the flue gas would be 13,000 × 0.003, or 39 lb/hr, which checks remarkably well with the values of 31.2 and 36.8 lb/hr obtained in the emission tests.

The sulfur oxide emissions have also been confirmed by material balance calculations. Thus, for an oil-firing rate of 13,000 lb/hr at a sulfur content of 2.7 wt%, the calculated SO_x emissions would be (13,000 × 0.027 × 64/32), or 702 lb/hr. This is in excellent agreement with the test values of 660 and 680 lb/hr.

The performance efficiencies for the multitube type of cyclone collector seems reasonable for the duty imposed. A particulate size analysis determination was not within the scope of the current test program. Actually, there have been serious doubts expressed regarding the validity of size analysis data for oil-fired flyash obtained by air elutriation methods because of the pronounced agglomerative characteristics of this material. Even if a more reliable size distribution technique is adopted, such as microscopy, there is some confusion as to the relevancy of these data to the performance characteristics of cyclone collectors. Therefore, in this particular test program, the cyclone collector of the multitube type was judged to be operating satisfactorily at the observed efficiencies when compared with results obtained for similar applications.

SAMPLING AND ANALYTICAL PROCEDURES

The gas stream mass-flow values were determined with a standard type pitot tube, having a correction factor of 1.0, and an inclined manometer; see Fig. 4-1. Two complete sets of readings were taken for each of the collector inlet and outlet ducts. The duct configurations made it necessary to take readings at 18 traverse points on the inlet duct and 15 on the outlet duct. The location of both testing stations ensured relatively smooth flow, as reflected by the velocity head measurements. An iron-constantan thermocouple was used for all temperature determinations.

The particulates sampling train was similar to that shown in Figure 4-4, with Alundum thimble filters (Fig. 4-5). Samples were collected isokinetically at an average gas sampling rate of 1.3 acfm.

The gas sampling train comprised four wet impingers; the first containing distilled water for condensate collection, the following two a weak caustic solution, and the fourth remaining empty to act as a deentrainment chamber. Total sulfur oxides, expressed as SO_2, were analytically determined by the barium sulfate precipitation method; see Table 4-5.

Both the particulates and gaseous emission values were confirmed by material balance calculations, based on the fuel-oil firing rate. Excellent agreement was obtained.

APPENDIXES

Observed data and calculations are contained in this section, as follows:
Appendix I: Pitot Tube and Particulate Sampling Data
Appendix II: Gas Flow and Particulates Loading Calculations
Appendix III: Sulfur Oxides Data and Calculations

Appendix I. Pitot Tube and Particulate Sampling Data
(for test of 12/11/69 only)

Plant: Commerce City Municipal Power Plant
Date: Dec. 11, 1969
Unit tested: Collector for Boiler No. 3
Sampling locations: Station 1, collector inlet; Station 2, collector outlet
Static pressure: Station 1, 6.0 in. w.g. Station 2, 3.4 in. w.g.
Barometric pressure: 29.7 in Hg
Pitot tube factor: 1.0
Sample probe: Dia. = 0.285 in.; area = 0.000443 ft^2

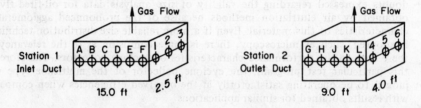

(see Table I-1 for Pitot tube readings)

Appendix II. Gas Flow and Particulates Loading Calculations
(for test of 12/11/69 only)

Station 1, Inlet Duct

Duct area = 37.5 ft^2
Avg. velocity head = 0.25 in. w.g.; avg temp = 650°F
Total sampling period = 90 min
Net dust sample wt; using Alundum thimble = 0.256 g
Sampling abs press = 29.7 + (6.0/13.6) = 30.1 in. Hg
    Average gas velocity from velocity head values, Table 4-1, for velocity head of 0.25 in. w.g. at 650°F,

$$\text{Velocity} = 48.3 \text{ fps}$$

Total flow in duct at 650°F,

$$\text{Flow} = 48.3 \times 60 \times 37.5 = 109,000 \text{ acfm}$$

Total flow in duct at standard conditions,

$$\text{Flow} = 109,000 \times (530/1110) \times (30.1/29.9) = 52,200 \text{ scfm}$$

Total gas sampled at 650°F,

$$\text{Vol} = 49 \text{ fps} \times 60 \times 0.000443 \times 90 = 117 \text{ acf}$$

Table I-1. Pitot tube readings

Sample Point	Time, P.M. Start	Time, P.M. Stop	Diff., min	Duct Temp., °F	Vel. Head	Sample Point	Time, P.M. Start	Time, P.M. Stop	Diff., min	Duct Temp., °F	Vel. Head
A-1	1:15	1:20	5	615	0.20	G-4	4:00	4:07	7	645	0.20
A-2	1:20	1:25	5	690	0.30	G-5	4:07	4:14	7	660	0.38
A-3	1:25	1:30	5	685	0.25	G-6	4:14	4:21	7	660	0.26
B-1	1:30	1:35	5	610	0.22	H-4	4:21	4:28	7	620	0.24
B-2	1:35	1:40	5	660	0.32	H-5	4:28	4:35	7	645	0.40
B-3	1:40	1:45	5	675	0.21	H-6	4:35	4:42	7	655	0.24
C-1	1:45	1:50	5	640	0.18	J-4	4:42	4:49	7	655	0.26
C-2	1:50	1:55	5	670	0.34	J-5	4:49	4:56	7	660	0.45
C-3	1:55	2:00	5	67C	0.23	J-6	4:56	5:03	7	655	0.13
D-1	2:00	2:05	5	620	0.15	K-4	5:03	5:10	7	610	0.18
D-2	2:05	2:10	5	655	0.45	K-5	5:10	5:17	7	650	0.37
D-3	2:10	2:15	5	670	0.15	K-6	5:17	5:24	7	650	0.39
E-1	2:15	2:20	5	620	0.15	L-4	5:24	5:31	7	650	0.24
E-2	2:20	2:25	5	670	0.48	L-5	5:31	5:38	7	650	0.26
E-3	2:25	2:30	5	670	0.12	L-6	5:38	5:45	7	650	0.38
F-1	2:30	2:35	5	620	0.20	Total			105		
F-2	2:35	2:40	5	680	0.40	Average				650	0.28
F-3	2:40	2:45	5	680	0.15						
Total			90								
Average				650	0.25						

Total gas sampled at standard conditions,

$$Vol = 117 \times (530/1110) \times (30.1/29.9) = 56.5 \text{ scf}$$

Dust loading = $0.256 \times (7000/454) \times (1/56.5) = 0.070$ grain/scf
Flyash mass rate = $0.070 \times (52,200/7000) \times 60 = 31.2$ lb/hr

Station 2, Outlet Duct

Duct area = 36.0 ft^2
Avg velocity head = 0.28 in. w.g., avg. temp. = $650°F$
Total sampling period = 105 min
Net dust sample wt; using Alundum thimble = 0.093 g
Sampling abs press = $29.7 + (3.4/13.6) = 30.0$ in. Hg

Average gas velocity from velocity head values, Table 4-1, for velocity head of 0.28 in. in. w.g. at $650°F$ = 51.1 fps

Total flow in duct at $650°F$

$$Flow = 51.1 \times 60 \times 36.0 = 110,000 \text{ acfm}$$

Total flow in duct at standard conditions,

$$Flow = 110,000 \times (530/1110) \times (30.0/29.9) = 53,000 \text{ scfm}$$

Total gas sampled at $650°F$

$$Vol = 51 \text{ fps} \times 60 \times 0.000443 \times 105 = 142 \text{ acf}$$

Total gas sampled at standard conditions,

$$Vol = 142 \times (530/1110) \times (30.0/29.9) = 68.4 \text{ scf}$$

Dust loading = $0.093 \times (7000/454) \times (1/68.4) = 0.021$ grain/scf

Flyash mass rate = $0.021 \times (53,000/7000) \times 60 = 9.5$ lb/hr

Appendix III. Sulfur Oxides Data and Calculations

(for test of 12/11/69 only)

Sampling Data

Meter pressure = -4 in. Hg
Meter temp = $65°F$
Volume sampled = 42 ft^3
Water condensed = 51 g
Mole vol at std. cond. = 387 ft^3

Gas Moisture Content

Volume of water condensed at standard conditions,

$$Vol = \frac{51}{454} \times \frac{387}{18} = 2.4 \text{ ft}^3$$

Volume of gas sampled at standard conditions,

$$Vol = 42 \times \frac{530}{460 + 65} \times \frac{29.7 - 4.0}{29.9} = 36.7 \text{ scf}$$

Gas moisture content (approx.)

$$Content = \frac{2.4}{36.7} \times 100\% = 6.5 \text{ vol } \%$$

SO_X Content

Laboratory analysis reports 3450 mg SO_X as SO_2

$$SO_2 \text{ concn} = \frac{3450}{454,000} \times \frac{387}{64} \times \frac{1}{36.7} \times 10^6 = 1250 \text{ ppm}$$

$$SO_2 \text{ emissions rate} = \frac{1250}{10^6} \times 53,000 \times \frac{64}{387} \times 60 = 660 \text{ lb/hr}$$

REFERENCES

1. *Velocity Tables.* Research-Cottrell, Inc., Bound Brook, New Jersey.
2. *Psychrometric Tables and Charts,* 2d ed., Zimmerman and Lavine, Industrial Research Service, Inc., Dover, N. H., 1964.
3. P. Muller, "Beitrag zur Frage des Einflusses der Schwefelsäure auf die Rauchgas–Taupunkt Temperatur," *Chemie Ing. Technik,* Vol. 31, No. 5 (1959), pp. 345–351.
4. W. C. L. Hemeon and A. W. Block, "Stack Dust Sampling: In-Stack Filter or EPA Train," *J-APCA,* Vol. 22, (1972), p. 516.
5. H. Devorkin, R. L. Chass and A. P. Fudurich, *Air Pollution Source Testing Manual,* Air Pollution District, Los Angeles County, Calif., Dec., 1972.
6. W. H. Walton, R. C. Faust, and W. J. Harris, "A Modified Thermal Precipitator for Quantitative Sampling of Aerosols for Electron Microscopy," Porton Tech. Paper No. 1, Ser. 83, March 1947.
7. P. Drinker and T. Hatch, *Industrial Dust,* 2d ed. McGraw-Hill Book Co. Inc., New York, N. Y., 1954.
8. T. Hatch, "Determination of 'Average Particle Size' from the Screen Analysis of Non-Uniform Particulate Substances," *J. Franklin Inst.,* Vol. 215 1933, p. 27.
9. J. L. Mills, R. T. Walsh, K. D. Luedtke, and L. K. Smith, "Quantitative Odor Measurement," *J-APCA,* Vol. 13, No. 10 (1963), pp. 467–475.
10. "Compilation of Air Pollution Emission Factors–AP42," U.S. Environmental Protection Agency, Office of Air Programs, Research Triangle Park, N.C., February 1972.
11. R. E. George, J. S. Nevitt, and J. A. Verssen, "Jet Aircraft Operations Impact on the Air Environment," *J-APCA,* Vol. 22, No. 7 (1972), p. 507.
12. M. Katz, *Measurement of Air Pollutants–A Guide to the Selection of Methods,* World Health Organization, Geneva, 1969.

Chapter 5

PARTICULATE POLLUTANTS
CONTROL EQUIPMENT

5.1. PROBLEM DEFINITION

Problem Statement

Particulate and gaseous pollutants were identified and classified in Chapter 2. The effluent being discharged to the atmosphere from any process may contain any one, or a combination, of these particulate and gaseous contaminants. Once the pollutant is identified, its discharge rate known, the waste gas process conditions determined, and the pertinent regulations defined, then the selection of the optimum type of control equipment may be considered.

In the definition of any air pollution problem, complete information regarding the process effluent is required. A sample data sheet, listing the major process specifications for a fictitious aggregates plant, is shown in Table 5-1.

Data Evaluation

These emissions data must be critically assessed and their relationship to the pertinent control regulations established before equipment selection can be considered. Because the definition of an air pollution problem is a very deceptive area, care must be taken to certify the types and emission levels of the pollutants. Although the Rock City Sand Company data do not indicate a need for gaseous emissions control, such a possibility should be questioned. If the aggregates drier is oil-fired, then the possibility of sulfur oxides emissions from sulfur-bearing fuel oil should be examined. A heavy residual fuel could contain up to 4% sulfur, equivalent to an emission concentration of about 3000 ppm of sulfur oxides. This value would be in violation of most regional codes. On the other hand, if gas firing is practiced, an examination of the nitrogen oxides level should be made. Should either gaseous pollutant pose a serious emissions problem, then the control equipment selection basis would be considerably broadened.

The gas flow decides the control equipment sizing and is therefore a most important variable. It is usually expressed as cubic feet per minute (cfm), either at actual conditions (acfm) or standard (scfm) at $70°$ F and 14.7 psia. Flows are usually determined by Pitot tube measurements, as explained in Chapter 4. Values representative of maximum, minimum, and average plant capacity conditions should be recorded.

Because of the importance of dust-loading values to control equipment perfor-

Table 5-1. Effluent Stream Process Specifications

Company name: Rock City Sand Company
Plant location: Commerce, Georgia
Plant type: Aggregates plant
Plant capacity: 30 tons/hr
Effluent source: Dryer

Pollutant classification: Particulates
Pollutant identification: Silica
Gas type: Flue gas
Gas flow: 32,000 acfm
Gas temperature: 450 °F
Moisture content: 18 vol %
Dust loading: 2.1 grains/acf

Dust size analysis:

Micron Size	Weight %
+10	2.4
+5–10	5.8
+2.5–5	19.2
+1.5–2.5	42.6
−1.5	30.0

mance, their validity must be confirmed. Dust loadings can be expressed in many units: lb/hr, mg/ft^3, grain/acf, and grain/scf, being only a few of them. In the interpretation of dust loadings, care must be exercised to determine their relationship to plant capacity, material being processed, and the cyclical nature of the process. It is most important that the specified loadings be representative of 100% plant capacity. If source testing was performed at the Rock City Sand Company while the plant was operated at, say, 80% capacity, then (based on the approximate loading-to-capacity square law) the actual dust loading might be as great as 5.3 grains/acf at the 30 tons/hr plant production level. The type of product being processed also has a considerable influence on the dust loading.

In the mythical aggregates plant we have chosen as our example, as many as six different aggregate mixes of varying particle size and moisture content might be handled. The more coarse products will yield relatively light loadings at equivalent processing rates, as will a partially dried product. A washed sand is also known to emit light concentrations of coarse particulates. The cyclical nature of various processes should also be considered in evaluating dust loadings. As an example, in processing secondary aluminum, the cycle period is approximately 2 hr. At the start of the operation, aluminum chloride loadings may be on the order of 1 to 2 grains/scf. However, as the reaction proceeds, the chlorine input is made increasingly available for reaction with the molten aluminum so that, at the completion of the cycle, loadings may reach a peak of 20 to 30 grains/scf. In the selection and sizing of equipment to meet specific regulations, those plant operating conditions must be chosen as represented by maximum gas flows, maximum dust loadings, and minimum dust particle size.

The Regulations

Having been assured as to the validity of the emissions data, the pertinent pollution control regulations must be considered. The location of the Rock City Sand Company brings it under the jurisdiction of the State of Georgia Regulations [1]. The 1967 Georgia Air Quality Control laws cite particulate emission limitations from manufacturing processes under Section 6.1 and 6.2. Under Section 6.1, an emission rate of 40 lb/hr is allowed, based on a process weight rate for an existing plant of 30 tons/hr. In addition, under Section 6.2, there is imposed an emissions concentration requirement not to exceed 0.07 grain/scf at a source-gas volumetric flow of 18,700 scfm, equivalent to 32,000 acfm at 450° F. A Ringelmann or opacity restriction does not appear to apply to manufacturing process in this version of the rules.

To determine the degree of emissions control required by the regulations, it is necessary to convert the emission values in Table 5-1 to those units specified by the regulations. The dust-loading value of 2.1 grains/acf is equivalent to an emissions rate of 575 lb/hr and an emissions concentration of 3.6 grains/scf. Based on the allowable emissions rate and emissions concentration requirements of 40 lb/hr and 0.07 grain/scf, there would be required particulate collection efficiency values of 93.0 and 98.1%, respectively. According to the interpretation of the regulations, the more severe performance value of 98.1% would prevail.

5.2. PROCESS EVALUATION

Before the selection of control equipment is seriously considered, the process should be examined to determine its amenability to emissions control. This is the system approach to air pollution control which is rapidly gaining wide acceptance. A simple representation of the relationship of process emissions, required control equipment performance, and emissions limitations imposed by the regulations is shown in Fig. 5-1.

The process emissions specified for the Rock City Sand Company were moderate,

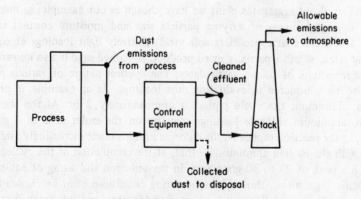

Figure 5-1. Dust emissions control sequence.

so the collection efficiency demands on the control equipment were not too severe. However, in some aggregates plants, loadings as high as 20 grains/acf have been experienced, which required collection efficiencies on the order of 99.8 to 99.9%. This efficiency level is difficult to attain with a single collector and therefore some investigation of the process parameters is warranted. One simple approach is to compare the design capacity of the process with the operating level. Very approximately, doubling the plant capacity could quadruple the particulate emissions. Duct design velocities might also be a factor responsible for unusually high dust loadings. Should operation of the plant beyond its rated capacity be responsible for the high dust loadings, then the costs of extending the operating day must be traded off against the increased capital investment requirement for the control equipment necessary at the expanded production level.

In systems for the control of particulate emissions at conveyor, screen, elevator, and bin transfer points, the hood design is very critical [2]. Should the ventilation velocities be too low, inadequate emissions capture will result. However, if hood design velocities are too high, then the dust pickup rate will be excessive, thereby imposing a heavy load on the control equipment performance.

One rather interesting industrial experience can illustrate the importance of ventilation velocities to emission control. [3]

> An indirectly heated steam-tube rotary drier for processing minerals was operated without control equipment in the hills of North Carolina in 1962. The need for control equipment was soon realized, more as a function of product recovery than the need to satisfy regulatory demands at that time. It was decided to install a wet scrubber, with the liquor effluent being returned to the process. In the interests of economy, a source test was not performed. Instead, an estimated drier air-flow rate of "about" 6000 acfm, as furnished by the processor, was used to design the scrubber, fan, and ductwork. However, at start-up, the draft created at the discharge end of the drier was so severe that as much as 25% of the dried product was sucked up in the ventilation air stream and drawn through the fan and scrubber. Estimated dust loadings in the neighborhood of 30 grains/acf were indicated. Further experimental and testing efforts indicated that the actual drier air flow should have been 2200 acfm. To reconcile these two flows, an atmospheric damper was installed downstream of the drier and ahead of the fan. A flow of 3800 acfm of air was constantly bled into the system at this point, thereby satisfying the drier ventilation requirement of 2200 acfm while maintaining the scrubber design flow at 6000 acfm. At this reduced ventilation rate, the drier-discharge loadings fell off to about 4 grains/acf.

This experience would indicate the need for confirming the definition of the emissions problem. It also introduces the use of the atmospheric damper, which permits process gas-flow variations while satisfying the control equipment design rate so necessary for optimum performance.

In batch operations, the dust loadings and gas flow rates are usually variable, peaking to a maximum value some time during the cycle period. Although the majority of emissions regulations do attempt to allow for these peak values, the designer of control equipment must reduce these maximum values within some economic framework. Where such peak loadings might represent a tenfold increase over the average emission rate, consideration must be given to the introduction of surge capacity into the system or lengthening of the operating cycle. The latter approach, involving process modifications, represents a systems approach to the problem and must be subjected to economic analysis.

One other common area involving process evaluation is the question of optimizing emission sources. In a metallurgical process such as aluminum reduction, which utilizes multiple operations, the question arises as to whether each individual source or the total building should be ventilated. Ductwork, equipment sizing, operating procedures, and plant hygienic standards are some of the factors to be considered in this decision. In one such systems study, the excessive building ventilation rates required to maintain acceptable in-plant pollutant levels discouraged the use of building monitor control. The fabric filter equipment requirements were economically unrealistic. Instead, individual equipment ventilation with a multiple wet scrubber system was recommended.

In the various foregoing examples an expedient solution to the emissions problem could have been accomplished by the application of suitably sized equipment to do the job. However, the capital equipment requirements and operating costs would have been considerable. The systems approach requires the evaluation of the overall process in terms of utility requirements, production costs, and air pollution control needs, and must establish the basis for the design of an integrated system that will provide for compliance with the most stringent air pollution control regulations at minimum annual costs.

5.3. EQUIPMENT SELECTION

With the establishment of the various process parameters involved in the problem, the regulations defined, and the process evaluated from a system viewpoint, the selection

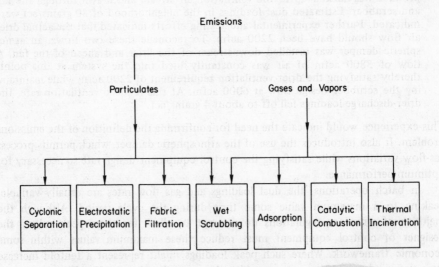

Figure 5-2. Particulate and gaseous pollutants control equipment.

of optimum control equipment can be undertaken. Just as the plant type and plant capacity serve to establish the allowable emissions rate as specified by the regulations, so do the plant type, effluent source, and pollutant classification indicate to the experienced environmental engineer the type of control equipment to recommend. For the aggregates plant-effluent conditions described in Table 5-1, reference can be made to Fig. 5-2, which discloses that for particulates abatement, a cyclonic separator, electrostatic precipitator, fabric filter, or wet scrubber can be applied [4].

The optimal choice of equipment for any single particulates collection problem is determined by many variables. A general control equipment selection guide for various industries based on the plant effluent source is shown in Table 5-2. Such factors as performance requirements, process considerations, and economic factors must be balanced in making a selection. There follows a discussion of these variables and their influence on the evaluation of particulate control equipment generally and, where applicable, to their effect on the problem facing the Rock City Sand Company specifically.

Collection Efficiency

The collection efficiency for a specific item of control equipment in a particular application is a measure of that equipment's performance. It is defined as the percentage of the entering dust that has been collected and removed from the effluent stream. Care must be taken to use consistent units, usually pound per hour, or grains per standard cubic foot. The required efficiency values for the Rock City Sand Company (see Section 5.1) were calculated as follows:

Process Weight Basis	
Process emissions, lb/hr	575
Allowable emissions, lb/hr	40
Concentration Basis	
Process emissions, grains/scf	3.6
Allowable emissions, grain/scf	0.07
Collection Efficiency	$[(3.6 - 0.07)/3.6]\,100 = 98.1\%$

A simplified calculation that improves slide rule accuracy would be: $100 - (0.07/3.6)100 = 98.1\%$.

Air quality regulations specify emissions loadings either as grains per standard cubic foot or pounds per hour. Suppliers of control equipment usually express the performance characteristics of their equipment in terms of collection efficiency. Although the supplier is usually willing to guarantee [6] discharge loadings from his equipment, consistent with the regulatory requirements, such promises are made contingent on the inlet loadings as specified by the process operator.

Various types of particulate control equipment are associated with different collection efficiency ranges. The cyclonic separator will operate at efficiencies in the range of 70 to 90% with most industrial dusts, and therefore is usually installed in series with one of the other collector types. The electrostatic precipitator, fabric filter, and wet scrubber can be effective in the range of $99^{+}\%$.

Once again it must be emphasized that equipment performance must meet a specific emissions discharge, and therefore its efficiency capabilities must take into

Table 5-2. Control Equipment Selection for Various Processes

| | Dust Properties | | Collector Types | | | | |
| | | | Cyclonic Separator | | | | |
Process, and Source	Loading	Particle Size	Cyclone	Multitube	Electrostatic Precipitator	Fabric Filter	Wet Scrubber
Coal Mining							
Material handling	Moderate	Medium	Seldom	Occasional	Never	Frequent	Frequent
Bunker ventilation	Moderate	Fine	Occasional	Frequent	Never	Frequent	Occasional
Dedusting	Heavy	Medium	Frequent	Frequent	Never	Frequent	Occasional
Drying	Heavy	Fine	Seldom	Occasional	Never	Never	Frequent
Rock Products							
Material handling	Moderate	Medium	Seldom	Occasional	Occasional	Frequent	Frequent
Dryers/kilns	Moderate	Med. coarse	Frequent	Frequent	Occasional	Seldom	Frequent
Cement kiln	Heavy	Med. fine	Seldom	Frequent	Frequent	Never	Seldom
Cement grinding	Moderate	Fine	Seldom	Seldom	Seldom	Frequent	Never
Power plant							
Coal firing, chain grate	Light	Fine	Never	Seldom	Never	Never	Never
Coal firing, stoker	Moderate	Coarse-fine	Seldom	Frequent	Seldom	Never	Never
Coal firing, pulverized	Heavy	Fine	Seldom	Frequent	Frequent	Never	Never
Foundry							
Shake-out	Moderate	Fine	Seldom	Seldom	Never	Seldom	Frequent
Sand handling	Moderate	Med. fine	Seldom	Seldom	Never	Seldom	Frequent
Tumbling mills	Heavy	Med. coarse	Never	Never	Never	Frequent	Frequent
Abrasive cleaning	Mod. heavy	Med. fine	Never	Occasional	Never	Frequent	Frequent

Flour and Feed Mills						
Grain handling	Light	Medium	Frequent	Occasional	Never	Frequent	Seldom
Grain dryers	Light	Coarse	Never	Never	Never	Never	Never
Flour dust	Moderate	Medium	Frequent	Frequent	Never	Frequent	Occasional
Feed mill	Moderate	Medium	Frequent	Frequent	Never	Frequent	Occasional
Steel Mills						
Electric furnace	Light	Fine	Never	Never	Seldom	Frequent	Frequent
Blast furnace	Heavy	Coarse-fine	Frequent	Seldom	Frequent	Never	Frequent
Open hearth	Moderate	Coarse-fine	Never	Never	Frequent	Seldom	Seldom
Ferrous cupola	Moderate	Coarse-fine	Seldom	Seldom	Occasional	Occasional	Frequent
Wood Products						
Woodworking machines	Moderate	Coarse-fine	Frequent	Occasional	Never	Frequent	Seldom
Sanding	Moderate	Fine	Frequent	Occasional	Never	Frequent	Occasional
Waste conveying	Heavy	Coarse-fine	Frequent	Seldom	Never	Occasional	Occasional
Rubber Products						
Mixers	Moderate	Fine	Never	Never	Never	Frequent	Frequent
Batch rolls	Light	Fine	Never	Never	Never	Frequent	Frequent
Grinding	Moderate	Coarse	Frequent	Frequent	Never	Frequent	Frequent

Source: Kane, Ref. 5.

101

consideration the process discharge loading. As an example, consider a process loading of 20 grains/scf with the allowable emissions discharge set at 0.05 grain/scf. This would demand an overall collection efficiency of 100 − (0.05/20) 100%, or 99.75%. This could be accomplished by the application of a cyclonic separator in series with an electrostatic precipitator. Assuming that the dust particle characteristics are suitable to predict collection efficiencies of 75 and 99% for the cyclone and electrostatic precipitator collectors, respectively, when connected in series, then the overall collection efficiency and discharge loading would be calculated as shown in Table 5-3.

As for the Rock City Sand Company, the required maximum collection efficiency of 98.1% does not present too formidable a problem. However, from the foregoing, it is doubtful that a cyclonic separator alone can perform the required duty, so that the selection can be narrowed to the precipitator, fabric filter, or wet scrubber. As an alternative, the cyclonic separator may be selected for operation in series with any one of these three collectors.

Gas Flow

The gas flow essentially fixes the size of the collection equipment. Average superficial velocity values for the four particulate collector types is shown in Table 5-4. These velocities can be directly applied to the gas flow rate in actual cubic feet per minute (acfm) in order to obtain the approximate cross-sectional area of the cyclonic separator, precipitator, and fabric filter. For the wet scrubber the equivalent saturated gas flow must be used to calculate the equipment area.

An increase in gas flow through the cyclonic separator and wet scrubber generally provides an increase in the collection efficiency, whereas in the precipitator and fabric filter the performance is impaired at increased throughputs.

The precipitator and fabric filter are of modular construction so that essentially single collectors comprising multiple modules can be provided for large gas flows. However, because of limitations in the particle collection mechanism, the cyclonic separator and wet scrubber are limited in capacity so that multiple units must be considered for the higher flow rates. Very large flows from numerous emission sources can be broken into several smaller systems for any of the four basic collector types, to provide increased flexibility and minimize the complexity of the ductwork. Electro-

**Table 5-3. Combination Collector
Efficiency Calculation**

Cyclonic Separator	
Inlet loading, grains/scf	20.0
Collection efficiency, %	75
Discharge loading, grains/scf	5.0
Electrostatic Precipitator	
Inlet loading, grains/scf	5.0
Collection efficiency, %	99
Discharge loading, grain/scf	0.05

Table 5-4. Design Superficial Velocities for Particulate Collectors

Collector Type	Design Velocity, fpm
Cyclonic separator	
Single cyclone	600–1000
Multiple tubes	1200–2400
Electrostatic precipitator	300–600
Fabric filter	1–30
Wet scrubber	200–800

static precipitators are very seldom recommended for gas flows less than 50,000 acfm because of the economic burden imposed by the basic electrical accessory equipment.

The gas flow specified for the Rock City Sand Company of 32,000 acfm would appear to eliminate the electrostatic precipitator as a collector candidate for this application.

Gas Properties

A number of gas properties must be weighed because of their influence on the selection and design of the various collectors [7]. Among these, the most important are temperature, moisture, and gaseous contaminants. The effect of each is described below.

Temperature. If the gas stream temperature is in excess of 600° F, the fabric filter is not recommended because of limitations imposed by the fabric. The cyclonic separator, of mild steel construction, is usually limited to operating temperatures of 750° F. With castable internal parts, the limiting temperature can be increased to about 2500° F. Increased temperature ratings for the electrostatic precipitator are obtained by the use of stainless steel or other high-temperature metal internals.

Moisture. Gases with a moisture or other condensible content in excess of about 20 vol % are liable to condensation inside the collector. The fabric filter is totally inoperative under these conditions because of the formation of mud on the collection surfaces. The cyclonic separator and electrostatic precipitator are similarly adversely affected.

Gaseous Contaminants. These fall into three major categories: combustible gases, such as hydrocarbons, that impose a fire or explosion hazard with the use of the three dry type collectors; corrosive gases, which, if maintained above their dewpoint in the dry collectors, do not constitute a corrosion problem, but which must lead to the use of expensive materials of construction in the wet scrubber; acidic or alkaline-based gases that must be present, although in very minute concentrations, to ensure the effective performance of the electrostatic precipitators.

In addition to the influence of these gaseous contaminants on equipment selection and design, there is the very important consideration that if the contaminant gas concentrations are in excess of the allowable regulation values, a combination particulate and gaseous pollutant control problem may develop. If this is the case, then either the wet scrubber alone, for the removal of both pollutants, or a dry collector in series with gaseous control equipment must be considered.

These various gas properties have very little influence on our continuing selection of suitable equipment for the Rock City Sand Company. The gas temperature of 450° F and moisture content of 18 vol % are both within the design limitations for the cyclonic separator, fabric filter, and wet scrubber. Furthermore, we will assume that the gaseous contaminants are not present.

Dust Loadings

The greater the dust loading, the more demanding is the required collection efficiency to meet a particular emissions regulation. In the previous discussion of collection efficiency, it was demonstrated that, at an emissions loading of 20 grains/scf and an allowable discharge of 0.05 grain/scf, a collection efficiency of 99.75% would be required.

The cyclonic separator and wet scrubber are not too sensitive to dust-loading levels, but the precipitator and fabric filter are influenced by this factor. In the precipitator the power input to the various sections must be variably controlled so as to be inversely proportional to the diminishing dust loadings as the gas stream passes from the inlet to the discharge section. In the fabric filter, high dust loadings demand increased filtration area and/or cleaning frequency.

The dust-loading value of 2.1 grains/acf specified in Table 5-1 is very nominal and will not be a relevant parameter in our efforts to determine whether a cyclonic separator, fabric filter, or wet scrubber should be chosen for this application.

Dust Particle Size. One of the most important considerations in the selection of particulate collection equipment is the dust particle size. Whether a dust particle is 5 μ or 0.5 μ in diameter makes a great difference in the type, complexity, and operating cost of control equipment at any particular performance level. The cyclonic separator can collect dusts having particle sizes of 5 μ and greater at reasonable efficiencies. The medium-energy wet scrubber can control effluents containing dust particles in the size range of 5 to 1 μ. Below 1 μ, high-energy scrubber operating at increased pressure-drop levels are required. The dust particle size is an important variable in the design of the electrostatic precipitator because the power input causing migration of the dust particle to the collection plate is inversely dependent on the particle diameter. The fabric filter operates at performance levels that are not too dependent on the dust particle size.

With these general criteria, let us examine the dust size analysis of the particulates to be collected for the aggregates plant, Table 5-1. The −5 μ fraction is 91.8% of the total size distribution. Because of the collection efficiency versus particle size limitation of the cyclonic separator (i.e., essentially ineffective at less than 5 μ), this collector type will not perform at the required 98.1% collection efficiency. Therefore, the choice must be either the fabric filter or the wet scrubber. As a promising alternate to the fabric filter, a cyclonic separator/wet scrubber combination can also be considered, for reasons explained below.

Dust Particle Properties. Dust particle properties other than size have an important bearing on control equipment choice. Particle properties such as specific gravity, shape, abrasiveness, resistivity, viscidity, flammability, and solubility influence equipment performance. The effect of these is described as follows:

1. Specific Gravity: Because it depends on inertial forces to separate the dust particles from the gas stream, the performance characteristics of a particular cyclonic separator is a function of the specific gravity of the dust particle. Thus, for the same operating conditions, a greater collection efficiency can be expected for silica dust (sp gr 2.5) than for sawdust, which has a specific gravity of about 50% of this value.

2. Shape: Plugging of collector passages is sometimes caused by the dust particle shape. Wood shavings, fabric lint, flat char, feather particles, and grain hulls are some of the dust particle types that pose a problem in their collection and transport. Because of their high void and low bulk weight characteristics, such materials are not recommended for handling in multiple cyclone collectors or electrostatic precipitators.

3. Abrasiveness: Dust loadings, particle size, and particulate shape contribute to the abrasive qualities of a specific material. The smaller the particle size, the greater the abrasive potential per unit concentration because of the increased surface area of the fine dusts. Careful design of ductwork, minimization of impact mechanisms in the collection devices, and reduction of equipment gas flow velocities will tend to reduce abrasion problems. The cyclonic separator and precipitator are not recommended for severely abrasive service.

4. Resistivity: The dust particle resistivity is most important to acceptable performance of the electrostatic precipitator. Resistivities in the range of 10^4 to 10^{13} ohm-cm are acceptable. Dusts with resistivities in excess of, or much less than, the values in this range will cause a reduction in the overall collection efficiency.

5. Viscidity: Dusts of a sticky nature will cause plugging in practically all four collector types. The multiple-tube type of cyclonic collector and the fabric filter are most susceptible to this type of problem. Hygroscopic dusts will cause difficulties in ductwork passages and dry collectors after shutdown because of the adsorption of atmospheric moisture. Insulation or a source of heat such as a simple gas or oil-fired burner can diminish this difficulty.

6. Flammability: Most materials in a finely divided state are potentially flammable. Flour dust, metallic fines, and coal and mineral dusts have all been known to form explosive mixtures in air streams at certain temperature levels and concentrations. The hazard is at a maximum for very fine particle sizes and high dust loadings. The obvious solution to this problem is the use of a wet scrubber.

7. Solubility: A common fallacy is in the belief that a soluble dust is more amenable to collection in a wet scrubber than one that is less soluble or even completely insoluble. This is not true because the controlling factor in the removal of particulates by a wet scrubber is not solubility but particle size. For example, aluminum chloride is violently soluble in water, yet this material as a fume is most difficult to transfer to the liquor phase in a wet scrubber.

The major influence of a soluble dust is the fact that it is more easily transferred from a wet scrubber, as a soluble liquor, once it has been collected. However, care

105

must be exercised at the various scrubber wet-dry interface areas, where temperature and concentration gradients tend to cause a buildup of such soluble materials.

None of the foregoing dust particle properties affect the conditions of the Rock City Sand Company problem.

Disposal Facilities

When considering the collected dust disposal facilities, the systems concept is being exercised once again. The nature of the process will influence whether dry or wet collection equipment should be employed. In a wet process the probable existence of liquor storage ponds would facilitate sluicing the collected dust rather than handling it dry with conveyors or other motorized equipment. Such a wet disposal system can be applied to either wet or dry collectors. If dry processing is involved, then (because of existing technology and expertise at the plant) dry collection and disposal would be favored.

It will be assumed that the aggregates plant in Commerce City, Georgia, has wet disposal facilities in connection with its gravel-washing operation. A quick check on the pond capacity indicates that there is sufficient water available for a wet scrubber installation.

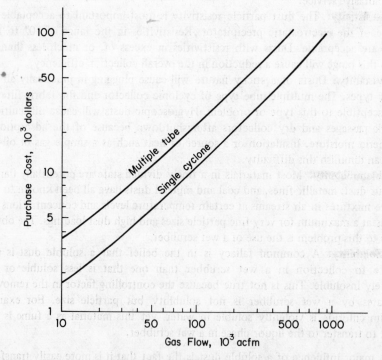

Figure 5-3. Cyclonic separator purchase costs. (Adapted from Edmiston and Bunyard, Ref. 8.)

Product Recovery

A process yielding a dry product usually favors dry collection equipment, with some credits accruing to the collected dust. Normally, the wet scrubber would be eliminated for such an application. In wet processing, it is sometimes possible to return a recoverable slurry to the process. In some industries the collected effluent, either wet or dry, may be reprocessed at an outside facility, with the return of a refined or partially refined product to the original manufacturing plant.

The possibility of product recovery at our Rock City Sand Company aggregates plant is real. Dry collection would be favored and therefore a cyclonic separator in series with a wet scrubber or a fabric filter would be the final candidate to be considered for this application.

Economic Factors

The economics of particulate collection equipment is based on capital equipment costs and operating costs. To present valid estimating figures, a systems approach should be employed, which would account for the accessory equipment as well as the primary collection equipment. Guides to the purchase costs for the four types of particulate collection equipment, on a flange-to-flange basis, are shown in Figs. 5-3 through 5-6.

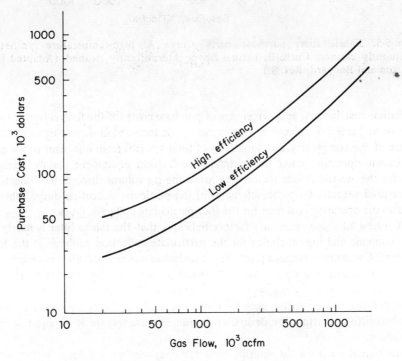

Figure 5-4. Electrostatic precipitator purchase costs. (Adapted from Edmiston and Bunyard, Ref. 8.)

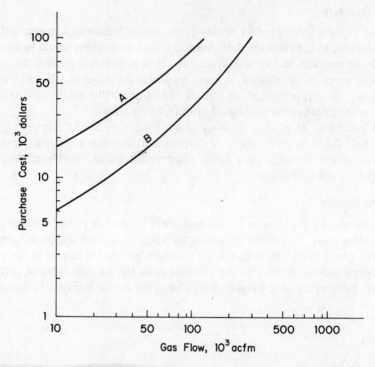

Figure 5-5. Fabric filter purchase costs. Curve A: high-temperature synthetics—continuously cleaned. Curve B: natural fibers—intermittently cleaned. (Adapted from Edmiston and Bunyard, Ref. 8.)

Installation cost factors, as a percentage of purchase costs for the four collector types, is shown in Table 5-5. Accessory equipment is not included in these estimating values because of the complexity and variability of total systems from one plant to the next.

Annual operating costs for particulate collection equipment mainly comprise those for the electric power required to move the gas volume through the collector at the required pressure-drop characteristics of the particular collection duty. Table 5-6 tabulates the operating cost data for the four particulate collector types.

A review of these economic factors indicates that the fabric filter is nearly the most economic and logical choice for the particulate collection problem at the Rock City Sand Company aggregates plant. This conclusion was reached after comparing the capital and operating costs for the cyclone-wet scrubber combination and the fabric filter. Based on design factors, which will be explained later in the chapter, it has been determined that the operating pressure drops required for the cyclone, wet scrubber, and fabric filter (to attain the desired collection efficiencies) are 3, 12, and 6 in. w.g., respectively.

The capital costs for the multiple-tube cyclone, medium efficiency wet scrubber, and intermittently cleaned fabric filter based on mild steel fabrication can be obtained from Fig. 5-3, 5-6, and 5-5 at $7,500, $11,000, and $13,000, respectively. Installation

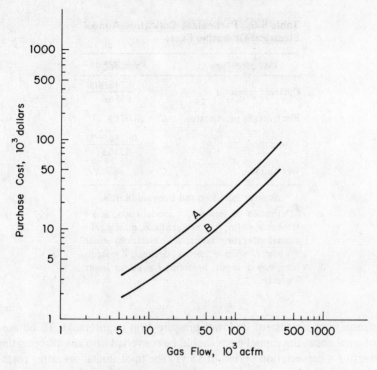

Figure 5-6. Wet scrubber purchase costs. Curve A: Medium and high efficiency. Curve B: low efficiency. Basis: mild-steel fabrication. (Adapted from Edmiston and Bunyard, Ref. 8.)

Table 5-5. Particulate Collector Installation Costs

Collector Type	Cost Range, % of Purchase Cost		
	Low	Typical	High
Cyclonic separator	135	150	200
Electrostatic precipitator	140	170	200
Fabric filter	150	175	200
Wet scrubber			
Low energy	150	200	300
High energy	200	300	500

Source: Edmisten and Bunyard, Ref. 8.

and operating costs are shown in Tables 5-5 and 5-6. Total installed costs and operating costs are tabulated in Table 5-7. The operating costs are based on 300, 24 hr operating days/yr and a power rate of $0.01/kwhr.

This comparison favors the fabric filter on costs alone. An additional factor, which will further favor the fabric filter, is the fan, which will cost considerably less

Table 5-6. Particulate Collectors Annual Electrical Operating Costs

Collector Type	Formula*
Cyclonic separator	$C = S\dfrac{0.746PHK}{6356E}$
Electrostatic precipitator	$C = S(JHK)$
Fabric filter	$C = S\dfrac{0.746PHK}{6356E}$
Wet scrubber	$C = (0.746)HKZ$

Source: Edmisten and Bunyard, Ref. 8.

*Definition of terms: C = annual costs, \$; S = flow rate, acfm; P = pressure drop, in. w.g.; H = annual operating hours; K = electricity costs, \$/kwhr; E = fan efficiency, decimal; Z = scrubbing power input, hp/acfm; J = power input, kw/acfm.

for the fabric filter because of its lower pressure-drop requirement. To obtain a true picture of total costs, the capital costs should be converted into annual operating costs on the basis of a depreciation period of 15 yr. The total annual operating costs would then be \$10,170 for the cyclone/wet-scrubber combination and \$6,430 for the fabric filter.

5.4. EQUIPMENT CHARACTERISTICS—CYCLONIC SEPARATOR

In this section each of the four particulate collection equipment types will be described under the following headings: operating principles, design parameters, mechanical design, installation features, applications, and sample design calculations.

Operating Principles

The cyclonic separator collects particulate matter from a gas stream by the action of centrifugal force. Cyclones are classified as either a single, large diameter unit usually associated with the control of woodworking shop effluents or the small highly efficient multiple-tube type. The single cyclone consists of a cylindrical shell, fitted with a tangential inlet through which the dust-laden gas enters, an axial exit pipe for discharging the cleaned gas, a conical base and a hopper for the collection and removal of the dust; see Fig. 5-7.

The three forces acting on individual dust particles are gravitational, centrifugal, and frictional drag. The gravitational force is defined by Stokes law as applied to freely falling bodies. The frictional drag on a dust particle is caused by the relative motion of the particle and the gas and opposes the centrifugal forces acting on the particle. The major force causing separation of the dust particles from the gas stream is the

Table 5-7. Collector Cost Comparison—Rock City Sand Co.

Collector Type	Cyclone/Wet Scrubber	Fabric Filter
Gas flow, acfm	32,000	32,000
Gas temperature, °F	450	450
Gas moisture, vol %	18	18
Inlet dust loading, grain/acf	2.1	2.1
Dust type	Silica	Silica
Dust size analysis, % less 5 μ	91.8	91.8
Operating pressure drop, in. w.g.	15	6
Delivered cost, $	18,500	13,000
Installation cost, $	33,300	22,800
Total installed cost, $	51,800	35,800
Power costs, $/yr	6,720	4,050

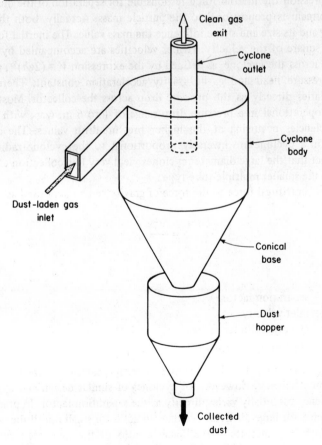

Figure 5-7. Cyclonic separator—cyclone type.

111

centrifugal force induced by rotation of the dust-laden gas stream within the collector. The gravitational, centrifugal, and frictional forces combine to determine the particle path and performance abilities.

The operating characteristics of the cyclonic separator can be determined by an examination of the formula for centrifugal force:

$$F = \frac{Mv^2}{r} \tag{5-1}$$

where

F = centrifugal force exerted on particle
M = particle mass
v = particle velocity
r = radius of dust particle path

In this expression the inertial force responsible for separation of the dust particle from the gas stream is proportional to the particle mass. Actually, both the particle specific gravity and its size and shape influence the mass value. The inertial force varies directly as the square of the velocity. Higher velocities are accompanied by increased pressure drops across the cyclone, as defined by the expression $V = (2gh)^{1/2}$, where h is the velocity pressure head and g the gravity acceleration constant. Therefore, the inertial force varies directly as the pressure drop across the collector. Most cyclonic separators are operational in a pressure drop range of 1 to 6 in. w.g., with increased collection efficiencies occurring at the higher pressure-drop values. The developed inertial forces in a cyclone are inversely proportional to the cyclone radius, which confirms the fact that the large diameter cyclones yield very low collection efficiencies as compared to the smaller multiple-tube type.

The ratio of centrifugal force to the force of gravity can be expressed as

$$S = \frac{v^2}{rg} \tag{5-2}$$

where

S = separation factor
v = inlet velocity
r = cyclone cylinder radius
g = gravitational constant

The separation factor [9], a dimensionless quantity, cannot be directly correlated to the collection efficiency. However, for cyclones of similar design and application, collection efficiency essentially varies directly as the separation factor. In practice, this factor varies from 5 for large diameter cyclones to 2500 for small multitube collectors. Thus, these two values indicate the efficiency ratio of the two rated cyclones to equivalent settling chamber designs.

Referring to the small tube type of cyclonic separator in Fig. 5-8, the dust-laden

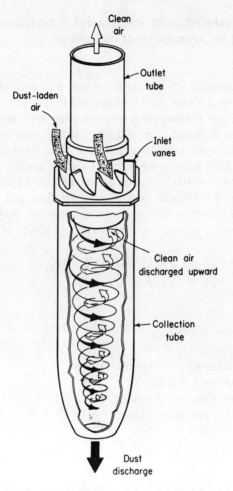

Figure 5-8.   Cyclonic separator—tube type.

gases enter the collection tube axially through the inlet vanes, which impart a swirling motion to their travel. The dust particles are forced toward the wall of this tube by centrifugal force. As they collide with the wall they lose their horizontal momentum and are carried downward by gravity and the continuous swirling action of the carrier gas. The dust is continuously removed from the bottom of the tube without disturbing the vortex action of the gas flow in the body of the tube. The clean gas reverses its flow and exits up through the outlet tube. The outlet tube may contain recovery vanes to convert some of the rotational energy into static pressure, thereby reducing the overall pressure drop across the tube. The action in the single cyclone type of collector is similar except that the gas enters the cyclone body horizontally, as shown in Fig. 5-7. In both designs the pressure drop across the cyclone is due to both frictional and dynamic energy losses. The frictional losses are determined by the cyclone surface roughness, gas velocity, and the physical properties of the gas and particulates. The

113

dynamic energy loss is inherent in the energy stored in the high-velocity rotating gas stream, part of which is lost as the gas leaves the collector.

Design Parameters

The major process parameters influencing the design of a cyclonic separator are gas flow rates and conditions, dust size, dust loadings, and the collection efficiency.

Both the large cyclones and individual collector tubes have capacity characteristics that relate a specific gas flow rate at conditions of temperature and pressure with the pressure drop across the collector. For example, a 24 in. diam cyclone is rated at a maximum capacity of 3000 acfm of air, having a specific gravity of 1.0 at 70° F and 14.7 psia. At these conditions this particular design will have a pressure drop of 1.5 in. w.g. From these data the pressure loss at any gas flow and conditions can be determined so that the collection efficiency can be predicted. Standard orifice design equations can be applied to express the relationship of two sets of conditions, so that

$$p_2 = p_1 \left(\frac{v_2}{v_1}\right)^2 \times \frac{M_2}{M_1} \times \frac{T_1}{T_2} \times \frac{P_2}{P_1} \tag{5-3}$$

where

p = pressure loss, in. w.g.
v = flow rate, acfm
M = molecular weight or specific gravity
T = temperature, degrees abs
P = abs pressure, psia or mm Hg, etc.
Subscript 1 = initial or known conditions
Subscript 2 = final or unknown values

Should it be required to utilize this 24 in. diam cyclone for flows and conditions other than those specified, then this equation could be used to determine the final pressure drop. Should the gas flow rate be increased, then the pressure drop would also increase as the square of the flow ratio. This, in turn, would improve the collection efficiency. The cyclonic separator is one of the particulate collection devices whose performance is improved by increasing the gas flow rate.

Large cyclones in the range of 18 in. to 84 in. diam have capacities of 1200 to 40,000 acfm. For flows in excess of these values, multiple cyclones are arranged in parallel. The superficial design velocity, assuming plug flow through the cyclone body, is about 1000 fpm. The inlet velocity is set at about 3000 fpm. With these two values, and other dimension criteria discussed under the mechanical design section, the overall cyclone dimensions can be established.

Collection tubes usually vary in size from ½ in. to 12 in. diam with corresponding capacities of 0.8 to 1800 acfm, equivalent to an average superficial velocity of about 1500 fpm. Because of the influence of both velocity and radius on the separation factor, the superior efficiency of the multitube type collector is quite evident. Collection tubes are mounted in parallel between tube sheets, as shown in Fig. 5-9. In some installations as many as 600 tubes are arranged in a single housing.

114

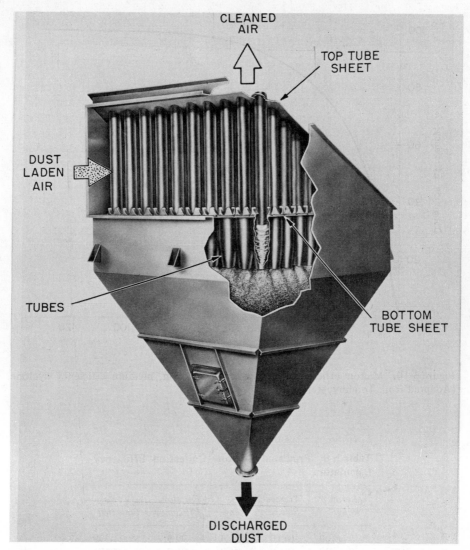

Figure 5-9. Multi-tube collector arrangement.

As illustrated by Eq. 5-1 for centrifugal force, the mass of the particle, and particularly its size and density, determine the cyclone's collection efficiency. Conventional single cyclones can provide high collection efficiencies for particles in the 40 to 50 μ size range. The collection efficiencies for the small-diameter cyclones or collection tubes can be quite high with dusts that are 5 μ and larger. In some instances, where the dust particles tend to agglomerate and/or increased dust concentrations are involved, high efficiencies can be attained with the multitube collector for agglomerated dusts having original particle sizes in the range of 0.5 to 3.0 μ. Because dust particle sizes are expressed as weight percentages for the different cuts or fractions, the

115

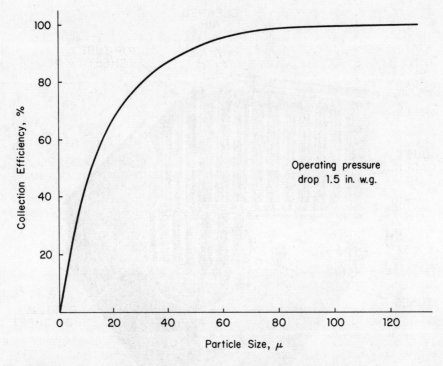

Figure 5-10. Micron efficiency curve for 24 in. diam, medium efficiency cyclone. (Adapted from Tomany, Ref. 3.)

Table 5-8. Predicted Cyclone Collection Efficiency Calculation

Micron Range	Average Size, μ	Wt % in Range	Fractional Efficiency	Wt % Collected
+100	100.0	21.4	99.9	21.379
−100 + 90	95.0	6.2	99.7	6.181
− 90 + 80	85.0	8.5	99.2	8.432
− 80 + 70	75.0	9.4	98.5	9.259
− 70 + 60	65.0	10.2	97.0	9.894
− 60 + 50	55.0	15.5	94.0	14.570
− 50 + 40	45.0	16.0	90.0	14.400
− 40 + 30	35.0	3.4	83.0	2.822
− 30 + 20	25.0	5.4	74.0	3.996
− 20 + 10	15.0	2.2	58.0	1.276
− 10 + 5	7.5	1.3	38.0	0.494
− 5 + 3	4.0	0.5	23.0	0.115
Total		100.0		92.818

116

collection efficiency characteristics for any type of cyclonic separator is expressed as a micron-efficiency curve. Such a curve is shown in Fig. 5-10 for a 24 in. diam, medium efficiency, cyclone collector. From this curve and a dust-sample particle size analysis, an overall collection efficiency can be calculated; see Table 5-8.

According to this calculation, with the dust having an analysis as shown and a known specific gravity, an expected collection efficiency of 92.8% is predicted at a pressure drop of 1.5 in. w.g. Considering the method of calculation as being rather approximate, the equipment manufacturer would make a performance guarantee somewhere in the neighborhood of 85 to 87%. Furthermore, any such guarantee would be contingent on the processor's furnishing a representative dust sample of the emitted pollutant. The collection efficiency value of less than 95% for this coarse dust underscores the reason why cyclones alone are usually not able to comply with emissions control regulations. However, as a primary collector they perform very well. In this role they decrease the load on the secondary collector, as we have seen in Table 5-3.

In utilizing the cyclone as a precleaner, it is important to realize that it preferentially collects the larger particles and permits the fines to continue to the final collector. For example, in Table 5-8, it is seen that the 3 to 5 and 5 to 10 μ range material entering the cyclone comprised 0.5 and 1.3% of the size distribution, respectively. However, because of the poor cyclone performance for this particle size, the dust leaving the collector contained 5.3 and 11.2% of these two fractions. This reduction in the size distribution imposes an extra performance burden on the secondary collector, whether it be an electrostatic precipitator or wet scrubber.

Cyclone separators can handle loadings up to about 10 grains/acf for the small tubes in the multitube design and as high as 100 grains/acf for the larger tubes up to 10 in. diam. The single cyclones are able to accommodate loadings as severe as 500 grains/acf. Performance usually improves with increasing dust loadings. This is probably due to the collision of particles at higher dust concentrations, with an attending loss of velocity followed by inertial separation. In those situations where there are very high inlet loadings and the regulations are very stringent, it is not uncommon to recommend the application of three collectors in series, a single cyclone followed by a multitube collector, and then a scrubber. For a particular problem where the inlet loading was 120 grains/scf and the allowable emissions were 0.05 grain/scf, efficiencies of 80, 85, and 99% were proposed for these three collectors, respectively. This combination produced an overall collection efficiency of 99.97%, equivalent to an emission of 0.036 grain/scf.

Prediction of collection efficiencies for the cyclonic separator can be accomplished directly from the micron-efficiency curve as illustrated in Table 5-8. In this example a relatively coarse dust permitted the use of a single cyclone, whose micron-efficiency curve is shown in Fig. 5-10. A similar curve is shown in Fig. 5-11 for a 6 in. diam tubular collector [3]. The influence of pressure drop across the tube on the collection efficiency corroborates the results predicted by Eq. 5-1, the centrifugal force equation. Reading directly from this curve, at a pressure drop of 1 in. w.g. across the collector, an efficiency of 70% is possible with 8 μ particles. By increasing the pressure drop to 4 in. w.g. the efficiency for this same size fraction can be increased to

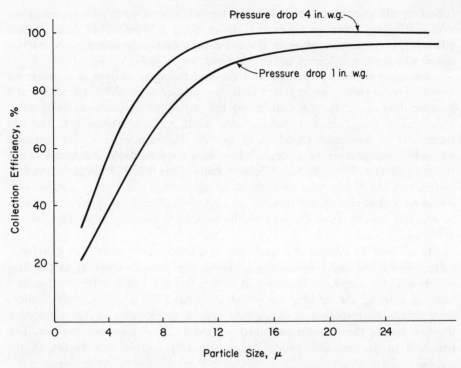

Figure 5-11. Micron efficiency curve for 6 in. diam tubular collector. (Adapted from Tomany, Ref. 3.)

87%. To illustrate the collection efficiency range for the 24 in. diam and 6 in. diam cyclones, the composite curve in Fig. 5-12 was constructed. This curve was plotted on the basis of dust size distribution characterized as that fraction less 10 μ. It shows that the collection efficiency for both collectors improves as the dust particle size increases, or as the fraction less 10 μ decreases. It also indicates that for the problem illustrated in Table 5-8, where the size distribution analyzed 1.8% less 10 μ, the 24 in. diam cyclone could deliver about 93% efficiency at 1.5 in. w.g. However, by the use of multiclones, the efficiency could have been increased to 98.5% at this same pressure-drop value.

In general, the cyclone collection efficiency varies directly with the dust particle density and particle size, dust loading, the square of the cyclone inlet velocity, cyclone body length, number of gas stream revolutions, and ratio of cyclone body diameter to outlet diameter. The collection efficiency varies inversely as the gas density, gas viscosity, cyclone diameter, gas outlet diameter, gas inlet duct width, and inlet area. In commercial practice an empirical design approach is used. Case histories of specific installations containing process conditions and collection efficiency data are the bases for handling fresh inquiries. This is particularly so recently because, with the existence of such demanding regulations, the emphasis is on performance. The collection efficiency guarantee is an essential component of equipment vendor proposals. In

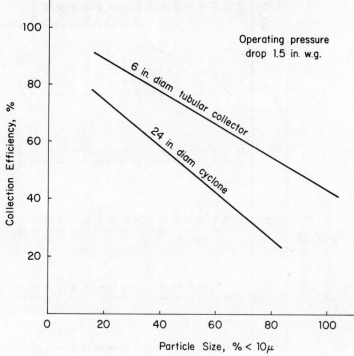

Figure 5-12. Cyclonic separator micron efficiency comparison.

familiar application areas such as coal-fired utility service, where operating experience is a major consideration, the guarantee statement is readily offered. However, in novel applications where the particulate properties are not too well understood, a performance promise usually falls short of a full guarantee. In such cases a demonstration test with a sample of dust provided by the client is recommended by the equipment vendor as a condition for a performance guarantee offer.

Mechanical Design

A discussion of the mechanical design of a cyclone separator is organized under the two types previously described: the cyclone in which individual collectors are used either singly or in a parallel arrangement, and the tubular collectors where multiple elements are mounted between tube sheets, with a common inlet and common discharge manifold.

Cyclone. The capacity range and approximate dimensions for a series of cyclones is shown in Fig. 5-13. These capacities are based on pressure-drop variations of 0.5 to 3.0 in. w.g. The dimension *A* shown in Fig. 5-13 is the effective diameter; therefore, if a ceramic or brick liner is required for high-temperature or corrosive service, the steel shell must be enlarged accordingly. The most common materials of construction are mild steel or stainless steel. Sizes from 18 in. to 42 in. diam can be fabricated in 10

Figure 5–13 Dimensions, Capacities & Weights for Single Cyclones

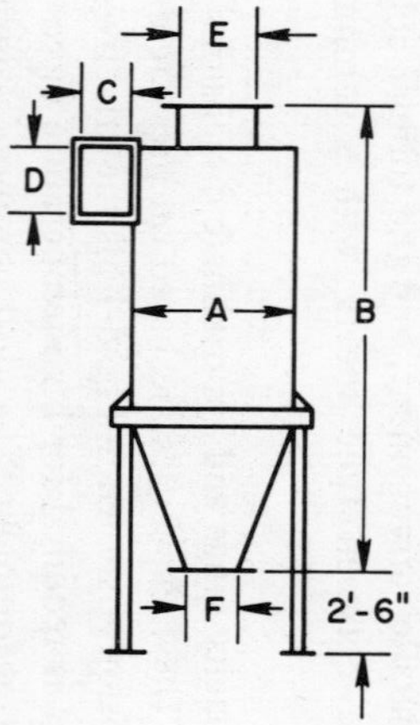

Dimensions, ft-in.

Nominal Capacity, acfm	A	B	C	D	E	F	Wt, lb
1,400–1,800	1–6	5–0	0–6	1–0	1–0	0–8	300
1,800–2,500	1–9	6–0	0–7	1–2	1–2	0–9	350
2,400–3,200	2–0	6–9	0–8	1–4	1–4	0–10	400
3,000–4,000	2–3	7–6	0–9	1–6	1–5	1–0	500
3,800–5,000	2–6	8–3	0–10	1–8	1–7	1–1	600
4,500–6,000	2–9	8–9	0–11	1–10	1–9	1–3	700
5,400–7,200	3–0	9–9	1–0	2–0	2–0	1–4	850
6,300–8,400	3–3	10–3	1–1	2–2	2–1	1–5	950
7,400–9,800	3–6	11–0	1–2	2–4	2–2	1–7	1,100
8,400–11,000	3–9	12–0	1–3	2–6	2–5	1–8	1,200
10,000–12,500	4–0	12–6	1–4	2–8	2–7	1–9	1,400
10,800–14,500	4–3	13–6	1–5	2–10	2–9	1–10	1,600
12,000–16,000	4–6	14–0	1–6	3–0	3–0	2–0	1,700
15,000–20,000	5–0	15–9	1–8	3–4	3–3	2–2	2,100
18,000–24,000	5–6	17–0	1–10	3–8	3–7	2–5	2,500
21,000–29,000	6–0	18–6	2–0	4–0	3–10	2–8	3,000
25,000–34,000	6–6	20–0	2–2	4–4	4–2	2–10	3,400
30,000–39,000	7–0	21–6	2–4	4–8	4–6	3–1	4,000

gage or 3/16 in. of either material. From 42 in. to 84 in. diam, the material could be 3/16 in. or 1/4 in. thick. The selection depends on the abrasive and temperature characteristics of the gas stream being handled.

The cyclone can be made in a variety of configurations. There are two general categories: a high-efficiency or a high-throughput design. The high-efficiency design is provided with a narrow gas inlet, which results in a shorter radial separation distance and a large cross-sectional area between the wall and the dust-laden gas vortex. The high-throughput cyclone sacrifices efficiency for capacity and is typical of large-diameter cyclones. A suggested relationship of the various dimensions for a high-efficiency cyclone is shown in Fig. 5-14. Because the collection efficiency is most dependent on the pressure-drop rating of the cyclone, this design has a resistance of 2 to 6 in. w.g.

In some designs, special high-efficiency inlet vanes or restrictions can be provided as an after-design feature to increase the collection efficiency of a large-diameter cyclone. With this feature the gas inlet scroll must reduce gas turbulence in the cyclone body to a minimum. Lower turbulence is accompanied by reduced resistance and a more effective efficiency/pressure-drop ratio. Higher gas velocities through the cyclone associated with improved performance causes an exponentially increased erosion effect. Maximum erosion occurs at an impingement angle of 20 to 30 deg. Erosion

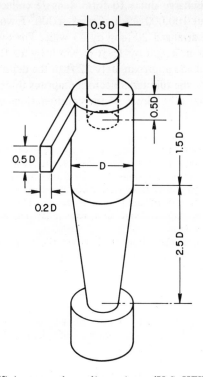

Figure 5-14. High-efficiency cyclone dimensions. (U.S. HEW publication, Ref. 10.)

effects may be minimized by the use of special alloys, abrasion resistant refractories, and rubber lining. Reinforcement of critical impingement areas is another method of overcoming abrasion. For some particulates having highly abrasive properties, economic trade-offs between the design gas velocities and their effect on performance and equipment life must be considered. For high-temperature service, external insulation is required. Usually, the equipment vendor furnishes the cyclone complete with external studs. Wire mesh and suitable insulation cements are then field-applied.

Tubular Collector. Tubular dust collectors are most commonly available in 2 in., 6 in., 9 in., 12 in., and 24 in. diam configurations with approximate flow capacities of 30, 220, 600, 1200, and 4500 acfm, respectively. The tubes are mounted between tube sheets, as shown in Fig. 5-9. In this configuration the dust-laden gas is shown entering horizontally, flowing downward and through the tubes where the flow is reversed, and then being discharged vertically to atmosphere. A lower horizontal tube sheet is used with a stepped upper-tube sheet to allow for the decreased flow of the incoming gas stream to the end row of tubes in the upper sheet. The dust is discharged from the multiple tubes into the single dust hopper. There are innumerable configurations available, such as horizontal in/horizontal out, angle in/angle out, or any combination of these. Both the gas flow arrangement and number of hoppers depend on the equipment arrangement at the plant site. The arrangement shown in Fig. 5-9 is a single module, and it can be combined with a number of others that have common gas stream inlet and discharge ducts to form a single collector. A single multitube collector to handle about 100,000 acfm of gases at 300° F would comprise about 450 six-inch diameter tubes arranged 20 long by 23 wide. The overall dimensions of this collector would be 14 ft in length by 15 ft in width by 12 ft deep. Two side-by-side collection hoppers would add approximately 12 ft to the depth.

As shown in Fig. 5-8, the tubular collector comprises three elements: the inlet, the tube body, and the outlet tube. Some of the considerations involved in the design of these components are listed below.

Tube Inlet: The function of the inlet section is to produce the highest possible rotational velocity of the gases to ensure maximum collection efficiency. This rotational velocity must be produced at the lowest possible pressure loss, since the resistance across the collector is a direct measure of the operating costs. Care must be taken to maintain critical velocities through the inlet section so as to prevent the precipitation of dust at this point. The inlet adds height to the collector. The extra capital cost involved must be regarded in the light of performance improvement.

The most common basic inlet designs are the tangential and vane types. The tangential device has for its principal advantage the ability to slowly increase the gas velocity from its value in the plenum chamber to that in the tube. This gradual velocity change reduces the entrance pressure losses so that the overall pressure drop is maintained at a minimum. The vane type comprises contoured blades that produce a more efficient spiraling of the gas. However, these blades are more susceptible to plugging than are the tangential type. As an alternate, a long curved vane inlet has been devised, which effects an efficient conversion of static pressure to rotational velocity. Because fewer of these blades are used around the tube periphery, larger inlet openings are produced and plugging is minimized.

Tube Body: The tube body diameter and length are the dimensions that are varied to gain optimum collection efficiency. The smaller diameters are necessary

122

to collect particles in the 5 to 10 μ size range. The tube length directly relates to performance. Increase in length will cause an increase in collection efficiency. However, with this length increase, the velocity at the dust-discharge end falls off and there is a tendency toward plugging. Various dust discharge shapes have been evaluated. A single open-end type of the same diameter as the tube body functions well. However, a conical section is favored to prevent gas flow recirculation between the tubes when a large number of tubes are installed in parallel. The function of the tube body is to contain the vortex produced by the tube inlet and outlet. To accomplish this, the tube body must be reasonably smooth internally, to minimize eddy currents, dust hold-up, and pressure drop.

Outlet Tubes: Two types of outlet tubes are commonly used. The first and most common is the straight tube. The second is similar but is provided with recovery vanes to reconvert some of the rotational energy of the gas to static pressure. Because these vanes perform the opposite function of the inlet vanes, a relationship exists between both sets. These recovery vanes reduce the overall pressure drop across the tube while maintaining optimum performance. In the use of these vanes the possibility of fouling by the residual dust leaving the tube must be considered.

The collection tubes are usually of cast metal, gray iron, white iron, or chrome-based alloys being chosen as a function of dust loadings and abrasive properties of the dust. Aluminum and plastic tubes, in the smaller 2 in. size range, are used for special light-duty applications. The collector housing is usually fabricated of 3/16 in. thick mild steel. The tubes are usually flanged and bolted to the housing. The dust hopper normally provides about 8 hr capacity. Depending on the process requirements,

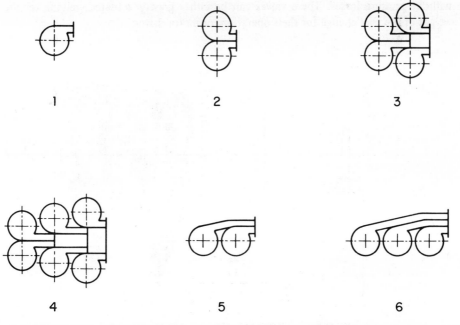

Figure 5-15. Typical cyclone parallel arrangement, schemes 1 through 6.

123

hoppers may be designed for as much as 24 hr capacity or for continuous discharge. The hopper walls are sloped at 50 to 60 deg from vertical for proper discharge of the collected dust.

Installation Features

The large-diameter cyclone is usually installed in a multiple, parallel arrangement. Six typical arrangements are shown in Fig. 5-15. The ductwork arrangement must be designed to obtain optimum gas distribution to the various individual cyclones. Both the cyclone and multitube collectors are usually installed in elevated positions to minimize the ducting from the process. The collected dust can then be gravity-discharged to suitable conveying equipment, such as a screw or belt type. Such an arrangement for an aggregates-processing plant is shown in Fig. 5-16.

The dust is discharged through solids-handling valves, which must usually operate under a negative pressure while maintaining a gas seal on the collector. Any air leakage at the base of either the cyclone or multitube collector upsets its operation and seriously affects performance. The dust valves are of two common designs, the rotary valve and the tipping valve. Both types are shown in Fig. 5-17. The rotary valve operates on the principle of the revolving door and discharges the dust continuously. Wiper blades, of soft rubber or plastic, are attached to the rotating vanes and maintain a seal against the casing walls so that individual pockets of dust are discharged. For high-temperature operation, more sophisticated designs are available; these utilize machined cast iron for the sealing material. The tipping valve is intermittently operated and presents two flap closures to the discharged dust stream. These operate alternately, thereby allowing the collected material to pass out of the cyclone hopper without pressure losses. These valves can be either gravity actuated, relying on the weight of the dust change for their operation, or motor driven.

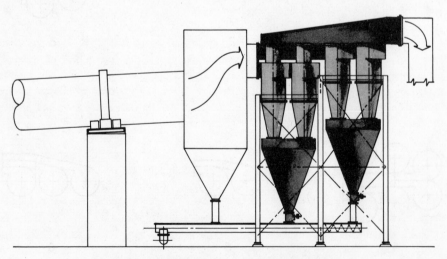

Figure 5-16. Rock industry cyclone arrangement.

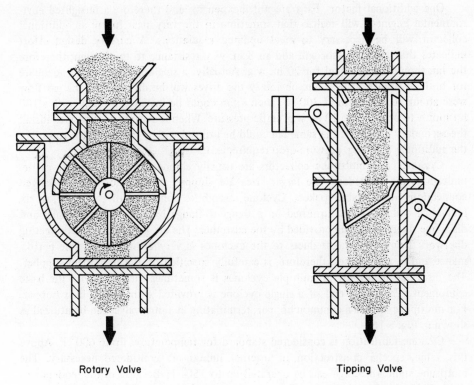

Rotary Valve   Tipping Valve

Figure 5-17. Dust discharge valves. (Courtesy of Western Precipitation Division.)

One of the major items of accessory equipment required by all pollution control equipment is the fan. Its selection, rating, and location in the system is of utmost importance. Because a cyclonic separator is almost invariably used as a precleaner, the fan is located downstream from it and ahead of the secondary collector. In this position the fan is protected from high dust loadings through the dust collection action of the cyclone. It is because of the fan location in the system that the majority of cyclonic separators are operated under negative pressure. When a fan is sized, the gas flow, gas conditions, and required pressure drop are the important design parameters. The pressure drop must usually be sufficient to draw the gases through the total process. For example, if a particular cyclone design requires a pressure drop of 6 in. w.g., a fan must be chosen to overcome this resistance.

In sizing the fan, the location of this cyclone and the duct arrangement must be determined so that the fan will have sufficient capacity to draw the gases from the process, through the collector, and discharge them to atmosphere. Before consideration of an emissions control system, the effluent was probably discharged through a stack, which might have drawn a negative pressure at the process of 2 in. w.g. If careful calculations indicate that the total ductwork both to the cyclone and from the cyclone to the fan to atmosphere would require an additional 4 in. w.g. resistance, then the fan will have to be designed for a total pressure drop of 12 in. w.g.

One additional factor: Fans are not inexpensive and therefore a farsighted environmental engineer will realize that sometime in the very near future an additional collector will be necessary to meet updated regulations. A cursory design effort indicates that this collector will add an 8 in. w.g. resistance to the system; therefore the fan should be specified for 20 in. w.g. Actually, a single fan size will be adequate for both resistance values, but obviously the drives will be different. If the gas flow were about 30,000 acfm at 300° F, then a fan could be installed initially with a 100 hp motor to provide the 12 in. w.g. static pressure. When it became necessary to install the secondary collector, the same fan could be utilized with a 150 hp motor to supply the additional 8 in. w.g. pressure-drop requirement.

Cyclones and multitube collectors are usually shipped assembled unless carrier limitations are exceeded. The larger sizes are shipped broken down, with flanged connections and suitable gaskets. Cyclone assemblies consisting of two, four, or six individual collectors are furnished on a flange-to-flange basis, with the gas inlet and discharge ductwork being provided by the customer. The design of the interconnecting ductwork from the process ducts to the cyclones is very critical to suitable performance and in many cases, therefore, is carefully specified by the equipment supplier. The discharge manifold for multiple cyclones is sometimes furnished with the basic equipment. The lower end of a single cyclone is provided with a small dust hopper. For multiple cyclones a common hopper, terminating in a dust valve, can be utilized as shown in Fig. 5-16.

Cyclone fabrication is considered standard for temperatures up to 600° F. Above this value, special construction or internal linings are considered necessary. The multitube standard design can be operated up to 750° F. Because of the tendency of the individual tubes to become plugged in some applications, a water-wash system is often furnished for the multitube collectors. Spray headers located between the tube banks maintain continuous or intermittent irrigation to wash down the collected dust. In continuous operation the use of wash water is marked by a considerable improvement in the collection efficiency.

Applications

The cyclonic separators, whether of the single cyclone type or multitube collector, are utilized for the collection of relatively large-size particulate matter, either solid or liquid. Because of their inability to effectively collect particulates less than 10 μ in size, these collectors are usually installed in series with secondary control equipment types such as the electrostatic precipitator, fabric filter, and wet scrubber. The normal pressure-drop range for the cyclone separators is between 1 and 4 in. w.g.

The application area for these collectors is usually restricted to free-flowing solid granular materials or liquids. Viscid or wet materials cause plugging, particularly in the multitube type of collector. The advantages to be considered in the case of this type of collector are:

1. Low capital cost.
2. Simplicity of operation.
3. No moving parts.
4. Moderate pressure drop.
5. Reasonable collection efficiencies with coarse dusts.

126

Limitations associated with the cyclonic separator are: (1) inability to collect fine dusts and (2) plugging difficulties with viscid materials.

One of the most outstanding applications areas for the multitube collector is the collection of coal- and oil-fired flyash by the utility industries. The aggregates industries involved in processing cement, crushed stone, sand, lime, talc, etc., also utilize the cyclone collector for the collection and recovery of the dry product. The chemical, paper, mining, steel, petroleum, and food industries all rely heavily on the cyclonic separator, usually as a primary collector. One notable exception is the petroleum industry, where by the use of specially designed in-series cyclones only, a collection efficiency of 99.9$^+$% is obtained. The process is a catalyst recovery system and the median particle size to be collected is about 35 μ.

A general rule for the application of control equipment is that the pyrometallurgical industries, utilizing high temperatures for processing metals and their oxides, produce submicron dust particles. As might be expected, the application of the cyclonic separator is extremely limited in such high-temperature operations as steel blast furnace, iron foundry cupola, glass-melt furnace, and sulfite recovery boiler. On the other hand, emissions from mechanical-handling processes such as crushing, grinding, screening, and conveying can be controlled at a reasonable level by the application of cyclonic separators.

There is one interesting application that recognizes the inability of the cyclonic separator to collect low-density dust particles. In the combustion of wood wastes, the effluent gases contain a low-density wood char type of dust, which presents a collection problem. Usually a collection of 75 to 80% would be expected, but the dust, because of its physical properties, tends to float in the upper section of the collection hopper. As the concentration builds up, the collected material becomes reentrained in the tubes and is then discharged to atmosphere. By evacuating a small gas flow from the hopper in the region of dust buildup, the rated collection efficiencies can be restored. In practice, a small side stream amounting to about 10 to 15% of the total gas flow is thus withdrawn and discharged through a secondary cyclonic

Table 5-9. Performance of Particulate Collection Equipment for Stationary Combustion Sources

Emissions Source	*Collection Efficiencies*			
	Med. Efficiency Cyclone	*Multitube Collector*	*Electrostatic Precipitator*	*Wet Scrubber*
Coal Fired				
Spreader stoker	80	90	99.5	99$^+$
Other stokers	85	90	99.5	99$^+$
Cyclone furnace	50	70	99.5	–
Pulverized coal	60	80	99.5	99$^+$
Oil Fired	60	80	75.0	–

separator. This "fractionating" operation produces an overall improvement in the collection efficiency at very little extra capital and operating costs.

To indicate the increasing role of the cyclonic separator as a precleaning device, reference is made to Table 5-9, which tabulates collection efficiency data for different types of particulate control equipment for various combustion processes.

The tabulated performance values in Table 5-9, when viewed in the light of the ever-tightening emissions control regulations, are the reason why both types of cyclonic separators will be relegated almost exclusively to the role of precleaner. In this position their application must be justified by two criteria: (1) the need for collecting a relatively coarse dry product, and (2) the requirements for precleaning as a function of the final collector performance. The wet scrubber for these applications represents a novel approach to stationary power-plant emissions control. A limited number of wet scrubbers has been installed and are presently being evaluated at various power plants for both sulfur oxides and particulates confinement.

Sample Design Calculations

To illustrate the sizing of equipment, a sample problem will be solved. The industrial application chosen will dictate that a cyclone separator be favored.

Problem Statement. Specifications were issued for an industrial boiler installation utilizing coal-spreader stoker firing. Particulate emissions control equipment was to be furnished to meet the existing regulations. Pertinent process data are as follows:

Duty	Flyash collection
Flue gas flow, lb/hr	480,000
Gas temperature, °F	400
Inlet dust loading, grains/acf	1.1
Dust size analysis	Not available
Maximum pressure drop, in. wg	2.5
Plant elevation	Sea level

Problem Solution. The allowable emissions rate for this size boiler, as of 1968 when this inquiry was processed, was 0.35 lb ash/1000 lb flue gas. The inlet dust loading, specified at 1.1 grains/acf must be converted to pounds of ash per 1000 lb of flue gas:

$$1.1 \text{ grains/acf} \times (1 \text{ lb/7000 grains}) \times (379 \text{ scf} \times 860° \text{ R/520}°) \text{ acf/mol}$$
$$\times (1 \text{ mol/30 lb gas}) \times 1000 \text{ lb gas} = 3.3 \text{ lb/1000 lb gas}$$

Required efficiency is

$$\text{Eff} = 100 - \left[(0.35/3.3) \times 100\%\right] = 89.4\%$$

Volumetric gas flow is

$$\text{Flow} = \frac{480,000}{30 \times 60} \times 379 \times \frac{860}{520} = 168,600 \text{ acfm}$$

Table 5-10. Typical Micron Analysis for Spreader-Stoker Firing

Size Range, μ		Average Particle Size, μ	Wt. % in Range
	+60	60	52.3
−60	+40	50	6.4
−40	+30	35	11.7
−30	+20	25	6.3
−20	+15	17.5	4.2
−15	+10	12.5	2.1
−10	+ 7.5	8.8	5.2
− 7.5		3.8	11.8
			100.0

Size Analysis and Efficiency Prediction: In the absence of size analysis data, some assumptions must be made. Considerable flyash analyses are available for various coal-firing techniques. For spreader-stoker firing, the typical micron analysis shown in Table 5-10 will be considered as representative.

The particulates emitted by a spreader-stoker boiler is a relatively fine dust for the centrifugal separator and therefore it is likely that a multiple tube collector will be required for this service. Still, an attempt will be made to meet the required 89.4% collection efficiency with either a large cyclone or a multiple tube collector. Rather than use a medium-efficiency design as rated for this service in Table 5-9, a high-efficiency cyclone will be considered. Layout requirements indicate that six large-diameter cyclones would be desirable, similar to the arrangement shown in Fig. 5-15, scheme 4. Based on the calculated volumetric flow of 168,600 acfm, each cyclone must have a nominal capacity of (168,600/6), or 28,100 acfm. From the data in Fig. 5-13, a 78 in. diam cyclone with a capacity range of 25,000 to 34,000 will be chosen.

To obtain performance characteristics for this size of cyclone, a curve similar to that in Fig. 5-10 must be plotted. Such a curve would represent performance data for a 78 in. diam, high-efficiency cyclone. Data for the multitube collector can be obtained directly from Fig. 5-11.

Using performance data for both collector types, at a 2.5 in. w.g. pressure drop, predicted overall collection efficiencies are calculated as shown in Table 5-11. A direct interpretation of these values would indicate that the high efficiency 78 in. diam cyclone would be adequate for this duty. However, in actual practice, to predict a 90.7% collection efficiency based on a "typical" dust analysis would be designing a little too closely. Because the allowable emissions will undoubtedly be reduced, the multiple tube collector would be recommended.

Design calculations for the multitube collector are as follows: At an allowable pressure drop of 2.5 in. w.g. (probably dictated by an existing or recently purchased fan) the capacity per tube is calculated by Eq. 5-3.

Table 5-11. Overall Collection Efficiency Calculations

				Collection Efficiencies			
				78 in. diam		*6 in. diam*	
Size Range, μ		*Average Size, μ*	*Wt. % in Range*	*Fractional Eff.*	*Wt. % Collected*	*Fractional Eff.*	*Wt. % Collected*
	+60	60	52.3	98	51.25	100.	52.30
−60	+40	50	6.4	97	6.21	100.	6.40
−40	+30	35	11.7	96	11.23	100.	11.70
−30	+20	25	6.3	94	5.92	99.5	6.27
−20	+15	17.5	4.2	92	3.86	98.2	4.12
−15	+10	12.5	2.1	89	1.87	97.5	2.05
−10	+ 7.5	8.8	5.2	80	4.42	87.0	4.52
−7.5		3.8	11.8	50	5.90	55.0	6.49
			100.0		90.66		93.85

$$p_2 = p_1 \left(\frac{v_2}{v_1} \right)^2 \times \frac{M_2}{M_1} \times \frac{T_1}{T_2} \times \frac{P_2}{P_1}$$

The 6 in. diam tube chosen for this design has a capacity of 190 acfm, for air at 1.0 sp gr, at 4 in. w.g. pressure drop, 70° F, and 30 in. Hg pressure. The capacity per tube for the conditions of the problem is

$$2.5 = 4.0 \left(\frac{v_2}{190} \right)^2 \times \frac{30}{29} \times \frac{530}{860} \times \frac{30.0}{30.0}$$

$v_2 = 188$ acfm,

Therefore

$$\text{Number of tubes} = \frac{168,600}{188} = 900$$

A configuration for the collector, similar to that shown in Fig. 5-9, must be determined. This arrangement is the most common with the collector inlet at an elevation, which places it level with the boiler discharge. The addition of an exhaust manifold converts the top discharge to a horizontal run that usually drops vertically to the fan inlet. Because of the fan size, weight, and operating dynamics, it is located at ground level.

The dust hopper(s) is sized to maintain at least an 8 hr holdup of the collected flyash. The hopper discharge must be located at an elevation to allow either truck or conveyor collection and disposal. For a 900 tube collector, one manufacturer's standard design would be:

35 tubes × 26 tubes
25'-0" wide × 18'-0" deep with four (4)-hoppers
Overall height, 24'-6"
Hopper volume, 450 ft^3 each

Dust load = 1.1 grains/acf × 168.600 acf/min × 1 lb/7000 grains × 60 min/hr
= 1590 lb/hr
Collected dust = 1590 × 0.94 = 1500 lb/hr

At a bulk density of 30 lb/ft^3

$$\text{Holdup period} = \frac{450 \times 4 \times 30}{1500} = 36.0 \text{ hr}$$

This is one of a number of designs and it must be checked against the client's plant arrangement. An 18 by 50 tube configuration might better suit the job requirements.

Weight and Pricing. Cyclone Collector: The overall dimensions of the 78 in. diam cyclone collector are shown in Fig. 5-13. The weight of the six parallel units would be

Total of six 78 in. diam cyclone = 20,400 lb
Ductwork = 2,800 lb
Total weight = 23,200 lb

For the 3/16 in. mild-steel fabricated, high-efficiency cyclone assembly, the cost on a delivered basis would be $16,000, equivalent to $0.095/cfm.

Multitube Collector: The approximate weight of the 25 ft wide by 18 ft deep collector, complete with steel tubes, 3/16 in. mild-steel casing, division plates, and hoppers is 85,000 lb. The cost on a delivered basis is $31,000, equivalent to $0.184/cfm.

5.5. EQUIPMENT CHARACTERISTICS—ELECTROSTATIC PRECIPITATOR

Operating Principles

The electrostatic precipitator collects particulate matter from a gas stream by imposing an electric charge on the particles. The charged particles migrate to a collection electrode consisting of an oppositely charged plate. The charge is neutralized and the particulates, either solid or liquid, are gravity-discharged from the precipitator. The process mechanism comprises a number of sequential actions. Refer to Fig. 5-18; these are:

1. Current is applied to the discharge electrode, a high-voltage corona is produced, and negative gas ions are generated.

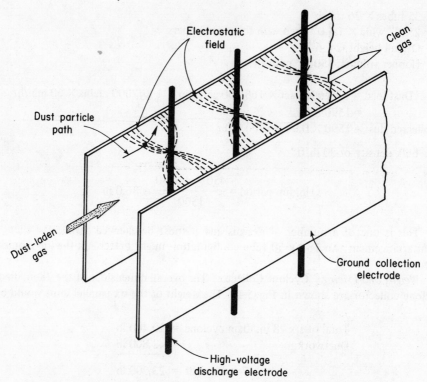

Figure 5-18.   Schematic arrangement of plate type precipitator.

2. The particles are charged by their exposure to the gaseous ions.
3. The charged particles, either solid or liquid, migrate through the electrostatic field to the ground collection electrode.
4. The particle charge is neutralized at the collection electrode and the particulates are gravity discharged.

The first commercial electrostatic precipitator was developed by Frederick G. Cottrell in 1907. It was applied to the collection of acid mist produced by a "contact" sulfuric acid plant. A rectified alternating current was used as the power source and the unit successfully collected acid mist from the 3 ton/day plant effluent stream. The first electrostatic precipitator patent, issued to Cottrell in 1908, was based on this successful installation.

Charged particles attract or repel each other, depending on the polarity of the charges, with like charges repelling and unlike attracting. The force of attraction or repulsion between two static charges can be expressed as

$$F = \frac{q_1 q_2}{DS^2} \tag{5-4}$$

where

F = force of attraction or repulsion between two particles, dynes
q_1, q_2 = charge on particles, coulombs
D = dielectric constant of medium surrounding particles, dimensionless
S = distance between particles, cm

Assume a theoretical vacuum condition where the dielectric constant D is 1. If the force is 1 dyne and the distance between the equally charged particles is 1 cm, then the basic electrostatic charge is defined at 1 coulomb. It is equivalent to approximately 2.08×10^9 electrons.

To explain the mechanism of electrostatic forces responsible for removing a particle from a gas stream, a definition of the electric field that produces the particle charge is necessary. The strength of this field may be written as the quotient of the force exerted on a charged particle, expressed in dynes divided by the magnitude of the charge in coulombs. The field strength may also be expressed as the potential difference divided by the distance between the charged particles or the charged particle and the collection electrode. Thus,

$$E = \frac{F}{q} = \frac{V_0}{S} \qquad (5\text{-}5)$$

where

E = field strength or electrostatic field gradient, v/cm
V_0 = electrostatic potential difference, v

Under typical operating conditions the charge imposed on a 10 μ particle is about 25,000 electrons, while the electrostatic field strength is 4500 v/cm. The coulomb force acting on the particle under these conditions, at a spherical particle density of 1, would be 700 times the force of gravity. Force values exerted on particles in the size range of 0.1 to 100 μ, expressed as multiples of gravitational units, is shown in Fig. 5-19 for these typical operating conditions, where for a 1 μ particle, the separation forces are 7000 times that of gravity. Thus, the curve in Fig. 5-19 illustrates the increased collection potential of the electrostatic precipitator for fine particle sizes. The mechanism utilized does not depend on exerting forces on the carrier gas, as does the cyclonic separator. Since forces in the electrostatic precipitator are exerted on the particles only, power consumption needed to attain the required field strength is minimal. The pressure-drop requirements are extremely low so that power inputs for transporting the gases through the collector are least of the four types of particulate control equipment.

Electrostatic precipitators require a discharge electrode of small cross-sectional area, such as a wire; see Fig. 5-18. The collection electrode is usually ground and offers a large surface area in the form of a plate or tube. The discharge electrode is usually negatively charged and the collection electrode is positively charged and at ground potential. A d-c high-potential field, on the order of 70,000 v is set up between the

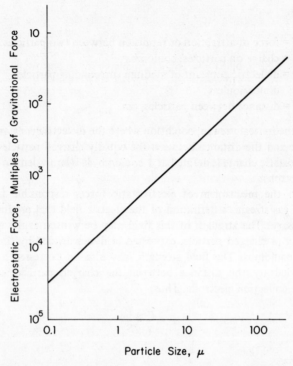

Figure 5-19. Electrostatic forces acting on dust particles. (Adapted from U.S. HEW publication, Ref. 10.)

discharge and the collection electrode while the gas to be cleaned passes between the collection electrodes and across the discharge electrode. As voltage is impressed on the discharge electrode, the air molecules are ionized in its vicinity, indicated by the formation of a faintly glowing corona. The negative ions move toward the positive collection electrode while the positively charged ions remain at the discharge electrode. As the neutral dust particles pass through the ionized field, they accept the negative charges. The suspended dust particles are charged in a few hundredths of a second as they enter the precipitator electrostatic field. Relatively high charging rates are accomplished, with a 1 μ particle acquiring about 200 electron charges and a 10 μ particle accepting up to 20,000 electron charges.

Because the negative ions move across the gas space toward the collection electrode, they contact a greater number of dust particles. Having received its maximum charge, the particle is under the influence of a force proportional to the product of the q on the particle and the strength of the electrostatic field E existing between discharge electrode and the collection electrode; see Eq. 5-5. Imposing a maximum charge on the particle and maintaining the field strength at a maximum while preventing sparkover between the electrode are the two major electrostatic precipitator design criteria.

The negatively charged dust particles are attracted to the positively charged collection electrode and migrate toward it, where they lose their charge. As they become neutral, their removal from the collection electrode is accomplished by gravity discharge, assisted by rapping the collection plates. In some instances a water wash-down system is employed to remove the collected dust.

Design Parameters

The major process parameters affecting the design of an electrostatic precipitator are collection efficiency, gas flow rate and conditions, and the dust characteristics. The general design concepts expressed in this section are for the single-stage plate type for the collection of solid particulates. Both the single-stage wire/plate and wire/tube types are shown schematically in Fig. 5-20.

The fundamental collection efficiency equation for a precipitator, as developed by Deutsch [11], can be expressed as follows:

$$E = 1 - e^{-A v_p/V} \qquad (5\text{-}6)$$

where

E = precipitator collection efficiency, %
e = base of natural logarithm, 2.7183

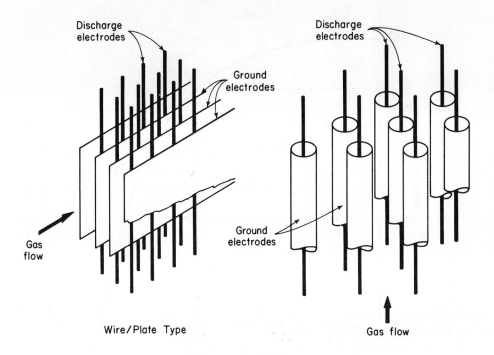

Figure 5-20. Schematic arrangement of wire/plate and wire/tube precipitators.

135

    A = collection electrode surface area, ft^2

    v_p = migration (drift) velocity of charged particles toward collection
        electrode, fps

    V = gas flow rate through precipitator, cfs

This equation gives the precipitator collection efficiency in terms of the three major design parameters: the available collection area, the gas flow rate, and the particle migration velocity normal to the gas flow. The expression Av_p/V is dimensionless. The drift velocity, as it is commonly called, may be expressed as a function of the particle and gas physical properties. Therefore,

$$v_p = \frac{r_p F_1 F_2}{2\pi\mu} \tag{5-7}$$

where

    r_p = particle radius

    F_1 = field strength at discharge electrode

    F_2 = field strength at collection electrode

    μ = gas viscosity

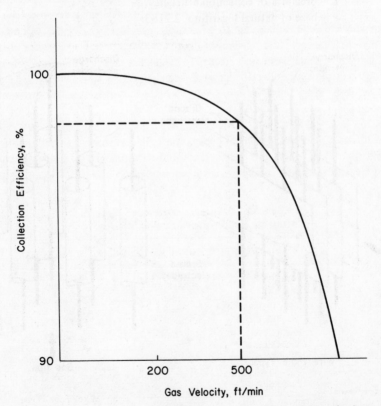

Figure 5-21. Precipitator performance as a function of gas velocity.

According to Eq. 5-6, the precipitator collection efficiency varies inversely as the gas flow rate. The greatest efficiencies are obtained in a precipitator having a long gas path and low gas velocities. At 30 to 40% of velocity values over the design rate, a substantial drop in efficiency occurs. Suggested design gas velocities are in the range of 300 to 600 fpm. A curve illustrating the typical relationship between gas velocities and precipitator collection efficiency is shown in Fig. 5-21. It shows that a precipitator which functions at its design gas velocity will exceed this performance level at lower gas flow rates. Higher velocities increase reentrainment and rapping losses considerably, and performance deteriorates rapidly above the design point.

In Eq. 5-7, the migration velocity is shown to be directly proportional to the particle size so that large particles are collected at a faster rate than the fine ones. However, there is no theoretical lower limit to particle sizes that can be collected. As shown in Fig. 5-19, the electrostatic forces responsible for particle migration and collection are greatest for the smaller sizes. Actually, the design parameter that is most important to the collection of particles, over the total size range, is retention time in the electrostatic field, as determined by the length of the gas path. Some of the various materials collected with the precipitator and their average particle size are shown in Table 5-12.

One source of difficulty with the collection of large particles, say, greater than 50 μ, is their tendency to become reentrained in the gas stream. The deposited dust will adhere to the collection plate up to a critical gas velocity. Beyond this velocity, dust-ridge formations are developed and then torn loose by the gas flow. For plain flat surfaces, critical velocities can vary from 2 to 10 fps for carbon black and cement-kiln dust, respectively. Special collection electrode configurations which will be discussed in the next section, "Mechanical Design," have been developed to reduce reentrainment.

Although the particle density is not included in Eq. 5-7, it is a factor to be considered in the overall collection efficiency. Light, fluffy dusts such as carbon black are more difficult to gravity-discharge from the collection electrode to the hopper. They tend to "float" and are more susceptible to carryover in the gas stream. Minimum electrode-rapping frequency is recommended for the collection of this type

Table 5-12. Various Materials Collected by the Electrostatic Precipitator

Material	Average Particle Size, μ
Coal flyash	150–1
Tar mist	0.3
Coal dust	10
Cement dust	100–5
Silica dust	5
NH_4 Cl fume	1–0.1
ZnO fume	0.05
Pigments	5–0.2
Flour dust	15

Source: Kirk and Othmer, Ref. 12.

137

of material. High dust-loading levels also affect performance because of the increased rapping efforts required to reduce overloads on the collection electrode. Too frequent discharge of the collected material by rapping the electrode affects the overall collection efficiency adversely. One solution, as discussed previously, is to utilize a cyclonic separator as a precleaner to reduce precipitator loadings.

Dust resistivity is one of the most important parameters influencing precipitator performance. When the charged dust particle contacts the grounded collection electrode, it is necessary that the particle lose its electric charge quickly. In most applications, the electrical properties of the dust allow the charge to readily leak off to ground. However, with some high-resistivity dusts the charge is not dissipated, and an electrical potential builds up on the surface of the precipitated material. Additional layers are deposited and the potential increases to the point where the potential difference between the two electrodes is lost. This loss is cumulative, and in time it reduces the corona discharge voltage to a point where the incoming dust particles fail to receive a maximum charge. This situation seriously reduces collection efficiency. As this condition advances, corona discharge takes place at the collection electrode, resulting in total malperformance of the precipitator.

At the opposite end of the resistivity range are dust particles of extremely low resistivity, such as those having high carbon content. The problem at the low end of the resistivity scale is that of holding the particles on the collection electrode when they are first precipitated. Because of their low resistivity, the carbon particle acquires a maximum charge very rapidly. However, because it is highly conductive, this charge is immediately lost as the particle reaches the ground electrode. This rapid loss of its negative charge leaves the particle with a weak net positive charge, which causes it to be attracted to the negative potential on the discharge electrode. It thus reenters the gas stream because of this negative ionic attraction. However, as it becomes negatively charged once again, it reverses direction and migrates to the ground electrode as before. This process is repeated indefinitely while the gas stream moves these unattached particles to the precipitator outlet, where they are discharged to atmosphere. Various types of baffled collection electrodes have been devised to minimize this effect.

Dust resistivity is normally expressed in ohm-centimeters and is defined as the resistance in ohms between two opposite faces of a cube, 1 cm on a side. The optimum resistivity range for effective precipitator performance is 10^9-10^{10} ohm-cm. Typical resistivity values for such dielectrics as porcelain and quartz are 10^{14} to 10^{17} ohm-cm. Resistivity of many dust types can be reduced to values dictated by good precipitator design practice, either by manipulation of the gas stream temperature or the addition of water vapor, ammonia, sulfuric acid, or other inorganic ion promoters.

Figure 5-22 shows a curve relating temperature and resistivity for a coal-fired flyash application. The values indicate that at a temperature level of either 285° F or 340° F the precipitator will yield an acceptable performance. At intermediate values poor performance would be expected. The normal operating temperature range for commercial precipitators collecting coal-fired flyash is about 250 to 300° F. However, although this low-temperature range is suitable for the required resistivity range and a high collection efficiency can be realized, there is a possibility that condensation will

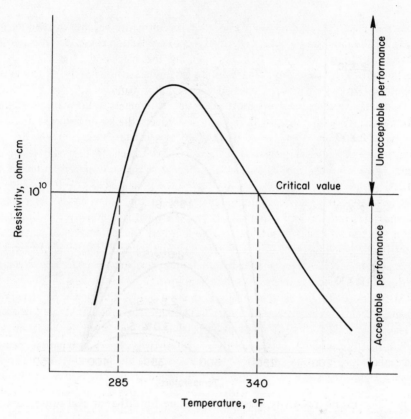

Figure 5-22. Dust resistivity variation with temperature.

take place with attending corrosion problems. To eliminate condensation and yet maintain suitable particle resistivity characteristics, some precipitators are being designed at temperature levels in the 600 to 800° F range.

In the combustion of bituminous coal, its sulfur content (varying from 0.2 to 4.0%) has a marked influence on particle resistivity. It is believed that the sulfur oxides resulting from the combustion of the sulfur-bearing coals are adsorbed by the flyash dust particles, thereby decreasing their resistivity value. Figure 5-23 illustrates the relationship between the resistivity of flyash and temperature for various bituminous coal sulfur concentrations. This correlation assumed great importance when air pollution legislation introduced measures to control gaseous emissions from power plants. In early 1969, New York City enacted a law limiting the sulfur content of coal to 1%, while demanding that the collection efficiency be maintained at 99%. Since that time the generally acceptable sulfur content of coal has been dropped to 0.5%. As might be expected from analysis of Fig. 5-23, this low-sulfur content increased the flyash resistivity values above those required for acceptable precipitator performance. A number of Western utilities, prior to this low sulfur regulation, were obtaining flyash

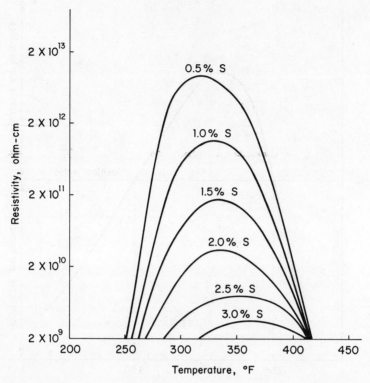

Figure 5-23. Flyash resistivity versus temperature for different coal sulfur contents.

collection efficiencies in the order of 99⁺%. When changing over to 0.5% sulfur coal, the efficiencies fell off to 50 to 60%.

It has been estimated that the reduction in the sulfur content of the coal at the 1% level has been responsible for a decrease in the SO_3 concentration to less than 5 ppm. Because SO_3 is considered to be the active oxide responsible for resistivity correction, one solution still being evaluated for the utility power industries is to bleed this gas into the flue gas stream, up to about 20 ppm, to reduce the flyash resistivity. At this writing, no commercial installations of this process have been undertaken. In fact, it is this sulfur oxides/flyash precipitator difficulty that is encouraging many of the utilities to consider wet scrubbing for the control of both pollutants; refer to Table 5-9.

The design of electrostatic precipitators is accomplished empirically. The most important term in the efficiency equation is drift (migration) velocity; see Eq. 5-6. The relationship of gas flows to collection efficiencies as a function of drift velocity is theoretically represented in Fig. 5-24. This curve, and Eq. 5-6, indicates that increased drift velocities at a constant gas flow rate will yield greater efficiency values. Conversely, for any single drift velocity value, the collection efficiency is inversely proportional to the gas flow. Actually, the values shown in this curve, as calculated

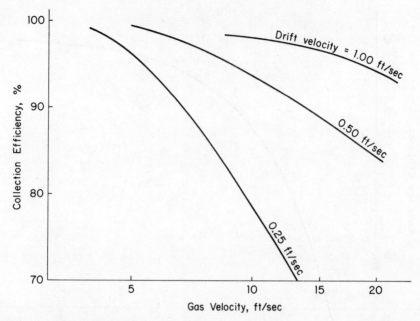

Figure 5-24. Theoretical precipitator efficiencies versus gas flows at different drift velocities. (U.S. HEW publication, Ref. 10.)

from Eq. 5-6, are about twice the values determined from actual efficiency measurements. As in the case of most particulate collection equipment, suppliers rely on parallel and almost duplicate experiences to size commercial precipitators. For new application fields, where experience is limited, pilot plant and prototype units are installed at the site to obtain basic process data.

The curve constructed in Fig. 5-25 is to be used with Eq. 5-6. In this working curve for the precipitator design, the collection efficiency has been plotted against Av_p/V. By choosing the desired efficiency, the exponential expression can be determined, and for known values of drift velocity and gas flow, the required collection electrode surface area can be calculated. The drift velocity value is dependent on the various parameters previously discussed. Some suggested values for various applications are tabulated in Table 5-13.

In the design of an electrostatic precipitator it must be realized that migration of the charged particle to the collection electrode is not the total mechanism. The particles must be removed from this plate and dropped into the dust hopper. The particle characteristics seriously influence this latter step. Both rapping and washdown systems have been devised to accomplish this function. The rapping cycle or washdown frequency, collection electrode designs, plate spacing, power input, and distribution are some of the mechanical details that convert the theoretical design into an operating reality.

To design an electrostatic precipitator for a specific application, the following information must be obtained [13]:

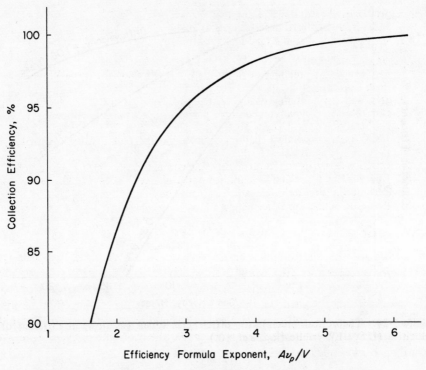

Figure 5-25. Precipitator efficiencies versus exponential function.

Table 5-13. Typical Precipitator Drift Velocities for Commercial Applications

Application	Drift Velocity, fps
Pulverized coal flyash	0.33–0.44
Kraft recovery boiler	0.25
Open hearth furnace	0.19
Secondary blast furnace	0.41
Gypsum kiln	0.52–0.64
Acid plant mist	0.19–0.25
Flash roaster	0.25
Multiple hearth roaster	0.26
Catalyst dust	0.25
Iron cupola	0.10–0.12

Source: Danielson, Ref. 9.

Plant Information
1. Plant name.
2. Plant location.
3. Plant type; i.e., power plant, steel, pulp and paper, etc.
4. Process to which precipitator will be applied.

5. Application area; boiler, kiln, melt furnace, etc.
6. Process raw material description and analysis.
7. Expected raw material feed-rate variations.

Operating Conditions
1. Gas volume at precipitator inlet: acfm, °F, pressure psia for both normal and peak conditions.
2. Gas moisture content, vol %.
3. Gas volume for which precipitator performance is to be guaranteed.
4. Gas analysis: CO_2, CO, O_2, H_2O, SO_x, NO_x.
5. Dust or liquid chemical analyses, specific gravity, bulk density, resistivity.
6. Particle size analyses.
7. Dust loading at precipitator inlet: continuous and peak rating, grains/acf.
8. Dust loading for which precipitator performance is to be guaranteed.
9. Barometric pressure or elevation at plant site.
10. Required collection efficiency, %.

Mechanical Design

Most of the previous discussion has been concerned with single-stage, dry, plate type electrostatic precipitators. The wire/plate and wire/tube configurations were shown schematically in Fig. 5-20. There is also a wet type precipitator wherein a constant water film uniformly flows down the inside of the collecting tubes so that the water surface constitutes the collection electrode. The mechanical design aspects of the electrostatic precipitator are discussed below under the two common types: wire/plate type dry and wire/tube type wet.

Single-Stage Wire/Plate Type Dry. In both precipitator types, there are four main components:

Electrode system
Precipitator casing
Collected material removal system
Power supply

The single-stage wire/plate precipitator is shown in Fig. 5-26. The gas flow enters through the perforated distribution baffle and flows through the parallel passages formed by the plate collection electrodes. The discharge electrodes, which may be round wire, square twisted rods, ribbons, etc., provide the corona for producing the electrostatic field. Round wires vary from 1/16 to 1/3 in. diam and are usually of steel alloy construction. Other materials such as stainless steel, silver, nichrome, aluminum, copper, and lead-covered steel have been used. The choice is a function of the corrosive service. The most common collection electrode in the wire/plate design is a smooth plate with vertical interlocking baffles. These baffles provide strength and also produce near-zero velocity conditions as the gas flows in a normal direction to them. Other special plate electrode configurations are rod curtains, zigzag plates, and various hollow electrodes with pockets on the outside surfaces for discharging the collected dust to the hopper from quiescent gas zones. For some particulates to be collected, perforated or expanded metal plates provide multiple closely spaced holes that hold the precipitated material while end baffles shield the perforated surfaces from the

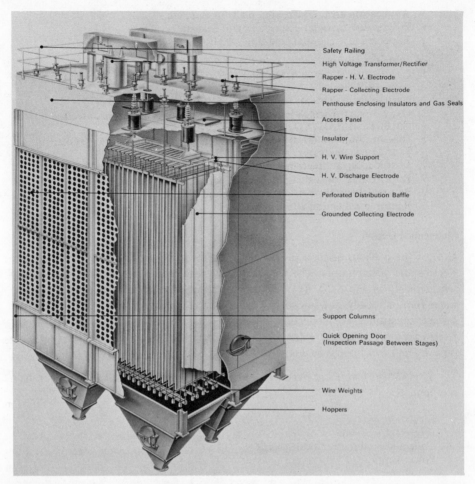

Safety Railing
High Voltage Transformer/Rectifier
Rapper - H. V. Electrode
Rapper - Collecting Electrode
Penthouse Enclosing Insulators and Gas Seals
Access Panel
Insulator
H. V. Wire Support
H. V. Discharge Electrode
Perforated Distribution Baffle
Grounded Collecting Electrode
Support Columns
Quick Opening Door
(Inspection Passage Between Stages)
Wire Weights
Hoppers

Figure 5-26. Typical plate type electrostatic precipitator. (Courtesy of Air Correction Division.)

scouring action of the gas. As in the case of the discharge electrodes, the plates are fabricated from a wide choice of construction materials, depending on the degree of corrosive service. The collection electrodes comprise the separating walls of the numerous parallel ducts that accommodate the gas flow. The high-voltage discharge electrodes are suspended vertically between each pair of collection electrodes. These wires are hung from a resiliently mounted, structural high-voltage frame, each wire being held in tension by a suitable weight. They are carefully centered between the collection electrodes to ensure proper corona gaps. Typical design parameters for commercial wire/plate precipitators are shown in Table 5-14.

The precipitator casing may be fabricated from a wide selection of materials: mild steel, lead-lined steel, acid-resisting brick, concrete, wood, rubber-lined steel, and tile are some of them. The inlet and outlet are arranged for horizontal flow through the

Table 5-14. Mechanical Design Parameters for Wire/Plate Electrostatic Precipitator

Design Parameter	Value Range
Plate spacing, in.	8–11
Gas velocity, fps	2–8
Plate height, ft	12–24
Plate length, ft	0.5–1.0 × height
Applied voltage, kv	30–75
Drift velocity, fps	0.10–0.70
Gas temperature, °F	700 max std; 1000–1300 special
Retention time, sec	2–10
Draft loss, in. w.g.	0.1–0.5
Efficiency, %	90–98 normal; 99.9 max
Corona current, ma/ft wire	0.01–1.0
Field strength, kv/in.	7–15

parallel paths formed by the collection electrodes. A perforated distribution baffle is provided at the inlet to ensure equal and parallel gas flows through the unit. The length of the precipitator may be comprised of several fields in the direction of gas flow, in some cases as many as four fields being required to obtain the desired performance. Each field section can be treated as a separate unit module so that additional fields can be easily tied into the system as the need for increased collection efficiencies arise. The entire casing, containing the internals, as shown in Fig. 5-26, is supported on a steel base that rests on support steel. Main support columns, attached to the steel base, support structural members near the top of the precipitator from which the grounded electrodes are suspended. The casing roof is also supported from these top structural members. The discharge electrodes are hung from high-voltage insulators located on and supported by the roof. These insulators, together with gas seals and rapper mechanisms, are enclosed in an attic space above the casing roof. The storage hoppers for the wire/plate precipitator are located under the collection electrodes and are a structural continuation of the casing. These hoppers catch and store the dislodged dust from the collection electrodes. The capacity of these hoppers is a function of dust loading and the type of application. The angle of repose for the particular dust being collected must be considered so as to determine the correct hopper slope. The dust removal system is similar to that for the cyclonic separators, the dust being transferred through a discharge valve to a conveyor.

There are two methods for removing the collected dust from the dry wire/plate type precipitator: rapping and washing. Rapping methods usually comprise a mechanically actuated "hammer," which is driven electrically or pneumatically. The motors are located in the precipitator attic, with connecting rods passing downward through suitable bushings in the casing and terminating at the point of impact on the electrode. The impact blow from the rapper (hammer) is transmitted vertically to the freely

145

suspended collection electrode so that the dust is released in the direction of gravity. The impact frequency must be so scheduled that successive blows will not interfere with the vibrations imparted by the previous rapping action. Rapping intensity is another variable that must be field-adjusted to suit the conditions. Overvigorous rapping can damage the electrodes. Frequently, improper rapping techniques may cancel out effective precipitation action, thereby impairing the overall collection efficiency. The ideal rapping system is one that will dislodge the dust from the collection electrode while avoiding the formation of a dust cloud in the gas paths between the electrodes. For flyash precipitation the dust buildup on the collection plates should attain a thickness of ¼ to ½ in. before the plate is rapped. When collecting dusts containing appreciable amounts of $< 10 \mu$ particles, rappers are provided for the discharge electrodes as well. In some cases where reentrainment is a serious problem, precipitators may be designed so that a number of sections may be closed during the rapping period by means of automatic dampers. As an alternate, the total precipitator may be deenergized while rapping is in progress. The necessity for these measures indicates the importance of effective electrode cleaning to the overall collection efficiency. In the case of some viscid dusts the removal of collected material can be accomplished by washing down the plate. Periodic water sprays are used, and by proper cycling of these sprays to maintain a wetted electrode, reentrainment can be completely eliminated.

The electric energy requirement for an electrostatic precipitator is that necessary to produce an effective corona. The power supply must deliver to the discharge electrodes a unidirectional negative current at a potential very close to that which will produce arcing across the electrodes. The value of the potential difference used in the single-stage precipitator is usually in the range of 20,000 to 100,000 v. The current delivered may vary from 20 to 400 ma. Power requirements are relatively small because, as stated previously, the dust only is treated rather than the total gas flow. For example, the power requirements to clean 500,000 cfm of gas at 95% efficiency, including the draft loss, is only about 70 kw. Alternating current is transformed to a high voltage and then rectified to convert it to unidirectional flow. Earlier precipitator designs utilized synchronous mechanical converters. These were later replaced by electronic tube rectifiers (circa 1920). More recently, solid state rectification is practiced. Selenium rectifiers provide reliable service with long life. At present, their life expectancy is estimated to be about 100,000 hr. Silicon rectifiers, which are hermetically sealed, seem to have almost unlimited life.

For the most ideal precipitating action, each discharge electrode should be individually energized to produce an optimum corona to satisfy the specific gas-and-dust process conditions in the vicinity of that particular electrode. Although this is not practical, most industrial precipitators are designed with sets of discharge electrodes tied into a common bus section. Each of these sets is separately energized by an individual rectifier set. The rectifier sets should be so arranged so that the precipitator inlet section is furnished with the lowest voltage input, because of the heavier loading. As the dust loading decreases along the gas path through the precipitator, the applied voltage may be increased so that maximum power can be impressed at the outlet to precipitate the smallest dust particle at the least loading value. For a precipitator with

146

two parallel flow sections and three sections deep, six rectifier systems would provide an ideal arrangement, with one system being available for each section. Recently, it has been shown that some controlled arcing or sparkover results in optimum performance. The frequency of arcing is a function of dust characteristics, dust loadings, and precipitator size. In general, maximum collection efficiency occurs at a sparkover rate of 50 to 100 arcs/min. In Fig. 5-27 there is shown the relationship between the sparkover rate and collection efficiency for a single section at a specific set of process conditions. To maintain such an arcing condition, automatic voltage regulation is necessary. The precipitator potential is maintained at the optimum value by a spark counter or current-sensing feedback circuit.

Single-Stage Wire/Tube Type Wet. The plate type precipitator is most commonly applied because of its ability to handle large volumes of gas per unit equipment volume. The single-stage wet wire/tube precipitator design, as shown in Fig. 5-28, is better suited to wet collection applications. This precipitator type is built in a cylindrical shell. The collection electrodes consist of nested pipes, which are connected and sealed to header sheets attached to the shell. The discharge electrodes are supported above the header sheet and are suspended axially in the collection electrode pipes. Water is introduced above the header sheet and flows over carefully leveled weirs at the tops of the pipes to form a water film on the inner walls of the pipes. The

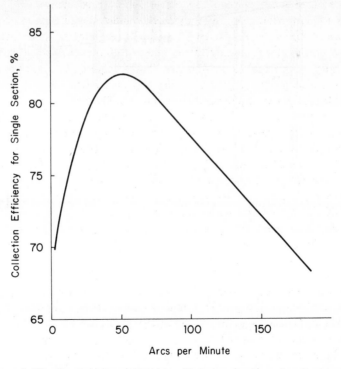

Figure 5-27. Precipitator collection efficiency versus arcing frequency.

147

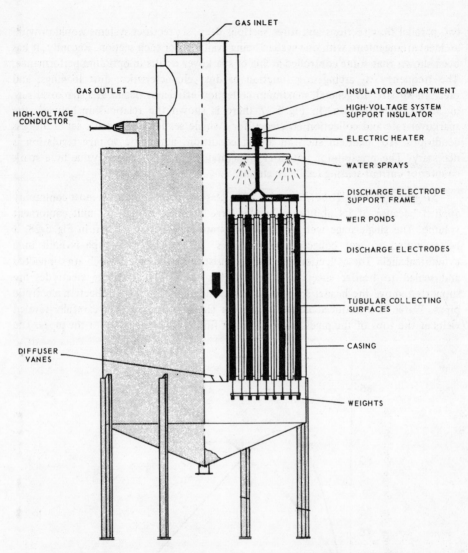

GAS INLET

GAS OUTLET

HIGH-VOLTAGE
CONDUCTOR

INSULATOR COMPARTMENT

HIGH-VOLTAGE SYSTEM
SUPPORT INSULATOR

ELECTRIC HEATER

WATER SPRAYS

DISCHARGE ELECTRODE
SUPPORT FRAME

WEIR PONDS

DISCHARGE ELECTRODES

TUBULAR COLLECTING
SURFACES

CASING

DIFFUSER
VANES

WEIGHTS

Figure 5-28.   Tube type precipitator. (Courtesy of Koppers Co., Inc.)

charged particles are collected in the water film, where their charge is neutralized and drained off from the bottom of the precipitator with the water. In some instances, the water is recirculated, after the solids have been removed. Alkaline salts are usually added to the water stream to make it conductive. This type of precipitator is often designed for high-pressure service. A typical design for cleaning blast furnace gases would have shell dimensions of approximately 20 ft diam by 24 ft high, and would contain one hundred and thirty 8 in. diam pipes, 15 ft long [14]. In this type of design, shown in Fig. 5-28, the gases enter the top inlet, passing vertically downward through a water-flushing system. The gas flow is reversed at the bottom of the shell and passes through the individual pipes, where precipitation and washdown are accomplished. The cleaned gas leaves the precipitator through the gas outlet located in the head of the shell. The wire/tube precipitator is commonly used for acid mist and tar fume collection. Because wet collection is involved, materials of construction are usually corrosion resistant.

The power supply is similar to that for the plate type precipitator. However, it is more simple in concept because the simple vertical gas flow pattern eliminates the need for sectionalization.

Installation Features

Plate type electrostatic precipitators are designed for gas flow rates in the range of 50,000 to 800,000 acfm. Because of the low gas velocities used, these units are large, and therefore their arrangement in the plant must be carefully considered. For example, for a gas flow rate of 500,000 acfm, approximate dimensions for a plate type precipitator collecting flyash would be 50 ft wide by 30 ft high by 20 ft deep. The dimensions defining the cross-sectional area can be considerably varied to suit field equipment arrangements. Some idea of the relative size of precipitators can be gained from Figs. 5-29 and 5-30. In Fig. 5-29 a typical plate precipitator is shown installed in series with multiple cyclonic separators. This system provides an overall collection efficiency of 99%, 80% for the cyclonic collectors, and 95% for the precipitator. The precipitator is located downstream from the air heater, where the temperature is about 280° F. In many recent installations, the collector is ahead of the air heater, handling gases in a temperature range of 600 to 700° F. These higher temperatures favor the flyash resistivity characteristics, thereby increasing the collection efficiency; refer to Fig. 5-22. Fig. 5-30 shows a plate type precipitator being installed. This unit handles about 500,000 acfm of boiler flue gases.

Because of its size, the plate type precipitator is shipped to the plant site unassembled. The structural support steel is erected and then the hoppers, collection electrodes, rapper assemblies, switch gear, inlet baffle, and walkways follow, all supported by the structural framework. In some advanced designs, a modular precipitator has been devised. Identical, interchangeable units can be bolted together in series or parallel to make up any size of precipitator having the required efficiency. Removal of any single, integrated module for repair or replacement—or, most importantly, for rearrangement of the units to meet more stringent regulations—can be easily accomplished.

The location, and particularly the elevation, of the precipitator must be carefully considered to accommodate the ductwork, fan, and dust discharge system. In Fig. 5-29 the precipitator has been located at the roof level to minimize ductwork, whereas in Fig. 5-30 it has been located at ground level. The other important accessory equipment item is the flyash conveyor. The selection of the final disposal method, whether it be pond storage, silo storage, or truck transfer, influences the location and elevation of the precipitator hopper discharge point.

In Fig. 5-29 the downstream location of the fan, relative to the precipitator, determines the design pressure for the latter. For this system, it is estimated that a negative pressure on the order of 12 to 16 in. w.g. exists in the precipitator. For a large

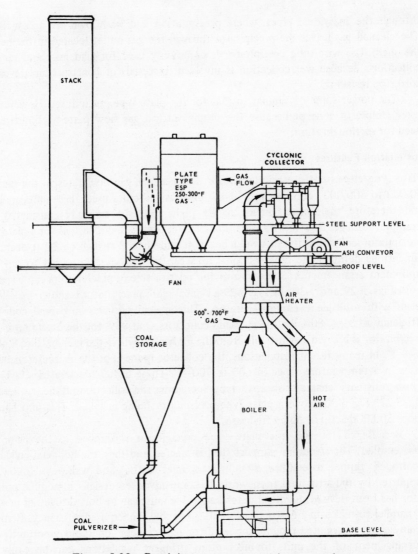

Figure 5-29. Precipitator arrangement in power plant.

precipitator (say, 50 ft wide by 30 ft high by 20 ft deep) handling 500,000 acfm, this design pressure is a most important consideration.

The ductwork path essentially decides the precipitator location. For a gas flow of 500,000 acfm, the ductwork cross-sectional area is over 100 sq ft, or 8 ft by 12 ft approximately. Notice in Fig. 5-29 the inlet duct transition section, which must reduce the duct gas velocity to that required through the precipitation (from 4000 to 400 fpm). Because of the importance of uniform and well-distributed gas flow through the precipitator, the use of Plexiglas scale models to study air flow patterns is a common practice. Gas flow distribution is considered uniform when at least 85% of the velocity

Figure 5-30.   Installation details of plate type precipitator.

readings at a point 5 ft ahead of the precipitator inlet are within ±20% of the average, with no single reading deviating more than 40% from average.

The power supply comprises two major components, the power unit and the control system. The former, containing a transformer and rectifier, is located at the precipitator so as to make the high-voltage leads as short as possible. The centralized control and instrument package is usually shipped as an integral unit, completely wired and contained in a steel cubicle, and ready for installation. Central control and instrument panels are becoming more accepted as a necessary component of the electrostatic precipitator. In major installations, readout of all process conditions at every point in the system is available on the central control and instrument panel. Automated self-correcting controls almost entirely eliminate human monitoring of the system.

Applications

The electrostatic precipitator has the capabilities of collecting particulates, both solid and liquids, over a broad particle size range from a comtaminated gas stream, at very low energy inputs. Operating pressure drops are in the range of 0.10 to 0.50 in. w.g. Some of the advantages in the use of the electrostatic precipitator are:

1. High collection efficiences.
2. Fine particle size collection capabilities.
3. Ability to handle solids and liquids.
4. Low pressure-drop requirements.
5. Low power requirements and operating costs.
6. Unique among particulate collectors in effective control of tar and acid mists.
7. Add-on features for increasing collection efficiency.
8. High gas flow capacity rating.

Some of the disadvantages of the precipitator are:

1. High initial costs.
2. Large space requirements.
3. Limited application to particulates in specific resistivity range.
4. Performance sensitive to upset process conditions.
5. Safety hazard presented by use of high voltage.

By far the most active application area for the electrostatic precipitator is the utility industry. Coal is still the most commonly used fuel for power production, and because of its ash content, its combustion results in the emission of particulates with the flue gas. The electrostatic precipitator is the established type of control equipment for the elimination of this type of emission. However, recent concern for the control of sulfur oxide emissions confuses the role of the precipitator at this time. A number of experimental prototype wet scrubbers have been installed for the simultaneous removal of both particulates and sulfur oxides. In this role, they are replacing the electrostatic precipitator.

In the cement industry, the precipitator has been used to treat the kiln effluent gases. The precipitator has been applied to both wet and dry cement processes. In the dry process the flue gas leaving the kiln is conditioned by the addition of water to lower the dust resistivity. As in the case of the utilities, the gas flow rates being treated in the cement industry are usually in excess of 100,000 acfm.

In the steel industry the precipitator handles gases from blast furnaces, reverberatory furnaces, sintering machines, coke ovens, scarfing operations, and open hearth furnaces, among others. Most of these processes involve unsteady-state conditions in which gas flows, temperatures, dust loadings, and particulate types vary considerably during any single run. According to industrial records, the steel industry is the second most active user of the electrostatic precipitator, following the utilities.

The use of the precipitator in the kraft paper pulp industry is an economic necessity in recovering particulates from the recovery furnace gases. In this furnace, spent black liquor, consisting of extracted lignin and a mixture of sodium sulfide and sodium carbonate, is oxidized to recover the sodium chemicals. This salt cake, leaving the furnace with the combustion gases, can be collected by either a high-energy wet scrubber or an electrostatic precipitator. In the case of a precipitator, collection efficiencies up to 99% have been obtained. The collected salt cake, equivalent to 150 lb of sodium chemicals per ton of pulp is returned to the process.

A tabulation of industries that utilize the electrostatic precipitator for the collection of particulates is shown in Table 5-15. Typical process data such as gas flows, temperatures, dust concentration, dust size, and efficiency are also listed in this table.

Sample Design Calculations

To demonstrate the sizing of a plate type precipitator, a sample problem will be solved. To compare the various particulate collection equipment types, the same basic problem stated in Section 5-4 for cyclonic separators will be used.

Problem Statement

Gas flow, acfm (avg)	168,600
Gas flow, acfm (peak)	210,000
Gas temperature, °F	400
Gas moisture, vol.%	8.5
Gas analysis, vol.%	
CO_2	14.2
H_2O	8.5
SO_2	0.2
Dust type	Flyash
Dust sp gr	2.5
Dust bulk density, lb/cu ft	30.0
Dust resistivity (estd.), ohm-cm	2×10^{10}
Dust loading, grains/acf (avg)	1.1
Dust loading, grains/acf (peak)	1.8
Barometric pressure, psia	14.71

Required Performance. A discharge loading of 0.03 grain/scf at average gas flow and dust-loading values is to be obtained. The peak-condition frequency is insufficient to demand compliance at these levels. The expected collection efficiency and emissions at peak flow and loading values should be indicated, however.

Problem Solution

$$\text{Inlet dust loading} = 1.1 \times \frac{860}{530} = 1.79 \text{ grains/scf}$$

Discharge dust loading = 0.03 grain/scf

$$\text{Required efficiency} = 100 - \left(\frac{0.03}{1.79}\right) \times 100\% = 98.3\%$$

Design Constants:

For 0.2% SO_2 flue gas, a flyash resistivity of 2×10^{10} ohm-cm, and a 98.3% efficiency requirement, design factors will be

1. Drift velocity, 0.40 fps

153

Table 5-15. Typical Application Data for Electrostatic Precipitators

Industry	Application	Gas Flow, thousand acfm	Temp, °F	Dust Concn, grain/acf	Dust Size, % < 10μ	Coll. Eff., %
Electric power	Flyash–pulverized coal boiler	50–800	270–600	0.4–5.0	25–75	95–99
Portland cement	Dust from kilns	50–1000	300–750	0.5–15.0	35–75	85–99+
	Dust from dryers	30–100	125–350	1–15	10–60	95–99
	Mill ventilation	2–10	50–125	5–25	35–75	95–99
Steel	Blast furnace gas cleaning	20–100	100–150	0.02–0.5	100	95–99
	Tar collection from coke oven gases	50–200	100–150	0.10–1.0	100	95–99
	Fume collection from open hearth and elect. furn.	30–75	300–700	0.05–3.0	95	90–99
Nonferrous metals	Fume from kilns, roasters and sinter machine	5–1000	150–1100	0.05–50	10–100	90–98
Pulp and paper	Kraft soda fume	50–200	275–350	0.5–2	99	90–95
Chemical	Acid mist	2.5–20	100–200	0.02–1	100	95–99
	Gas cleaning, SO_2 CO_2, etc.	5–20	70–200	0.01–1	100	90–99
Petroleum	Powdered catalyst recovery	50–150	350–550	0.1–25	50–75	99–99.9
Gas	Tar removal from gas	2–50	50–150	0.01–0.2	100	90–98
Carbon black	Carbon–black collection and agglomeration	20–150	300–700	0.03–0.5	100	10–35*

Source: Magill et al., Ref. 15.

*Precipitator agglomerates and cyclonic separator collects.

2. Plate spacing, 9 in.
3. Plate height, 30 ft
4. Four 6 ft fields (sections) in depth
5. Power input potential, 40 kv
6. Pressure drop of 0.5 in. w.g.

Plate Collection Area (A)
From Eq. 5-6, Eff $= 1 - e^{-x}$.

$$e^{-x} = 1 - 0.983 = 0.017$$
$$x = Av_p/V = 4.1$$
$$A = 4.1 \times \frac{168,600}{0.40 \times 60} = 28,800 \text{ ft}^2$$

Number of Ducts in Width

Field depth = 4 fields $\times$ 6 ft = 24 ft
Field height = 30 ft

Since each duct has two surfaces,

$$\text{No. ducts} \times 2 \times 24 \text{ ft} \times 30 \text{ ft} = 28,800 \text{ ft}^2$$
$$\text{No. of ducts} = 20$$

Overall Dimensions

Width = 20 ducts $\times$ (9/12) ft = 15 ft
Depth = 24 ft
Height = 30 ft

Gas Velocity

$$\frac{168,600}{20 \times (9/12) \times 30 \times 60} = 6.2 \text{ fps}$$

Contact Time

$$\text{Time} = \frac{4 \text{ fields} \times 6 \text{ ft}}{6.2} = 3.9 \text{ sec}$$

Partial Efficiency at End of First Field

$$x = Av_p/V = \frac{28,800}{4} \times \frac{0.40}{168,600} \times 60$$
$$x = 1.02; \quad e^{-x} = 0.36$$
$$E = (1 - 0.36) \times 100 = 64.0\%$$

Efficiency at Peak Gas Flow Rates

$$x = 28{,}800 \times \frac{0.40}{210{,}000} \times 60 = 3.3$$

$$\bar{e}^{-x} = 0.037; E = 96.3\%$$

At peak conditions

$$\text{Emissions} = 1.8 \times \frac{860}{530} \times 0.037 = 0.11 \text{ grain/scf}$$

Estimated Power Demand

In a single field, the number of plates is 20. Therefore,

Plate area $= 20 \times 6 \text{ ft} \times 30 \text{ ft} \times 2 = 7200 \text{ ft}^2$

At 25 ma/1000 ft^2

$$\text{Current} = \frac{7200}{1000} \times 25 = 180 \text{ ma/field}$$

$$\text{Power output} = 40 \text{ kv} \times \frac{180 \text{ ma}}{1000} = 7.2 \text{ kw}$$

Hence, the total power required is

$$\text{Total} = 4 \text{ fields} \times 7.2 = 28.8 \text{ kw (d-c)}$$

Weights and Pricing. Based on calculated weights of the various plate thicknesses, support steel, electrodes, etc., the total weight for this design is 230,000 lb. The cost, on a delivered and unerected basis, is $121,000, equivalent to a unit cost of $0.72/cfm. Included in this price are the following items;

Casing and structural members
Discharge electrodes
Collection electrodes
Division plates
Perforated baffle
Electrode support frames
Rapper assemblies
Hoppers
High-voltage substation
Control station

5.6. EQUIPMENT CHARACTERISTICS—FABRIC FILTER

Operating Principles

The fabric filter collects solid particulates by retaining them on fabric bags as the gas continues through the bag or envelope. The fabric filter is capable of providing high collection efficiencies for particles as small as 0.5 μ and will remove some particles down to 0.01 μ in size. Since the fabric filter is a positive, or "total," collector, collection efficiencies are usually measured at the 99⁺% level. Operating pressure drops across the filter lie in the range of 2 to 8 in. w.g.

The fabric filter comprises multiple bags or envelopes, which are either tubular or flat, suspended inside a housing. The housing, usually referred to as a "bag house," may contain as many as several thousand bags. Dust is removed from the fabric surfaces by mechanical or pneumatic means. A typical fabric filter is shown in Fig. 5-31.

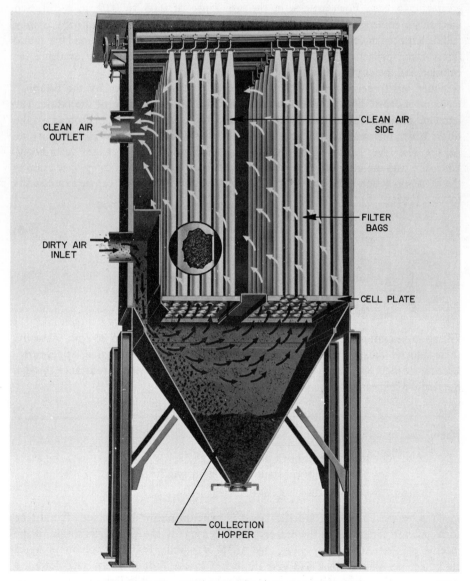

Figure 5-31. Tubular type fabric filter baghouse.

Air flow in fabric filtration is usually laminar. As the particle conveyed by the gas stream approaches a single fabric fiber and comes within a distance equivalent to the radius of the particle, it is intercepted by and captured by this fiber. This direct interception mechanism applies to relatively inertialess particles in the submicron size range. When the particle has appreciable inertia, it will not follow a streamline path because the gas flow is deflected by the individual fibers. The inertia and size of the particle and the size of the fiber will determine whether collision will occur between particle and fiber. This impingement or inertial impaction is predominant for particles greater than 1 μ. For particles in the size range of 0.05 to 0.01 μ, a diffusion mechanism controls. These very small particles exhibit a Brownian activity as they collide with the molecules of the carrier gas. Lower filtration velocities and large fabric fibers favor collection by diffusion because they increase the time available for collision and thereby increase the deposition rate.

Filter resistance comprises the resistance to air flow presented by the filtering medium or fabric plus the layer of dust particles on the surface of the fabric. The latter resistance is the major factor responsible for the overall pressure drop across the filter. Where intermittent cleaning is practiced, the filtration process is a cyclical one. As the dust layer builds up on the fabric, the pressure drop increases accordingly. When the bags are cleaned, the pressure drop falls off and then builds up once again as the filtration action is repeated. Total resistance through the filter can be expressed as follows [16] :

$$R_t = R_f + R_d \qquad (5\text{-}8)$$

where

R_t = total resistance
R_f = resistance through fabric
R_d = resistance through dust layer

Filtration velocities are relatively low, being in the range of 1 to 30 fpm. Actually, these velocity values represent the ratio of gas flow to fabric filter area, or acfm/ft^2. Because of their low value, streamline flow takes place so that the resistance through the fabric is a linear function of velocity. Therefore,

$$R_f = K_o v \qquad (5\text{-}9)$$

where

R_f = resistance through fabric, in. w.g.
K_o = permeability constant of fabric, in. w.g.ft^2/ft^3-min
v = filtration velocity, fpm

The permeability of a specific fabric is experimentally determined. It must be corrected for variations in the viscosity and density of the gas. A definition of the specific permeability of any fabric, by ASTM standards, is the air volume in actual cubic feet per minute that will pass through a square foot of clean new cloth at a pressure differential of 0.5 in. w.g. At the velocities used in industrial filtration, the resistance offered by the fabric only is practically negligible. At a filtration velocity of about 3 fpm, the pressure drop is less than 0.10 in w.g.

The resistance through the accumulated layer or mat of collected dust particles is given by

$$R_d = K_d \frac{Ctv^2}{7000} \qquad (5\text{-}10)$$

where

R_d = resistance through the dust mat, in. w.g.
K_d = dust resistance coefficient, in. w.g./lb dust/ft^2 fabric
  area/fpm filtration velocity
C = filter inlet dust concentration, grain/acf
t = filtration period, min
v = filtration velocity, fpm

The value of K_d depends on the viscosity and density of the gas as well as the density, porosity, particle size, and bulk density of the dust. The value of this resistance coefficient varies inversely as the square of the particle size so that the pressure drop across the filter is greater for the collection of fine particles. The value K_d also varies exponentially inversely as the mat porosity. Fine particles are more easily compacted by the filtration action than are the larger ones, so that the deposited mat is less porous. Because of this porosity factor the resistance across the filter for fine particles is greater than for the larger ones.

From Eq. 5-10 it can be seen that the resistance increases directly as the dust loading and the filtration period. For any given operating pressure drop, the greater the inlet loading, the shorter is the time required to build up the mat to the thickness equivalent to this pressure drop.

Although the resistance through the fabric (R_f) in Eq. 5-9 is a linear function of velocity, it is an exponential function of the mat thickness. This is due to the increasing thickness of the dust buildup. High velocities tend to compact the mat and decrease porosity, thereby causing the resistance to increase. If the fabric surface were continuously cleaned and the mat thickness were not permitted to build up, then the value of R_f would approach zero. Thus, the filtration velocities used are dependent on the bag-cleaning method. For continuous cleaning, velocities in the range of 10 to 20 fpm. can be tolerated, thereby decreasing the filtration area requirements and the cost of the fabric filter. One limit on the velocities to be used is the possibility that the fabric may be distorted or ruptured, thus adversely affecting performance and increasing maintenance costs.

A more practical approach to pressure drop in fabric filtration can be taken when it is realized that the resistance of clean new fabric can never be attained again, once the fabric has been used. Therefore, the resistance of the fabric-residual dust mat combination can be considered and the resistance values expressed as follows [17] :

$$R_f' = K_0' v \qquad (5\text{-}11)$$

where

R'_f = basic (used) cloth resistance, in. w.g.
K'_0 = resistance factor, in. w.g./fpm
v = filtration velocity, fpm

The magnitude of K'_0 depends on the characteristics and quantity of dust remaining in the cloth immediately after the cleaning cycle. It therefore depends on the cloth, the dust, and the cleaning method. Values of K'_0 for different types of dust and applications are shown in Table 5-16. For any particular application, the total resistance R_t can be taken as the sum of R'_f in Eq. 5-11 and R_d in Eq. 5-10. Values of the dust resistance coefficient K_d are tabulated in Table 5-17. Equations 5-8 through 5-11 cannot be directly applied to the design of fabric filters, but they can be used as guides, to be modified according to experience for any particular application.

Table 5-16. Basic Cloth Resistance Factors (K′₀)

Collector Type and Dust Source	Cloth Area, ft²	K'_0
Flat Bag Collectors:		
Stone crushing	250	0.83
	250	0.49*
	500	0.83
		0.78
		0.75
		0.74
	2250	0.79
	9000	1.01
Synthetic abrasive crushing	–	0.80
Clay crushing	500	1.60
Cloth Tube Collectors:		
Stone crushing	2150	0.47
		0.45
		0.60
	4300	0.45
		0.37
	1500	0.40
Stone chiseling	400–1000	0.17–0.27
Electric welding fume	10	0.70
Iron cupola fume	–	2.50
Foundry dust core handling	5200	0.28
		0.25
		0.58
Shot–blast room ventilation	2350	0.63
		0.39
		0.39
Clay crushing	500	0.60

*Same as first operation but with pneumatic vibrator.
Source: Hemeon, Ref. 17.

Table 5-17. Dust Resistance Coefficient K_d

	K_d (in. w.g./lb dust/ft² /fpm velocity) for Particles Sized at Less than:						
	20 Mesh*	140 Mesh	375 Mesh	90 μ	45 μ	20 μ	2 μ
Graphite	1.58	2.20	–	–	–	19.8	–
Foundry	0.62	1.58	3.78	–	–	–	–
Gypsum	–	–	6.30	–	–	18.9	–
Feldspar	–	–	6.30	–	–	27.3	–
Stone	0.96	–	–	6.30	–	–	–
Lampblack	–	–	–	–	–	–	47.2
Zinc oxide	–	–	–	–	–	–	15.7†
Wood	–	–	–	6.30	–	–	–
Resin (cold)	–	0.62	–	–	–	25.2	–
Oats	1.58	–	–	9.60	11.0	–	–
Corn	0.62	–	1.58	3.78	8.80	–	–

Source: Williams et al., Ref. 18.

*U.S. Standard sieve

†Flocculated material; size actually larger.

In normal filtration, particles considerably smaller than the voids in the fabric will be trapped by impingement on the fine fibers that span the openings formed by the main threads. After a period of operation, a loose cake or mat is built up on the filter fibers. This mat may have voids as great as 80 to 90%, which provide additional filter media and yield very high collection efficiencies. It is during this period of mat buildup that the fabric filter attains its maximum efficiency. However, to avoid excessive pressure drop, the fabric must be periodically cleaned. Theoretically, after cleaning, the filter will operate at a reduced efficiency until the mat is built up again. Therefore, a balance must be made between the cleaning effectiveness and frequency and the attainment of maximum collection efficiency.

Both the intermittent and continuous cleaning methods are available with industrial fabric filter designs. The intermittent method utilizes mechanical shaking devices, while the continuous method depends on a pneumatic reverse-purging air flow. When intermittent cleaning is used, the fabric filter is limited to low operating velocities of 1 to 6 fpm. High velocities tend to compact the dust too greatly between cleaning cycles and build up the operating pressure drop. Frequency of cleaning must be economically traded off against the greater fabric surface area required at the very low velocity values. With continuous cleaning, a wool felt is usually chosen as the filtering medium. Because of the close texture, a mat of collected dust is not required for effective performance, and the dust is almost completely and continuously removed. Allowable design velocities for continuously cleaned filters are in the range of 10 to 20 fpm.

Design Parameters

The process parameters influencing the design of a fabric filter are gas flow, temperature, moisture content, dust concentration, dust characteristics, gas properties, pressure drop, and collection efficiency.

Table 5-18.   Filtration Velocities for Various Dusts and Bag Cleaning Methods

DUST	Filtration Velocity, cfm/ft²			DUST	Filtration Velocity, cfm/ft²		
	Shaker Collector	Pulse Jet	Reverse Air Collapse		Shaker Collector	Pulse Jet	Reverse Air Collapse
Alumina	2.5–3.0	8–10	—	Leather dust	3.5–4.0	12–15	—
Asbestos	3.0–3.5	10–12	—	Lime	2.5–3.0	10–12	1.6–2.0
Bauxite	2.5–3.2	8–10	—	Limestone	2.7–3.3	8–10	—
Carbon black	1.5–2.0	5–6	1.1–1.5	Mica	2.7–3.3	9–11	1.8–2.0
Coal	2.5–3.0	8–10	—	Paint pigments	2.5–3.0	7–8	2.0–2.2
Cocoa, chocolate	2.8–3.2	12–15	—	Paper	3.5–4.0	10–12	—
Clay	2.5–3.2	9–10	1.5–2.0	Plastics	2.5–3.0	7–9	—
Cement	2.0–3.0	8–10	1.2–1.5	Quartz	2.8–3.2	9–11	—
Cosmetics	1.5–2.0	10–12	—	Rock dust	3.0–3.5	9–10	—
Enamel frit	2.5–3.0	9–10	1.5–2.0	Sand	2.5–3.0	10–12	—
Feeds, grain	3.5–5.0	14–15	—	Sawdust (wood)	3.5–4.0	12–15	—
Feldspar	2.2–2.8	9–10	—	Silica	2.3–2.8	7–9	1.2–1.5
Fertilizer	3.0–3.5	8–9	1.8–2.0	Slate	3.5–4.0	12–14	—
Flour	3.0–3.5	12–15	—	Soap, detergents	2.0–2.5	5–6	1.2–1.5
Graphite	2.0–2.5	5–6	1.5–2.0	Spices	2.7–3.3	10–12	—
Gypsum	2.0–2.5	10–12	1.8–2.0	Starch	3.0–3.5	8–9	—
Iron ore	3.0–3.5	11–12	—	Sugar	2.0–2.5	7–10	—
Iron oxide	2.5–3.0	7–8	1.5–2.0	Talc	2.5–3.0	10–12	—
Iron sulfate	2.0–2.5	6–8	1.5–2.0	Tobacco	3.5–4.0	13–15	—
Lead oxide	2.0–2.5	6–8	1.5–1.8	Zinc oxide	2.0–2.5	5–6	1.5–1.8

NOTE: Values tabulated are based on light to moderate loadings of granular dust having particle size and shape characteristics typical of the specific material. Ratios will normally be less when dust loading is very heavy, temperature is elevated, or particle size is smaller than commonly encountered.

Source: American Air Filter Co., Inc.

162

The gas flow, expressed in actual cubic feet per minute, determines the fabric filter size by application of the optimum filtration velocity. This filtration velocity, or filter ratio, is defined as the ratio of gas filtered in cubic feet per minute to the area of the filtering media in square feet. Table 5-18 tabulates values of filtration velocity for various industrial pollutants. In sizing fabric filters, these values are empirically determined and are based on prior experience. The value represents the average velocity with which the gas passes through the fabric surface area. It is actually a superficial face velocity, as it does not truly represent the actual velocity through the fabric openings. Excessive filtration velocities cause high pressure drops, reduced collection efficiency, and high maintenance costs due to fabric wear. For common dusts such as sand, fertilizer, and grinding dusts, velocities in the range of 2.0 to 5.0 fpm are recommended for an intermittently cleaned filter. For fumes comprising submicron particulates, which tend to bind the fabric, velocities less than 2 fpm should be chosen. The choice of design velocity depends on the type of fabric. For example, for durable Orlon bags the cleaning frequency can be increased and higher velocities utilized because of the resistance of this fabric to wear. In one application involving the collection of metallic fumes from an electric furnace, a velocity as high as 2.5 fpm has been used. From an operational point of view, minimum filter ratio values are favored. However, overdesign must be avoided and an optimum value must be chosen to reflect an economic balance between equipment capital costs and troublefree performance.

A variety of fabrics are available for the filter design. The major criteria in their selection are temperature and dust characteristics. The various fabric types, with their temperature limitations and chemical resistance ratings, are shown in Table 5-19. The use of wool felt is still common for the collection of metallurgical fumes at temperatures up to 240° F. Continuous jet-cleaning devices favor the use of felt because of its excellent permeability and tendency to form an easily removable cake. The superior temperature rating of such synthetic fibers as Nylon, Orlon, and Dacron—plus their smooth surface and chemical resistance—ensures their selection for many high-temperature chemical dust-collection applications.

Table 5-19. Industrial Fabric Materials

Fabric	Temp. Range, °F	Recommended Max. Temp., °F	Chemical Resistance	
			Acid	*Alkali*
Cotton	160–190	180	Poor	Fair
Dynel	150–180	175	Good	Good
Wool	180–240	220	Good	Poor
Nylon	200–290	220	Good	Poor
Orlon	200–350	275	Good	Fair
Dacron	250–350	275	Excellent	Good
Glass	500–700	550	Excellent	Excellent

Source: Danielson, Ref. 9.

By far the most favored fabric for high-temperature operation is glass. As indicated in Table 5-19, the maximum recommended operating temperature with this material is 550° F. In those applications where temperatures of 2000 to 2500° F are reached, dilution air bleeds, water jackets, water addition, and radiant heat-transfer surfaces are employed to reduce the temperature to an acceptable operating level. However, due to the physical weakness and low abrasion resistance of glass, cleaning techniques are severely limited. Vigorous shaking must be avoided and extreme care must be exercised even in the shipment and storage of this material. Low filtration velocities must be used to avoid binding. In most applications a graphited silicone coating is applied to reduce the inherent fragility of the glass.

The moisture content of the dust-laden gas is extremely important to the application of the fabric filter. For some effluent control problems where the moisture content exceeds 20 vol %, care must be exercised to maintain the operating temperature above the dewpoint. Any condensation of water vapor would blind the filter fabric and the collector would then become inoperable. Even at lower moisture contents, temperature controls must be provided to prevent condensation. One method of control consists of a burner that introduces hot flue gases into the dust-laden gas stream to maintain the temperature level above the dewpoint. The burner operation is controlled by temperature-sensing elements located in the gas stream at the filter exit.

The filter resistance varies directly as the dust concentration so that high loadings require the most frequent cleaning. Since the cleaning operation, especially by mechanical means, is one of the principal causes of bag wear, minimum cleaning is one of the criteria for good design. In cement kiln operation, dust loadings may vary from 5 to 60 grains/acf. At loadings above 20 grains/acf, the use of a cyclonic separator ahead of the filter can be considered as an alternate to reducing the filtration velocity to an economically unacceptable level. For a typical application involving the collection of cement or limestone dust, a filter ratio of 25 fpm would be permissible at dust loadings of 2 to 3 grains/acf. However, for a dust loading of 40 grains/acf, a practical filter-ratio value of about 15 fpm would be chosen. Both the nature of the dust and loading values are the most important variables affecting the filter ratio.

Such dust characteristics as particle size, shape, viscidity hygroscopicity, and dielectric properties influence the fabric selection and filter ratio. As mentioned previously, the filter resistance varies inversely as the particle size, so that lower filter ratios must be chosen for fine dusts. With some dusts, electrostatic forces contribute to the cake buildup rate. For these dusts, synthetic fibers such as Orlon, Dacron, or Dynel are superior to natural fibers. For electrically conductive dusts such as those of inorganic salts, electrostatic effects are negligible and wool fibers are suitable. Certain viscid and/or hygroscopic dusts strongly adhere to the fabric and therefore smooth, glossy bag materials are necessary to facilitate cleaning.

The chemical composition of both the gas stream and particulates must be considered in selecting the proper filter fabric. Such corrosive gaseous contaminants as SO_2, NO_2, and HCl, among others, dictate the use of chemically resistant fabrics. Both the gas density and viscosity affects the resistance through the fabric and dust mat.

The filter pressure drop or resistance is affected by the physical properties of the dust and fabric, the filter ratio, and the cleaning frequency and it is usually specified as an average value for a "commercially clean" filter. At startup, the pressure drop across the filter is low before the collected dust builds up, and only after hours or perhaps days of operation will steady-state conditions be attained. Pressure-drop values for different filter ratios are shown in Fig. 5-32. In these curves, the resistance values for three different fabric materials are illustrated for a 60 min cleaning cycle. Each curve demonstrates that, at constant dust loadings, the resistance generally varies as the square of the filtration velocity (filter ratio). This relationship was defined in Eq. 5-10. The filter resistance is influenced by the frequency and completeness of bag cleaning. It is important that complete removal of the dust cake be avoided because of the effectiveness of this residual dust as a filtering medium. In such intermittent bag-cleaning methods as mechanical shaking, an optimum frequency must be selected to avoid too great a pressure-drop variation between cleaning cycles.

In continuous cleaning, the resistance is usually maintained at a constant and minimum level, depending, of course, on the dust loading and filter ratio. Figure 5-33 shows the effect of continuous versus cyclical cleaning on the filter pressure drop for a

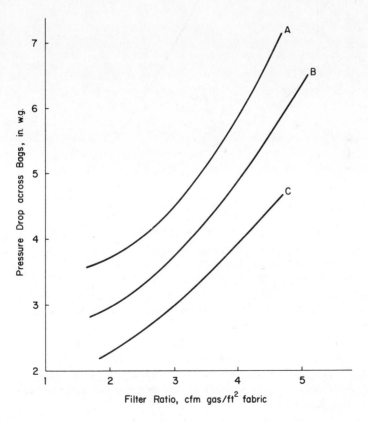

Figure 5-32. Pressure drops at various filter ratios on a 60 min cleaning cycle. Curves A, C: siliconized glass fabric; B: siliconized Dacron. (Adapted from Danielson, Ref. 9.)

165

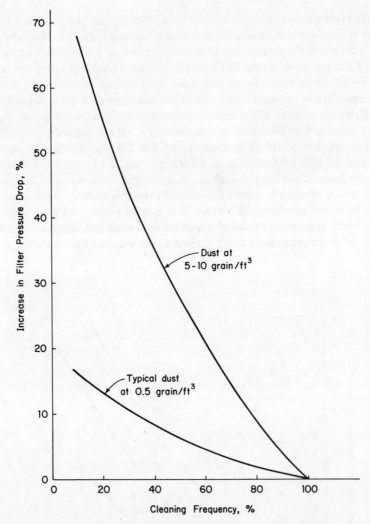

Figure 5-33. Filter pressure drop versus cleaning frequency for reverse air-jet operation. (Adapted from Danielson, Ref. 9.)

reverse air-jet type of cleaning mechanism. At the lower dust-loading value of 0.5 grain/acf, this curve indicates that operating the cleaning mechanism only 30% of the time causes an increase in pressure drop of only 10%. Because of the wear on most fabrics imposed by cleaning, these values would indicate the desirability of accepting the minimal pressure-drop increase while reducing bag maintenance. At 5 to 10 grains/acf, continuous or near-continuous cleaning is recommended because of the greater pressure-drop variation.

The collection efficiency as commonly evaluated for other types of particulate collection equipment is not so significant a factor with the fabric filter. Fabric filters are usually rated at collection efficiencies in the order of 99 to 99.8%. Most important, this performance can be accomplished without extraordinary energy inputs, as is necessary for the wet scrubber and to a lesser degree for the electrostatic precipitator. An effective performance can be obtained by the expenditure of fixed costs for purchase of the equipment and maintenance charges. Dust losses from a fabric filter occur primarily from leaks that are the result of fabric wear. Thus, minimum filter velocities equivalent to maximum investment values and maintenance are the major costs associated with the fabric filter.

Before the equipment can be sized for a fabric filter dust collection system, the proposed application must be fully described. Based on the data furnished by the user, the equipment manufacturer will provide an optimum design to satisfy the user's performance requirements. As in the case of the cyclonic separator, electrostatic precipitator, and wet scrubber, the fabric filter designs are empirically determined. Actual industrial performance records for specific applications provide design criteria, and when these are lacking, pilot plant investigations are undertaken.

Both pertinent plant information and process-operating conditions must be obtained. The data requested for the design of the electrostatic precipitator (see Section 5.5) would be generally applicable to the fabric filter. Emphasis should be placed on the three items of data most important to the design of the fabric filter. These are (1) gas temperature, particularly the maximum value; (2) gas moisture; and (3) cyclical nature of process.

Mechanical Design

Although the basic design of the fabric filter is well established, there are a number of mechanical variations in their construction and operation. The two common industrial filters are of the tubular or bag and the envelope types. In the tubular fabric filter, shown in Fig. 5-31, the dirty air enters at the side and flows downward toward the collection hopper. On reversing its flow, the heavier dust particles are inertially separated and collected in the hopper. The gas stream continues up through the inside of the tubes, where the remainder of the dust is removed. The clean gases pass through the tube fabric and are discharged to atmosphere. In this particular design the collected dust, on the inside surface of the tubes, is shaken loose intermittently by oscillation of the bag support frame. The dislodged dust falls into the collection hopper, from where it is discharged into a disposal system.

Overall dimensions for a series of tubular-type baghouses are shown in Fig. 5-34. The housing is of modular design so that, for large capacities, two or more units can be connected in parallel. The bag dimensions are 6 in. diam by 10 ft in length. The cleaning mechanism used in this series is a motor-operated shaker system. During the cleaning cycle the compartment being cleaned is removed from the system, and during this period it is important that the collection capacity not be exceeded. For example, consider a filter design to handle 1800 acfm at a filter ratio of 2:1 so that the nominal fabric area is 900 ft^2, assuming the filter has three compartments, each with 300 ft^2 of area. At any single instant, two compartments will be on stream while the third is being cleaned, so that the actual filtration area is 600 ft^2, which is equivalent to a filter ratio of 3:1. For a realistic design, a four-compartment design should be chosen, each with 300 ft² of fabric area. With three compartments constantly in operation, then the true filter ratio is 1800/(3×300), or 2:1.

In this design the bags are attached to thimbles located on a tube sheet at the bottom of the collector, fastened in place by steel bands. Extra care must be taken to ensure that a tight seal is made at this connection to prevent leakage. In some applications involving the collection of fine dusts, losses up to 10% of the inlet dust loading have been experienced because of inadequate sealing at this point. The top of the bag can be similarly connected to thimbles or spring-loaded caps. In Fig. 5-31, the bags have been sewn closed and hung over hooks that are connected to the movable support frame. The clearance between the bags must be sufficient to prevent contact between adjoining bags. Such contact could result in wear as the bags rub against each other during mechanical shaking. A minimum clearance of 2 in. has been adopted for

Figure 5-34 Tubular Type Baghouse Dimensions—Shaker Cleaning Models

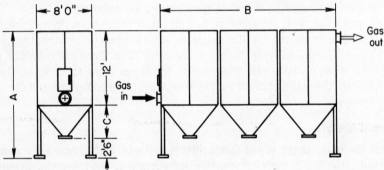

Capacity Ratio, cfm			Cloth Area,	No. of	No. of	Dimensions, ft-in.		
2:1	2½:1	3:1	ft.²	Bags	Hoppers	A	B	C
1,880	2,350	2,820	940	60	1	20-0	4-3	5-6
2,500	3,125	3,750	1,250	80	1	20-0	5-6	5-6
3,120	3,900	4,680	1,560	100	1	20-0	6-9	5-6
3,740	4,675	5,610	1,870	120	1	20-0	8-0	5-6
4,360	5,450	6,540	2,180	140	1	21-6	9-3	7-0
4,980	6,225	7,470	2,490	160	1	22-6	10-6	8-0
5,600	7,000	8,400	2,800	180	2	20-0	11-9	5-6
6,220	7,775	9,330	3,110	200	2	20-0	13-0	5-6
6,840	8,550	10,260	3,420	220	2	20-0	14-3	5-6
7,460	9,325	11,190	3,730	240	2	20-0	15-6	5-6
8,080	10,100	12,120	4,040	260	2	21-6	16-9	7-0
8,700	10,875	13,050	4,350	280	2	21-6	18-0	7-0
9,320	11,650	13,980	4,660	300	2	22-6	19-3	8-0
9,940	12,425	14,910	4,970	320	2	22-6	20-6	8-0
10,560	13,200	15,840	5,280	340	3	20-0	21-9	5-6
11,180	13,975	16,770	5,590	360	3	20-0	23-0	5-6
11,800	14,750	17,700	5,900	380	3	20-0	24-3	5-6
12,420	15,525	18,630	6,210	400	3	20-0	25-6	5-6

bags of average length. Bag dimensions are usually 6 to 12 in. diam with lengths varying from 6 through 20 ft.

Envelope-shaped bags must be mounted on a support frame, usually fabricated of wire. The major advantage in the use of envelopes is that maximum filter area can be obtained for minimum filter size. An envelope type fabric filter baghouse is shown in Fig. 5-35. However, inspection and maintenance of this type of bag is somewhat more complicated. Wear is increased because of the friction between the filter cloth and the wire support frame. Glass fabric, because of its fragility, cannot be used in this design configuration. If a particular service demands severe mechanical shaking for cleaning the bags, then the use of envelopes is not recommended. Because of the necessity of using an inside support frame, dust is collected on the outside of envelope bags.

Two basic types of cloth are used in fabric filters: woven cloth and felted cloth. Woven fabric acts as a support on which a layer of dust is deposited to form a microporous mat capable of removing additional particles. Cake filtration on woven cloth is accomplished when the new filter cloth becomes thoroughly impregnated with dust. When a new woven fabric is placed in service, performance may suffer until the dust mat is built up [20]. This deposition of mat may require a period of a few hours or as much as several days, depending on dust loading, filter ratio, and nature of the dust. Clean felted fabrics are more efficient dust collectors. The dust particles are trapped on the interstices formed by the felt fibers so that a dust precoat or mat is not required.

Woven fabric filters usually employ filter ratios of 1:1 to 5:1, depending on the fabric permeability. The permeability or porosity can be varied by using different materials and weave patterns. The basic weaves used for filtration are plain, twill, and sateen. The plain weave has a simple "one up" and "one down" pattern, which allows a high impermeability if woven tightly. If the count per square inch is decreased, an open weave can be obtained and the permeability can be increased. The plain weave is used with certain cotton ducks and many synthetic fibers. The twill weave is characterized by a sharp diagonal "twill" line and has fewer interlacings than the plain weave and is therefore more porous. Cotton and synthetic filter twills are commonly used. The sateen weave has still less interlacings and is spaced widely, thereby providing a smooth surface with increased porosity. Cotton fabrics in this weave are commonly known as "sateens" and are used more than any other fabric for filtration at ambient temperature. Woven fabric bags are made from cotton, wool, Dacron, Nylon, Orlon, Nomex, polypropylene, Teflon, and fiberglass. Fiberglass fabric bags are treated with silicone, mixtures of silicone and colloidal graphite, and Teflon lubricants to provide protection against abrasion and flexing failures.

Felted fabric bags, used with reverse-jet cleaning devices, can be operated at filter ratios of 5:1 to 20:1. Felted bags are more expensive than woven bags. Felted fabrics can be made with wool and a limited number of synthetic fibers. Felted fabrics are complex, labyrinth-type masses of randomly oriented fine fibers. The thickness of the bag wall provides maximum dust impingement targets to trap small dust and fume particles. In some applications, felted bags do not function well because of the difficulty in removing very fine fumes that have become embedded in the felt.

As dust is continuously collected on the fabric bag, the pressure drop increases

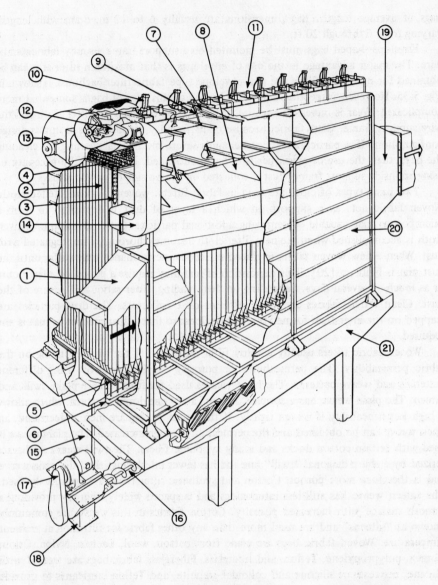

1. Gas inlet manifold
2. Fabric-pocket filter panel
3. Pocket retaining spirals
4. Retaining bar
5. Guide rods
6. Guide rod channels
7. Upper damper

8. Lower damper
9. Connecting rod
10. Pneumatic operating cylinder
11. Electrical solenoid valves
12. Rotary air-valve
13. Scavenge air pressure fan
14. Scavenge air suction duct

15. Dust hopper
16. Dust conveyor
17. Conveyor drive unit
18. Dust outlet
19. Gas outlet manifold
20. Casing
21. Support structure

Figure 5-35.   Envelope type fabric filter baghouse. (Courtesy Prat-Daniel, Ltd.)

until a maximum design value is reached. The filter elements must then be cleaned. This can be done either intermittently or continuously by manual, mechanical, pneumatic, or sonic means. Fully automatic systems can be incorporated into the design, which will actuate the cleaning device when the pressure drop across the filter reaches a preset value. For small fabric filters having collection areas up to about 500 ft^2, manual cleaning is sometimes used. A manually operated handle transmits an impact to the framework supporting the bags. This impact, or "rap," shakes the dust loose. This is an ideal arrangement for batch processes where the need for cleaning coincides with the completion of a process operation. The manual action can be easily converted to a motor-driven system, as shown in Fig. 5-31. Regardless of the cleaning method used, a manometer should be provided to indicate the pressure drop across the filter. In any application the frequency of cleaning is a function of the pressure drop, and therefore a pressure-sensing instrument is essential.

In motor-driven shaking systems the rotary motion imparted by the drive is converted to oscillations by a cam or eccentric. The bags may be shaken either horizontally or vertically. There must be "zero" pressure inside the bags to ensure their complete collapse when cleaning takes place. Pressures as low as 0.05 in. w.g. can prevent adequate cleaning because of the force still exerted on the adhering particles by this pressure. The use of butterfly or louver dampers alone to seal off a section being cleaned is not sufficient to maintain "zero" pressure conditions. Either the blower moving the dust-laden process gas must be shut down or, as an alternate, a small amount of reverse air flow must be introduced into the isolated section being cleaned.

In the various mechanisms available for cleaning fabric bags, both the frequency and amplitude of the energy input must be considered. In one method of pneumatic cleaning, compressed air is used to operate an air motor that imparts a high-frequency vibration to the bag suspension framework. Although the frequency is high, the amplitude is low; thus the overall transferred energy is low. This particular method is not too effective for materials that adhere strongly to the bags. This system is commonly used with fiberglass filter bags. Either a pneumatic motor or cylinder oscillates the bag suspension framework. The action is very gentle because of the strength limitations of the glass fibers. To supplement this system, a small reverse flow of air is introduced to the bags, causing their collapse. By close control of this reverse air flow, a gentle collapse and inflation of the bag can be accomplished without damaging the fabric. Short air pulses (about three, each of 1 sec duration) sweep down the filter tube, thereby creating a gentle shaking action. Another method for cleaning fiberglass bags is to install several rings inside the bag so that when the air is reversed, the bags will collapse but glass-to-glass fabric contact will be avoided.

Envelope type baghouses usually depend on reverse air for cleaning. The dust is collected on the outside of the fabric and, during cleaning, purge air is introduced into the inside of the envelope. A number of bags are thus cleaned at one time, thus reducing the unavailable filtration area to a minimum. A separate air blower is used to provide the cleaning air.

Another reverse-jet cleaning system utilizes an air supply ring that travels vertically outside and around the bag. The cleaning air is directed at the bags through a narrow slot in the ring as it travels up and down the tubular bag. The force of the jet causes the bag to be flexed at the point of air impingement, thus effecting the cleaning process by a combination reverse air and flexing action. Filter bags utilizing this cleaning mechanism are usually made of felted fabric. Felted Orlon or Dacron are used for high-temperature and chemically resistant service. Because these felted materials do not require residual dust for effective filtration action, this system provides total and continuous cleaning. As a consequence, relatively high filter-ratio values can be used. However, some reports have indicated considerable fabric wear and operating difficulties with the moving-ring mechanism, thereby contributing to high maintenance costs.

Recently, sonic cleaning has been utilized on a very limited scale. Sonic generators are used to provide additional fabric vibration for applications involving the collection of such fumes as zinc oxide and carbon black. The sonic generators are relatively expensive and some means must be employed to deaden the sounds associated with their operation. This cleaning method has still not been fully evaluated.

Cleaning systems can be simply manual or totally automatic, depending on the service, labor demands, and plant design sophistication. In simple manual systems the shakers are activated by push-button control or valve operation. Interlocks are usually provided to shut down the fan before the cleaning cycle is initiated. Such a system is suited to the operations of small baghouses, particularly when used with intermittent processing where the filter must be cleaned at the completion of each process batch. In a fully automatic system, the complete filtering/cleaning cycle is programmed to the operation. When the cleaning cycle is initiated, one compartment of the baghouse is isolated by actuation of appropriate dampers and a small volume of reverse air is injected into that compartment to ensure collapse of the filter bags. The isolated section is then cleaned by one of the systems previously discussed. After completion of the cleaning cycle, the compartment is returned to service by rearrangement of the dampers, and the next compartment isolated for cleaning. Each compartment, in turn, is cleaned in a similar manner. As discussed previously, this system of one compartment's being consistently unavailable for service must be considered in sizing the filter. The optimum filter ratio must be applied to the gas flow rate being treated, based on one compartment being out of service; this was discussed at the beginning of this section.

Installation Features

Of the four types of particulate collection equipment, the fabric filter requires the most plant space. At a filter ratio of 5:1, a gas flow of 500,000 cfm would require a baghouse having dimensions of 40 ft wide by 32 ft high by 80 ft long, equivalent to a volume of slightly greater than 100,000 ft^3. The precipitator sized for this same volumetric gas flow occupied 30,000 ft^3 as calculated in Section 5.5. A tubular type fabric filter at an aggregates plant designed to handle approximately 40,000 acfm is shown in Fig. 5-36. The filter is connected in series with and located downstream from a cyclonic separator. The gases leaving the cyclones enter the fan, which is located at ground level. The fan discharges the partially cleaned gas stream up to the side inlet of the fabric filter. The completely cleaned gases are released to atmosphere

Figure 5-36. Tubular fabric filter installation. (Courtesy of Western Precipitation Division.)

from the top of the filter. The overall collection efficiency for both collectors is in excess of 99.9⁺%.

The inlet gas temperature for this application is about 250° F, so that a treated wool-felt fabric is suitable. In furnace and kiln effluent-control applications the gases must be cooled to protect the fabric. This can be done by indirect cooling, dilution with ambient air, or evaporative cooling. Indirect cooling requires either gas/water heat exchangers or gas-ambient air radiation surface. In cooling the gases, an allowance of at least 50 to 75° F above the dewpoint must be made to prohibit condensation on the fabric surfaces. The use of dilution air usually involves relatively large amounts and therefore the increased investment for the increased capacity of the filter and fan is considerable. Evaporative cooling is about the least expensive solution, but sophisticated control instrumentation is required to control water flow through the sprays. A balance must be made between the added moisture and final reduced temperature, to avoid condensation.

The fabric filter must be located conveniently with respect to the collected dust-handling and disposal system. Ground-located filters are usually provided with support legs that allow a 3 ft clearance between the hopper discharge opening and ground level. The hoppers are emptied through a dust valve into a screw or belt conveyor. In some plants the filter hoppers are located sufficiently high to permit discharge through a dust valve, directly into a dump truck or tote bin. The installation

Figure 5-37.   Bag installation procedure. (Courtesy of Envirotech Corp.)

arrangement must also allow for easy replacement of the bags. Usually, monthly inspection is recommended as one phase of an overall preventive maintenance schedule. An average bag life is approximately 18 months. However, the majority of operators do not wait for leaks to occur but replace the bags on a fixed schedule, the frequency being a function of the service. For one application, one-quarter of the bags are replaced every three months so that every bag is changed once each year.

The position of the fabric filter with respect to the cyclone precleaner and fan, as shown in Fig. 5-36, is typical. The precleaner ahead of the fan lessens the dust loading, thereby reducing fan wear. In this position the filter is under negligible pressure, and in some climates the bags are not enclosed in a housing. This open construction allows hotter gases to be treated because of the cooling achieved as the gases pass through the fabric bags. Construction of an open fabric filter is considerably simplified and therefore entails lower capital cost requirements. An added advantage is ease of bag inspection. With the fan located on the discharge side, the total negative pressure necessary to draw the effluent gas through the train is imposed on the filter. Therefore, an airtight housing must be provided. Ability to withstand negative pressures, sometimes as high as 15 in. w.g. (equivalent to a pressure of about 80 lb/ft^2) must be built into the housing.

Shipment and erection procedures for the fabric filter are similar to those for the electrostatic precipitator. The unit is shipped unassembled and is then field-erected. The support structure, housing, hoppers, and walkways are installed first. The individual tubular or envelope elements are then fastened and suspended. Installation of a tubular type fabric filter is illustrated in Fig. 5-37.

Applications

The fabric filter is essentially a positive collector. It can be designed to handle gas flows in the range of 10,000 to 500,000 cfm. It can remove dust particles, over the complete size range, from effluent gas streams at efficiencies of 99$^+$%. Some of the advantages in its use are as follows:

1. Almost 100% collection efficiency
2. Fine particle size collection capabilities
3. Relatively low pressure-drop requirements
4. Pollutant material collected in dry state
5. High gas flow capacity rating.

The disadvantages in the use of the fabric filter are:

1. High initial costs
2. Operation sensitive to gas moisture content
3. Process gas temperatures limited to 550° F
4. Space requirements considerable
5. Flammability hazards high for a number of dusts

Descriptions of some of the more common applications follow.

Cement Kilns. The collection of dust from rotary cement kilns is a difficult problem to solve economically because of the large volumes of flue gas to be handled and the heavy dust loadings. Gas flows on the order of 100,000 to 200,000 acfm at temperatures of 1100 to 1600° F are typical for a dry process cement kiln with lifters or chains. The moisture content of the gas is 5 to 12 vol % and dust loadings up to 55 grains/acf can be attained. A cyclone precleaner is installed ahead of the fabric filter. The gases are cooled to about 500° F by water-jacketed heat exchangers. An envelope type of fiberglass bag design can be applied for this service, with filter ratios of 2 to 3:1. A continuously operated reverse air-manifold system is utilized to clean the bags. The overall collection efficiency for the cyclone and fabric filter collectors is in excess of 99.9%.

Foundry Emissions. Emissions from foundry cupola are marked by very variable process conditions. The gas flow, temperature, and dust loadings fluctuate widely during any single batch cycle. The gases from the cupola are usually "burned off" in a combustion chamber so that the temperature of the flue gas to be treated is at about 1800° F. However, the use of heat recuperators reduces this temperature to about 500° F while recovering heat for combustion makeup air. With inlet dust loadings as high as 15 grains/scf and an atmospheric emissions requirement of 0.03 grain/scf, a collection efficiency on the order of 99.8 is required. In one specific installation designed to handle 60,000 acfm of gases at 500° F, a baghouse was provided having dimensions of 15 ft wide by 50 ft high by 60 ft long with 30,000 sq ft of cloth area. The housing contained 460 silicone-coated fiberglass bags, each 12 in. diam by 23 ft long. Although exact inlet dust loadings are unavailable, measured emissions from this filter have proved its compliance with local codes, which demanded that loadings did not exceed 0.03 grain/scf.

Carbon-Black Plants. These plants have almost unanimously accepted the use of the fabric filter for final cleanup of the process furnace effluent. Furnace temperatures are maintained at about 2500° F and the effluent gas temperatures are reduced to 450 to 550° F by evaporative cooling. Because of the extremely fine particle size in the range of 0.01 to 0.5 μ, the fluffy nature of carbon black dust, and the extreme low resistivity values, the only applicable particulate collection equipment is the fabric filter. Fiberglass bags are used, with cleaning being accomplished by bag collapse and some supplementary gentle vibration from sonic horns. Filter ratios of about 1.5:1 are used at process gas flows of about 50,000 acfm.

Oxygen Furnace Emissions. Emission controls for the basic oxygen furnace effluent have usually been electrostatic precipitators or high-energy scrubbers. In Europe, a number of fabric filters has been installed for this service. The major problem, which is common to all pyrometallurgical processes, is that of temperature control. The gases leaving the oxygen converter reach a maximum of 2100° F halfway during the blow period. Even with regenerative cooling, it is sometimes found necessary to introduce minimal amounts of dilution air to bring the peak temperatures within the operating limitations of the fiberglass bags. As with most steel mill processes, gas flows and dust loadings as well as temperatures are cyclical, with the total blow period averaging 20 min. Process values for a 220 ton furnace operation are 270,000 scfm at 1800° F when leaving the furnace and at 350° F when entering the filter. The average inlet dust loading was 9 grains/scf. The fabric filter is a glass-fiber envelope type utilizing continuous reverse air blowing for cleaning the bags. A total filter surface area of 115,000 ft^2 was used, equivalent to a filter ratio of about 3.5 fpm. The average filter emissions were measured at 0.005 grain/scf, equivalent to an average collection efficiency of 99.95%.

Grain Processing Emissions. The major sources of grain processing emissions are cleaning, grinding, blending, and truck loading. Usually, medium efficiency cyclonic separators do an effective job, but once again because of the upgraded regulations, a final collector for the cyclone effluent cleanup is required. The fabric filter is preferred for this duty, with reported collection efficiencies of 99.9[+]% for the 1 to 5 μ particles. Mechanically shaken, woven cotton bags are used with filter ratios of about 5:1. More recently, felted fabrics have been used with the attainment of the same high level of collection efficiencies, but at filter ratios as high as 15:1.

Sample Design Calculations

The following problem statement and solution will be based on the same process conditions defined for the cyclonic separator and electrostatic precipitator. Pertinent data for the design of the fabric filter follow.

Problem Statement

Gas flow, acfm (avg)	168,600
Gas flow, acfm (peak)	210,000
Gas temperature, °F	400
Gas moisture, vol %	8.5
Gas analysis, vol %:	
CO_2	14.2
H_2O	8.5
SO_2	0.2
Dust type	Flyash
Dust sp gr	2.5
Dust bulk density, lb/cu ft	30.0
Dust loading, grain/acf (avg)	1.1
Dust loading, grains/acf (peak)	1.8
Barometric pressure, psia	14.71

Required Performance. Atmospheric emissions of 0.03 grain/scf at average gas flow and dust-loading values are to be obtained. The peak conditions occur too infrequently to demand compliance at these levels. The expected collection efficiency and emissions should be specified at peak conditions.

Problem Solution

Laboratory Data. The collection of flyash with fabric filters is a relatively new industrial experience. Therefore a filter-ratio value will be determined experimentally from the following laboratory test data.

Inlet dust concentration, C: 1.1 grains/acf
Filtration velocity, v: 3.1 fpm
Filtration period, t: 60 min
Initial resistance of fabric/residual dust, R'_f: 2.7 in. w.g.
Final total resistance: 6.3 in. w.g.
Resistance of dust mat, R_d: 6.3 − 2.7 = 3.6 in. w.g.
Collected dust: 20 g/ft^2; W = 0.044 lb/ft^2

Resistance Coefficients. The total resistance across the filter can be expressed as

$$R_t = R'_f + R_d$$

The fabric/residual dust resistance, from Eq. 5-11, is

$$R'_f = K'_0 v$$

The dust mat resistance, from Eq. 5-10, is

177

$$R_d = \frac{K_d C t v^2}{7000}$$

Solving for K'_0

$$K'_0 = \frac{R'_f}{v} = \frac{2.7}{3.1} = 0.87 \text{ in. w.g./fpm}$$

Referring to Table 5-16, the experimentally determined value of K'_0 for flyash is relatively conservative, being a little in excess of the maximum value for stone crushing operations. Solving for K_d,

$$K_d = \frac{R_d}{vW} = \frac{3.6}{3.1 \times 0.044} = 22.0 \text{ [in. w.g./(ft/min)} \times \text{(lb/ft}^2\text{)]}$$

Pressure Drop and Filter Ratio. Because this is a power plant application, the maximum pressure drop across the fabric filter is critical. In evaluating the filter, its pressure drop will be compared with that of the precipitator at 0.5 in. w.g. and of the cyclonic separator at 2.5 in. w.g. An optimum total resistance value for the fabric filter will be chosen at 3.2 in. w.g.

Therefore, substituting the process data for this application into Eqs. 5-8, 5-11, and 5-10, the filter ratio can be determined as

$$R_t = K'_0 v + \frac{K_d C t v^2}{7000}$$

$$3.2 = 0.87v + \frac{(22.0 \times 1.1 \times 60 \times v^2)}{7000}$$

$$3.2 = 0.87v + 0.207v^2$$

$$v = 2.4 \text{ fpm}$$

Therefore, when operating at a filter ratio of 2.4 acfm/ft^2 at an inlet gas flow of 168,600 acfm and an inlet dust concentration of 1.1 grains/acf, a maximum pressure drop of 3.2 in. w.g. across the filter can be expected, with cleaning of the bags to be performed every 60 min.

At the peak gas flow of 210,000 acfm, equivalent to a filter ratio of 3.0 fpm, at the same cleaning rate, the pressure drop would increase to about 4.5 in. w.g. The bags would have to be cleaned about every 20 min to maintain the pressure drop at 3.2 in. w.g. for a gas flow of 210,000 acfm.

Collection Efficiency. The collection efficiency can be assumed to be 99.5%. Actually, the laboratory test data indicated collection efficiency values in excess of this. Therefore, the emissions at both filter inlet loadings of 1.1 grains/acf average and 1.8 grains/acf peak would be 0.006 and 0.009 grain/acf, respectively. Correcting for temperature conditions, the standard values would be 0.009 and 0.015 grain/scf, respectively. Both values are below the allowable regulations emissions of 0.03 grain/scf.

Cleaning Cycle. At the average dust loading of 1.1 grains/acf and a cleaning cycle of 60 min, the resistance of the filter can be limited to 3.2 in. w.g. To maintain this value as a maximum at the peak dust loading of 1.8 grains/acf, the cleaning cycle must be decreased. From the composite equation,

$$3.2 = 0.87 \times 2.4 + \left[\frac{22.0 \times 1.8 \times t \times (2.4)^2}{7000}\right]$$

Therefore, $t = 34$ min.

In summary, at peak loadings, the filter collection efficiency will be unaffected and compliance with the regulations will be accomplished. To maintain the design pressure drop across the filter at peak loadings, bag cleaning must be performed about every 30 min rather than 60 min for the average loading. Automated pressure-control cleaning can easily vary the cleaning cycle to suit inlet load conditions.

Fabric Filter Sizing. The fabric filter will have fiberglass tubular bags to withstand the 400° F operating temperature. The bags will be cleaned by a reverse air-purge system, with the cleaning cycles to be automatically controlled.

$$\text{Fabric area} = \frac{168,600}{2.4} = 70,000 \text{ ft}^2$$

A six-compartment fabric filter will be chosen, with each section being programmed to be shut down in sequence for cleaning. Therefore, the total fabric area required is $(70,000 \times 6/5)$, or 84,000 ft^2.

The filter will contain 1400 bags, each measuring 12 in. diam by 20 ft long, and occupying an area of 87,000 ft^2. The filter dimensions are 40 ft wide by 30 ft high by 60 ft deep.

Weights and Pricing. The total delivered weight for the fabric filter is 240,000 lb. The cost on a delivered and unerected basis is $180,000, equivalent to $1.07/acfm. Included in the price are the housing with division plates and structural support members, walkways and handrails, fiberglass bags, hoppers, and a reverse air-cleaning system including internal dampers, compressor, control mechanisms, and instrumentation.

5.7. EQUIPMENT CHARACTERISTICS—WET SCRUBBER

The wet scrubber is the only particulate control device that introduces an additional process stream into the contaminated gaseous effluent to accomplish dust collection. The wet scrubber effects intimate contact between a gas stream and scrubbing liquid for the purpose of transferring solid and gaseous pollutants from the gas to the liquor. Because of the need for an effective gas/liquor contacting mechanism, there are a variety of basic wet scrubber designs. Most importantly, because of its ability to remove both solid and gaseous pollutants from a gas stream, the need for the wet scrubber in some application areas is mandatory.

Operating Principles

Besides the collection of solid and gaseous pollutants from an effluent gas stream, the wet scrubber can perform a number of additional process duties such as heat transfer, chemical reaction, evaporation, and distillation. In this section, only the transfer of particulates to a suitable scrubbing liquor to produce a clean gas will be considered.

The principal mechanism utilized by wet scrubbers to accomplish dust collection is to condition the individual particles so as to increase their size and thus permit them to be removed from the gas stream more easily. A secondary mechanism involves the entrapment of the dust particles in a liquid film and their removal in the liquor stream. Both collection mechanisms are used in the various commercial wet scrubber designs. Therefore, practically all scrubbers comprise a liquid/gas contacting section followed by a deentrainment chamber where the wetted particles are removed by inertial forces. Conditioning of the dust particles is accomplished by bringing them into contact with liquid droplets to produce a particle-liquid agglomerate. This agglomeration action is achieved by forcing the collision of particles with liquid droplets. For example, when a 5 μ dust particle collides with a 50 μ liquid droplet, the mass of the combination is increased by a factor of 1000. Because of the increase in both mass and size, the resultant 50 μ particle can be easily removed from the gas stream by simple inertial separation. Such collisions are promoted in the various wet scrubber designs by gravitation, impingement, and mechanical impulsion of the liquid droplets. Spray towers, venturi scrubbers, and disintegrator scrubbers are examples of some of the more common commercial designs that utilize these three mechanisms, respectively.

Particle-liquid agglomerates can also be formed by liquid condensation on the dust particles. If the temperature differential between the gas stream and scrubbing liquor is sufficiently large, it causes the gas to drop below its dewpoint. The dust particles act as nuclei for the condensed liquid so that the effective size of the particle-liquid combination is increased. Condensation can remove only a relatively small amount of dust because of the thermodynamic limitations of the system to yield sufficient condensate. Therefore, this mechanism is effective for initially hot gases containing relatively small dust concentrations, say, less than 1 grain/acf.

Theoretical analyses of the mechanism of particulate collection, comparable to those developed for the cyclonic, electrostatic, and filtration particulate collection devices, have not been made for the wet scrubber. The selection and design of a scrubber is usually based on performance tests for the particular application being considered. It is recognized that the smaller the particle size to be collected in a wet scrubber, the more difficult is its removal. To transfer the smaller particles from the gas phase to the liquid phase requires increased energy inputs, most commonly expended in the form of greater pressure drops across the scrubber. As a corollary to this relationship, for a specific particle size, increased collection efficiencies can be obtained at greater pressure drops. This empirical interrelation of power requirements to particle size and collection efficiency applies to all wet scrubber types.

Low-pressure-drop scrubbers such as simple spray towers, operated at pressure drops on the order of 1 in. w.g., can perform at a collection efficiency of about 90% with coarse dusts in the size range of 2 to 5 μ. For this same dust size, a baffle plate

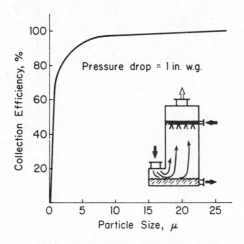

(A)   Spray Tower (see Fig. 5-44)

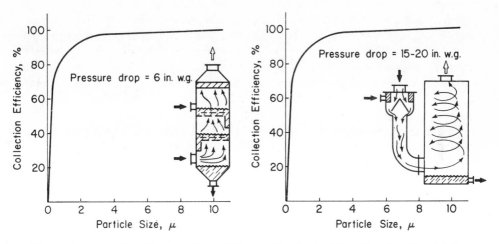

(B) Impingement Scrubber (see Fig. 5-48)   (C)   Venturi Scrubber (see Fig. 5-50)

Figure 5-38. Comparative micron efficiency curves for three wet scrubber types. Adapted from *Air Pollution Manual*, Ref. 21.)

type of impingement scrubber operated at 6 in. w.g. can achieve a 95% efficiency. A venturi scrubber can collect this identical dust at an efficiency of 99[+]% when operated at a pressure drop in the range of 15 to 20 in. w.g. Figure 5-38 illustrates these three types of scrubbers and their micron-efficiency curves. From any single curve, the dependence of the collection efficiency on the dust particle size can be observed. For example, the spray tower operated at about 1 in. w.g. can collect 5 μ dust particles at an efficiency of about 95%, whereas, at this same pressure drop, the collection efficiency for a micron particle would be 50%.

The energy requirement for any particular scrubber design to achieve a certain performance level is referred to as the "contacting power." In this concept it is assumed that in the gas/liquid contacting process, power is dissipated in fluid turbulence for both the gas and liquid phases. This energy expenditure, ultimately converted to heat, can be expressed as power per unit of volumetric gas flow rate. This value is the criterion of scrubber collection efficiency, and it has been designated as contacting power [22]. The contacting power must represent the effective or useful energy expended in causing intimate contact of the gas and liquid streams. It should not include mechanical losses in the gas and liquid transfer equipment or frictional losses imposed by the gas flow through the dry sections of the system. Thus, this net energy input value has been successfully correlated with collection efficiency for a wide selection of wet scrubber types.

As illustrated in Fig. 5-38, the most important factor in scrubber performance is the dust particle size. The contacting power required to attain a specific efficiency increases with a decrease in particle size. Generally, low contacting power equivalent to 6 in. w.g. pressure drop or less is required to accomplish almost 100% collection of particles greater than 5 μ. Moderate contacting power, equivalent to pressure drops of 10 to 12 in. w.g., will yield collection efficiencies of 90$^+$% for particles of 1 to 2 μ in size. For high-efficiency collection of submicron particles, much greater pressure drops are required. For example, for aluminum chloride fume particles measuring an average of 0.4 μ in diameter, a scrubber operated at a pressure drop of at least 35 in. w.g. is required to attain efficiencies of 90 to 95%.

The contacting-power concept is applicable to all types of wet scrubber designs. In the majority of them, the energy input is produced when the gas flow undergoes a pressure drop across the scrubber; in others, the contacting energy is imparted by a power-driven rotor, as in the disintegrator scrubber; still other designs employ high-pressure liquid pumps to atomize the scrubbing liquor. Values of contacting power can be calculated by application of the mechanical energy balance equation, as follows:

$$W = \int_{p_1}^{p_2} v \, dp + (\Delta z)\left(\frac{g_L}{g_c}\right) + \frac{\Delta u^2}{2g_c} + F \qquad (5\text{-}12)$$

where

W = shaft power input, ft-lb force/lb-mass
v = specific volume, ft^3/lb-mass
p = pressure, lb-force/ft^2
z = elevation, ft
g_L = acceleration of gravity, ft/sec^2

g_c =conversion factor, 32.17 lb-mass ft/lb-force sec^2
u =velocity, fps
F =friction loss, ft lb-force/lb-mass

Equation 5-12 has been written for the flow of a unit of 1 lb-mass of fluid. Actually, contacting power is computed by relating all power terms in this equation to the volumetric gas flow rate, and is usually expressed as horsepower per cubic foot per minute. If Eq. 5-12 is written for the gas stream, then the friction-loss term F is equivalent to the contacting power. In most wet scrubber applications, all energy for the contacting is derived from the gas stream so that the shaft-work term W is zero, and the changes in elevation and kinetic energy are negligible. Therefore, Eq. 5-13 can be written as a simplified version of Eq. 5-12, as follows:

$$F = - \int_{p_1}^{p_2} v\, dp = -v\, \Delta p \qquad (5\text{-}13)$$

If the energy per unit-volume of gas is used, then the contacting power is simply equivalent to the gas pressure drop Δp, expressed as in. w.g. Expressing the contacting power P_G as a function of the gas pressure drop,

$$P_G = 0.1575 \Delta p \qquad (5\text{-}14)$$

where

P_G = contacting power based on gas stream energy input, hp/1000 acfm
Δp = pressure drop across the scrubber, in. w.g.

Similarly for scrubbers, which utilize the energy furnished by the liquid stream, the contacting power P_L is expressed as

$$P_L = 0.583 p_F \, (q_L/q_G) \qquad (5\text{-}15)$$

where

P_L = contacting power based on liquid stream energy input, hp/1000 acfm
p_F = liquid inlet pressure, lb-force/in.^{2} gage
q_L = liquid feed rate, gpm
q_G = gas flow rate, cfm

To correlate contacting power with scrubber collection efficiency, the latter is best expressed as the number of transfer units. The relationship of transfer units to efficiency can be shown as

$$N_t = \ln\left(\frac{1}{1-E}\right) \qquad (5\text{-}16)$$

where

N_t = number of transfer units, dimensionless
E = fractional collection efficiency, dimensionless

When the number of transfer units is plotted against the contacting power on a logarithmic plot, a straight line is produced. Considerable empirical data are available in this form for the more common effluent control applications. Figure 5-39 shows two contacting power plots, one for the collection of open hearth furnace fume and the other for metallurgical fume. A high-energy venturi scrubber was applied for both

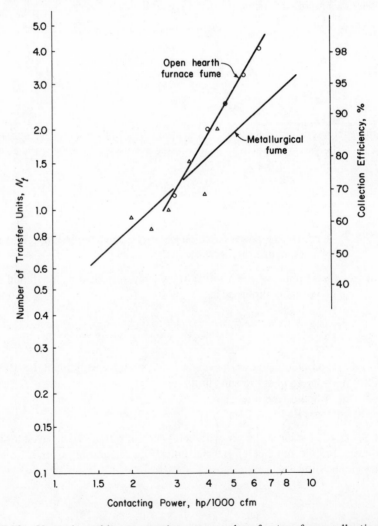

Figure 5-39. Venturi scrubber contacting power values for two fume collecting applications. (Adapted from Semrau, Ref. 22.)

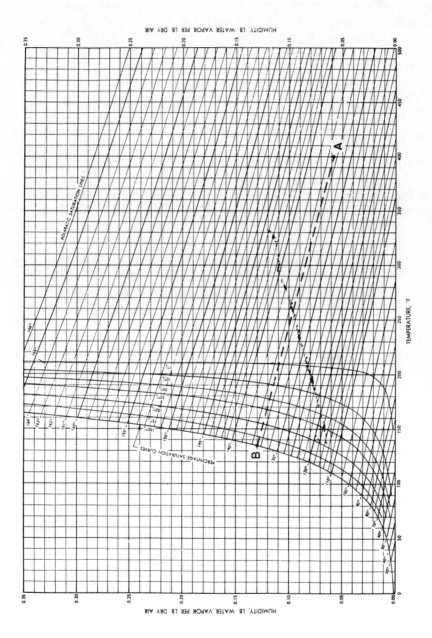

Figure 5-40.   Psychrometric chart. (From Zimmerman and Lavine, Ref. 23.)

collection problems. The considerable agreement of the individual test points indicates the reliability of this type of plot for the prediction of scrubber energy requirements for specific collection problems. The contacting power values for the open hearth and metallurgical fume duties at an efficiency level of 95% are 5.2 and 8.3 hp/1000 cfm, respectively. Because the major energy input for a venturi scrubber is the gas pressure drop, the values of Δp calculated from Eq. 5-14 are 33.0 and 52.7 in. w.g., respectively. The greater pressure-drop requirement for the metallurgical fume collection duty confirms the relatively finer particle size of this particulate material.

Design Parameters

The various process parameters influencing the design of a wet scrubber are gas flow, temperature, gas moisture content, dust particle size, dust type and concentration, gas properties, liquor flow rate, pressure drop, and collection efficiency. In a wet scrubber the gas flow must be corrected for its humidification by contact with the scrubbing liquor. This correction usually results in a reduction of the volumetric flow rate and is influenced by the temperature and moisture content of the incoming gas. Evaporative cooling of the incoming gases accounts for both temperature and volumetric reduction. For example, an inlet gas flow of 1000 acfm at 400° F, containing 8.5 vol % water would be reduced to 760 acfm, saturated at 134° F, and would contain 17.1 vol % water. The conditions of the gas entering the scrubber is designated by point A on the psychrometric chart in Fig. 5-40. Both heat and mass transfer effects are reflected by the adiabatic cooling line, with the gas being saturated at the conditions defined by point B. Calculations illustrating the volumetric reduction for these specific conditions are presented in the wet scrubber sample design calculations at the end of this section. Because this cooling action takes place almost immediately as the gas is contacted by the scrubbing liquor, all scrubber sizing is based on the saturated-gas volumetric flow.

As discussed previously, the relationship between the dust particle size and collector energy input is more relevant to the wet scrubber than to the other three particulate collection devices. With a definition of dust particle size and required collection efficiency, the energy requirements are empirically determined, most usually as a function of the gas pressure drop across the scrubber. Thus, to collect a dust analyzing 50% at $< 10\ \mu$ at an efficiency of 98% in a wet scrubber, a pressure drop of about 6 in. w.g. would be required. For the identical dust type analyzing 50% at $< 1\mu$, the pressure-drop requirement for this same collection efficiency would be 60 in. w.g. A simple baffle plate type of scrubber would be suitable for the low-pressure-drop duty and the more sophisticated venturi scrubber would be selected for the collection of the submicron particulates. Efficiency performance curves for both scrubber types are shown in Fig. 5-41.

The dust type normally has very little effect on wet scrubber collection efficiency. In some applications where the wetting characteristics of the dust are poor, chemical agents may be added, but this practice is not always too effective. In the other direction, highly soluble dusts such as aluminum chloride are not more easily collectible than sand. This is due to the fact that the liquid-contacting and agglomeration mechanisms, responsible for particulate collection, rely almost solely on the particle size.

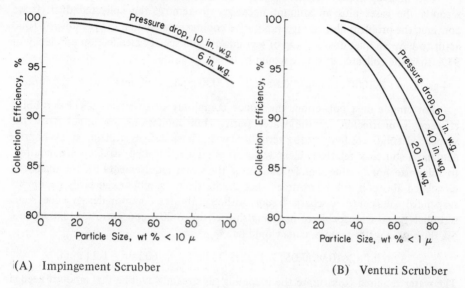

(A) Impingement Scrubber   (B) Venturi Scrubber

Figure 5-41. Performance curves for impingement and venturi type scrubbers. (Tomany, Ref. 24.)

Wet scrubber performance is essentially unaffected by the dust concentration in the incoming gas stream. At very high loading values, the probability of both "particle to particle" and "particle to water" collision is increased, and theoretically, therefore, the collection performance should be improved. However, any increase is usually negligible and is not considered as an important scrubber design parameter. Regardless, high dust loadings could cause plugging difficulties in certain types of scrubbers and must be compensated for by increasing the makeup water flow rates. For loadings in excess of 20 grains/acf, the need for a precleaner such as a cyclonic separator should be considered, to ease the scrubber load.

The presence of contaminant gases in the dust-laden air or flue gas stream influences the scrubber design considerably. If such gases as HCl, HF, and SO_x must be removed in the scrubber, then the optimum selection must be predicated on both particulate collection and gas absorption. Such a design approach usually results in the selection and sizing of a completely different scrubber type than that for particulate collection service only. Whether or not the gases are to be removed, there remains the problem of providing corrosion resistant materials of construction. Corrosion problems represent the most outstanding maintenance burden associated with the wet scrubber.

In some applications where the carrier gas is not air or flue gas, corrections must be applied to determine the true pressure-drop characteristic of these gases in the scrubber. For example, ammonia and carbon dioxide, with specific gravities of about 0.6 and 1.5, respectively, are two common gases for which pressure-drop values must be corrected. Simple orifice calculations can be applied for these corrections; see Eq. 5-3.

The makeup liquor required by a wet scrubber is that quantity required to saturate the gases plus an amount necessary to remove the collected dust. If we continue the problem illustrating evaporative cooling by assuming the gas flow of 1000 acfm to have an inlet dust loading of 4 grains/acf and specify a collection efficiency of 95%, then the collected dust rate would be

4 grains/acf $\times$ 1000 acf/min $\times$ 0.95 $\times$ 1 lb/7000 grains $\times$ 60 min/1 hr = 32.6 lb/hr

For simple dust collection, the liquor commonly used is water. The scrubbing efficacy is unaffected by the water purity, but limitations are imposed on the suspended solids content by the scrubber type. Those designs utilizing spray nozzles demand a source of relatively clean water. If pond water is used, strainers are required, to minimize nozzle plugging. To determine the water requirements for the dust load calculated above, it will be assumed that the particular scrubber can easily handle 5% suspended solids in the discharge liquor. Although this is a common design value, some scrubber types are capable of handling slurries up to 20% concentration. Based on a 5% slurry, the water requirements would be

32.6 lb/hr $\times$ (0.95/0.05) $\times$ 1 gal/8.33 lb $\times$ 1 hr/60 min = 1.24 gpm

The water required to saturate the incoming gas stream would be that amount needed to increase its moisture content from 8.5 to 17.1 vol %. Therefore, the gas saturation water requirements would be

$$1000 \text{ acf/min} \times \left(\frac{17.1 - 8.5}{100}\right) \times \frac{18 \text{ lb}}{480 \text{ acf}} \times \frac{1}{8.33} = 0.39 \text{ gpm}$$

The total makeup requirements for a flue gas flow of 1000 acfm entering the scrubber at 400° F, containing 8.5 vol % moisture and a dust loading of 4.0 grains/acf—assuming 95% collection efficiency and a discharged slurry concentration of 5%—would be (1.24 + 0.39), or 1.63 gpm. The gas flow leaving the scrubber would be 760 acfm, saturated at a temperature of 134° F, and would contain 17.1 vol % of moisture and a dust loading of 0.26 grain/acf.

Most scrubber designs require a recirculated liquor to achieve effective liquor/gas contact. The recirculation requirement is independent of the makeup liquor flow rate. For example, in a simple impingement type scrubber, which relies on impaction of the gas stream on a liquid reservoir, the recirculation rate is zero; therefore, the liquor entering the scrubber would be the makeup quantity only. However, in a venturi type scrubber, a liquid/gas ratio (L/G) of about 10 gal/1000 ft^3 is recommended. This recirculation flow can be provided from an outside source such as a settling pond, or can be pumped directly from the scrubber sump. Schematic flowsheets for both conditions, $L/G = 0$ and $L/G = 10$, are illustrated in Fig. 5-42. This L/G value is extremely important to collection efficiency in a number of scrubber types. Increased L/G values ensure higher droplet/particle collision probabilities and thus improve performance. A typical L/G versus collection efficiency plot for a venturi type scrubber is shown in Fig. 5-43.

Wet scrubbers can be operated over a wide range of collection efficiencies,

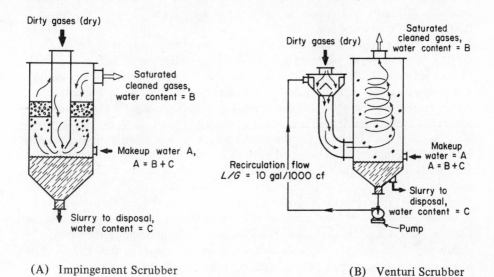

(A) Impingement Scrubber   (B) Venturi Scrubber

Figure 5-42. Schematic flowsheets for typical scrubber liquor systems.

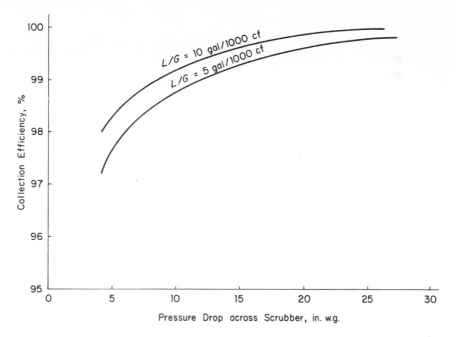

Figure 5-43. Liquid/gas ratio effect on collection efficiency for a venturi scrubber.

depending on the required performance level. The dust particle size and specified collection efficiency will decide the operating pressure drop. Although the empirical relationship of contacting power versus performance is generally applicable to all types of commercial scrubbers, each design has its own efficiency-pressure drop characteristics. Many classes of scrubbers are suitable for the collection of dusts at acceptable efficiencies in the 1 μ and large size range. However, for submicron particle collection, a different type of scrubber must be considered. For example, a multiple stage, countercurrent flow scrubber has been operated up to 30 in. w.g. for the collection of submicron particulates at near-acceptable collection efficiencies. A decided improvement in collection efficiency for the same application was noted with the use of a venturi type scrubber. when operated at the identical pressure drop. This observation gives credence to the theory that the contacting power must be effectively utilized since some bypass of the liquor and gas stream was apparently taking place in the countercurrent flow scrubber, thereby reducing the contact efficacy. Because pressure-drop requirements for the wet scrubber, whether on the gas side or the liquor side, can become economically burdensome, the selection of the optimum scrubber type to utilize the input energy effectively becomes most important. In steel mill practice where a waste gas stream of 500,000 acfm must be treated in a wet scrubber operated at a pressure drop of 60 in. w.g., an energy input equivalent to about 8000 hp is required.

Mechanical Design

Almost an infinite number of wet scrubbers are commercially available. Therefore, it is practical to describe only the basic components common to almost all wet scrubbers, and to define the contacting mechanism and physical configuration of some of the more common designs. The two major elements common to all wet scrubbers are the contacting section and the deentrainment section. In the former, intimate mixture of the gas and scrubbing liquor is accomplished, and in the latter the agglomerated droplets are removed by inertial devices. In practically all commercial designs the prevalent liquor/gas contacting mechanism involves impingement of the dust particles in the incoming gas stream on a liquid surface and/or against individual liquor droplets. The deentrainment of the liquor/dust droplets from the gas stream is achieved either by impaction against extended surface baffles or by centrifugal separation. A description of five commercial wet scrubber designs follows.

Figure 5-44, shows a gravitational open-spray tower. This type of scrubber is usually vertical and circular in cross section. The scrubbing liquor is sprayed down the tower, with the gas rising countercurrently and being discharged at the top. Removal of the larger dust particles is accomplished by impingement of the inlet gases against the liquor pool at the base of the tower. Final scrubbing takes place as the gases flow upward and the contained dust particles are impinged against the downcoming liquor flow. Liquor droplet entrainment is prevented by the use of low gas velocities through the tower, in the range of 100 to 250 fpm. With the advance of wet scrubber design technology, this type of scrubber is being limited to precleaning duty, saturating the hot gases and removing the larger dust particles in the range of 5 to 10 μ. Performance characteristics are defined in Fig. 5-38. Water is introduced through low-pressure

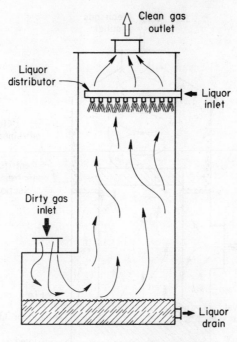

Figure 5-44. Open spray tower.

nozzles at a rate of 2 to 30 gal/1000 ft^3 of gas. Typical design parameters for the application of such a scrubber to the collection of iron ore dust from a roasting furnace are:

Gas flow, acfm	65,000
Gas temperature, °F	1,650
Water content, vol. %	20
Dust loading, grains/acf	12
Dust size, % greater than 5 μ	90
Scrubber dimensions, diam × hgt., ft.	12 × 22
Water flow, gpm	1200
No. of nozzles	90
Gas discharge temperature, °F	170
Pressure drop, in. w.g.	0.8
Collection efficiency, % (estimated)	80

There are innumerable medium-energy scrubber designs relying on impingement for the liquor/gas contacting mechanism. In these scrubber types, liquid reservoirs, wetted baffles, wetted packing shapes, and highly atomized liquor droplets are utilized to achieve capture of the dust particles by the scrubbing liquor. In Fig. 5-45 there is illustrated a medium-energy, fixed-orifice type of wet scrubber. In this design, the gases enter the side inlet nozzle and flow vertically downward through a restricted orifice, causing them to be impacted against a static liquid reservoir. The level of this reservoir is maintained in a fixed position, slightly below the bottom of the orifice, by

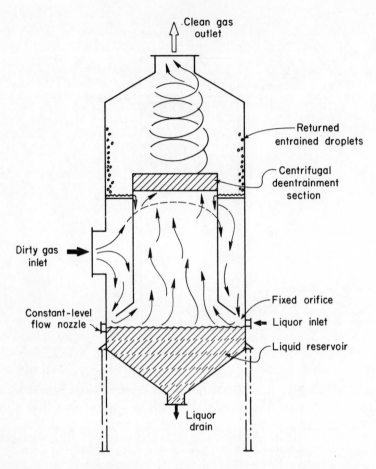

Figure 5-45. Liquor impingement scrubber. (Courtesy of Western Precipitation Division.)

an adjustable weir. After impingement against the liquid surface, the gas flow direction is reversed; as it travels upward, additional liquor/gas contact is accomplished when the liquor droplets formed at the highly turbulent liquor surface are entrained in the gas stream. The gas passes through a centifugal type deentrainment section where the agglomerated liquor-dust particle droplets are removed. Inertial separation is performed in this section by initially imparting a high-velocity swirling motion to the gas stream, followed by a severe reduction in velocity. The droplets are thus removed from the gas stream by their own momentum, striking the walls of the vessel and returning by gravity flow to the liquid reservoir. In this design, the fixed orifice is set for a specific pressure drop, depending on the required performance level. This classification of scrubber is operated at pressure drops in the range of 3 to 10 in. w.g. One advantage in its use is that the gas impingement mechanism depends on a fixed liquid reservoir and thus the recirculation of liquor is unnecessary. Superficial gas velocities, based on

Figure 5–46 Impingement Scrubber Dimensions
(Courtesy of Western Precipitation Division)

Capacity, cfm, sat	Dimensions, ft-in.					Weight, lb
	A	B	C	D	E	
3,000	2-9	11-6	3-0	1-6	1-3	1,200
4,000	3-3	12-0	3-6	1-9	1-6	1,500
6,000	4-0	12-9	4-3	2-3	2-0	1,800
9,000	4-8	13-3	5-0	2-9	2-6	2,500
12,000	5-6	14-0	6-0	3-0	2-9	3,000
20,000	7-0	15-0	7-0	3-8	3-4	5,000
30,000	8-8	17-0	8-3	4-2	4-0	7,000
40,000	9-4	18-6	9-3	4-10	4-6	8,000
65,000	12-0	23-3	12-0	6-6	6-0	12,000
100,000	16-0	29-3	15-0	8-0	7-6	27,000
130,000	18-0	32-0	17-0	8-6	8-0	30,000
200,000	22-0	39-3	21-0	10-3	9-6	45,000

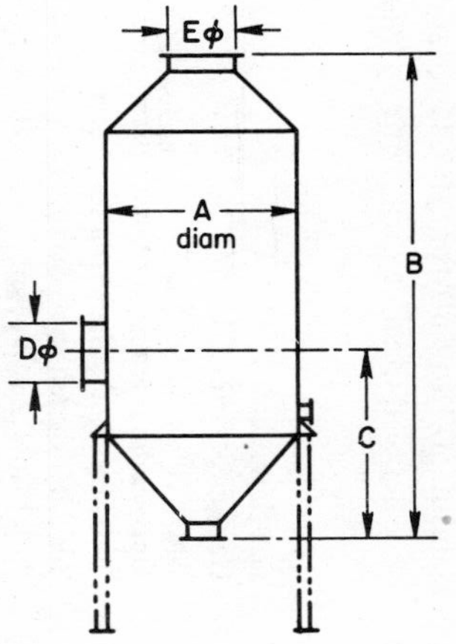

193

the shell diameter, are about 500 to 550 fpm. Process data for the application of this type of scrubber to the collection of silica dust from an aggregates plant operation are as follows:

Gas flow, acfm	75,000
Gas temperature, °F	325
Water content, vol %	14
Dust loading, grains/acf	8.5
Dust size, % less than 10 μ (est.)	50
Scrubber dimensions, diam × hgt., ft	12 × 25.0
Water flow makeup, gpm	220
Gas discharge temperature, °F	142
Pressure drop, in. w.g.	10.0
Collection efficiency, %	99

Figure 5-46 illustrates capacities, overall dimensions, and weights for a series of this type of medium-energy scrubber. The practical maximum capacity of 200,000 cfm, saturated basis, is predicated on the problem of maintaining uniform gas distribution for flows in excess of this value.

One of the most basic wet scrubber designs is the packed scrubber, which utilizes various types of packing to break up the countercurrent flows of gas and liquor. The packing materials vary from wood slats to complex ceramic shapes. The ideal packing is one that provides maximum wetted surface at minimum pressure drop for any specific recirculation rate. Recirculation flows, or L/G ratios, are usually expressed as gallons per 1000 ft³. Normally, countercurrent flow is utilized, but more recently an accumulation of considerable data indicates that a cross-flow configuration is equally effective with minimization of pressure-drop requirements. In this design the gas flows horizontally through the packed section of the scrubber while liquor is introduced vertically at the top. A conventional countercurrent-flow packed scrubber design is shown in Fig. 5-47. Deentrainment may be achieved by the use of labyrinth impact separator vanes.

Packed scrubbers are primarily applied for the cooling and/or absorption of gases. However, at very low dust loadings and with packing having a high percentage of voids, particulate collection can be realized. For any particular type of packing, data are available which define the pressure drop as a function of gas and liquor flow rates. Thus, at equivalent gas and liquor flow rates of 200 fpm and 5 gpm/ft², based on the open tower area, an average pressure drop of about 4 in. w.g. can be expected for a 20 ft depth of packing. Further treatment of packed scrubber design parameters will be undertaken in Chapter 6.

An interesting development in the design of wet scrubbers is directed at heavy particulate loading applications. Because of the tendency of fixed packings to plug, a mobile packing has been devised [25]. This packing, comprising 1 1/4 in. diam plastic lightweight spheres, is restrained between two grids; under the impetus of gas and liquor flows, the spheres move freely about in random motion. This turbulent action prevents the accumulation of solids on the packing so that a nonplugging operation is attained even at high dust loadings. One added advantage of this design is the ability to handle high flow rates of liquor and gas at minimal pressure drops. For example, a

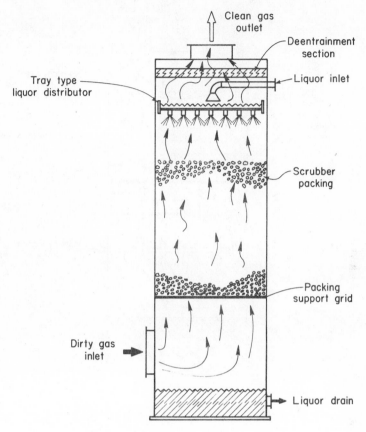

Figure 5-47.   Fixed-packing scrubber.

single stage, defined by a support and restraining grid spaced about 3 ft apart, can handle gas and liquid flow rates of 750 fpm and 20 gpm/ft^2 at a pressure drop of about 3 in. w.g.

A wet scrubber relying on the impingement of the gas stream on wetted baffle plates is shown in Fig. 5-48. This particular design is the Peabody scrubber, which contains one or more impingement stages, depending on the difficulty of the dust collection duty. Each stage or plate comprises a perforated sheet above which there is supported an impingement baffle grid structure. Each plate is flooded by the scrubbing liquor that flows across it, normal to the upward gas flow. As the gas passes through the numerous perforations at high velocity, it impinges against the baffle located above each orifice. This impingement action causes transfer of the dust particles from the gas stream to the liquor. The gas passes the baffle and bubbles up through the head of liquor, which is maintained at about a 1 in. depth on each plate, thereby providing additional intimate contact between the liquor and gas streams. The gas similarly passes through succeeding contact stages and then flows through the helical type

195

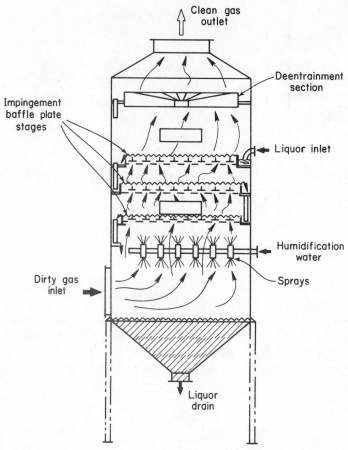

Figure 5-48. Baffle impingement scrubber. (Courtesy of Peabody Engineering Co.)

entrainment separator, where the agglomerated liquor-dust droplets are removed from the gas stream and returned by gravity to the contact stages.

The scrubbing liquor is introduced at the top stage, whence it flows across the plate and then is gravity-discharged through a downcomer and liquid seal to the stage below. With this arrangement, the recirculation rate L/G is essentially zero so that only a makeup liquor flow to saturate the gases and discharge the collected dust is required. Because this type of scrubber is frequently applied to the collection of dusts from high-temperature gas streams, humidification sprays are provided below the bottom stage. The introduction of a portion of the makeup liquor at this location promotes condensation upon the fine particles, thereby causing their agglomeration. Each stage of this type of scrubber is designed for pressure drops in the range of 1.5 to 4.0 in. w.g. The capacity of the scrubber is about 500 to 600 cfm/ft^2, based on the shell free cross-sectional area. Collection efficiencies of 97 to 99% can be expected for dusts down to 1 μ in particle size. Typical process data for this scrubber applied to the collection of particulates from an iron pyrites roasting operation are:

Gas flow rate, acfm	75,000
Gas temperature, °F	900
Water content, vol. %	9.0
Dust loading, grains/acf	20.0
Dust size, % less than 10 μ (estimated)	10.
Scrubber dimensions, diam × hgt., ft.	11.0 × 28.0
Water flow, gpm	550
Gas discharge temperature, °F	160
Pressure drop, in. w.g.	8.5
Collection efficiency, %	99.8

In the majority of commercial wet scrubbers, the contacting power is in the form of energy-input equivalent to the gas pressure drop across the scrubber. However, some scrubbers rely on the input energy being applied through high-pressure liquor sprays in the range of 100 to 300 psi. Such scrubbers offer very low pressure-drop resistance to the gas stream, and for many applications involving the processing of large gas volumes, this approach is very advantageous. Instead of requiring high pressure drop fans, reliance is placed on relatively inexpensive liquid pumps.

Figure 5-49 shows a specific design utilizing high-pressure sprays. The Elbair scrubber relies almost entirely on the formation of fine liquor droplets developed by the impact of high-pressure spray water against a rebound screen or plate. Considerable savings in total horsepower requirements are claimed for this system whereby the finely divided scrubbing liquor particles are developed directly as opposed to scrubbing systems relying upon scrubber pressure drop for this purpose.

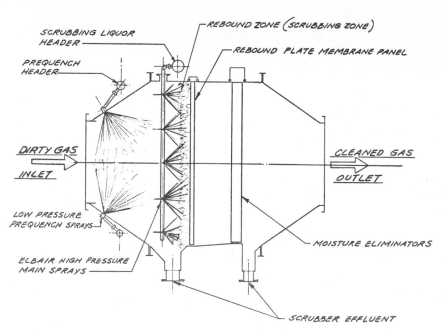

Figure 5-49. High-pressure spray scrubber. (Courtesy of Krebs Engineers.)

197

The dirty gas enters the scrubber horizontally and passes through a series of prequench spray nozzles. This prequenching humidifies the hot gases and causes some agglomeration of the dust particles by nuclei condensation. Following humidification, the gases continue into the contacting section where a series of high-pressure sprays direct the scrubbing liquor against a screen composed of vertically mounted, closely spaced parallel rods, or against a specially designed louvered rebound plate. Arrangement of the sprays, spray pressure, and distance of the nozzles from the screen are fixed so that a zone of rebounding, fast-moving, atomized liquor droplets is developed at the surface of the screen. The incoming gas passes through this zone and the contained dust particles are subject to high-velocity impact by the finely divided liquor droplets. On reaching the screen, the gas is accelerated through the narrow spaces between the vertical rods, where further scrubbing occurs.

The liquor and the major portion of the agglomerated liquor-dust particles are removed in the area of the contacting screen, being gravity discharged into a sump located at the base of this contacting section. Finer liquor-dust particles are removed from the gas stream by impaction against an extended surface-baffle type of eliminator. The cleaned gases are discharged horizontally from the scrubber. Liquor ratios on the order of 3 to 6 gal/1000 ft^3 of incoming gas are required. This scrubber can be arranged in multiple stages, with the gases passing horizontally through two or three spray-contacting zones. The gas flow pressure drop in each stage does not exceed 1 in. w.g. Operating data from a installation for the collection of coal flyash from a utility boiler flue gas stream are:

Gas flow rate, acfm	300,000
Gas temperature, °F	300
Water content, lb H$_2$O/lb dry gas	0.088
Dust loading, grains/dscf (dry std. ft^3)	3.4
Dust size, % less than 10μ	34.0
Scrubber dimensions, length × width × hgt., ft	20 × 30 × 16
Scrubbing water flow, gpm	1,200
Gas discharge flow rate, acfm	247,500
Gas discharge temperature, °F	136
Pressure drop, in. w.g.	2.5
Discharge dust loading, grain/dscf	0.046
Collection efficiency, %	98.7
Contacting power, hp/1000 acfm inlet gas	1.2

In view of the recent increased severity of air pollution control regulations, directed toward reduced particulate emissions, only those wet scrubbers with high performance capabilities are being considered. Because of the empirically proven contacting power theory, therefore, it can be expected that operating pressure drops for wet scrubber applications will be continuously increased to meet these revised particulate emissions regulations. The venturi type scrubber has long been proved capable of effectively converting contacting power to maximum collection efficiency levels. Figure 5-50 shows a schematic representation of a venturi type scrubber. There have been almost an infinite number of venturi designs developed recently by the various commercial scrubber suppliers in recognition of the continuously escalating emissions restrictions.

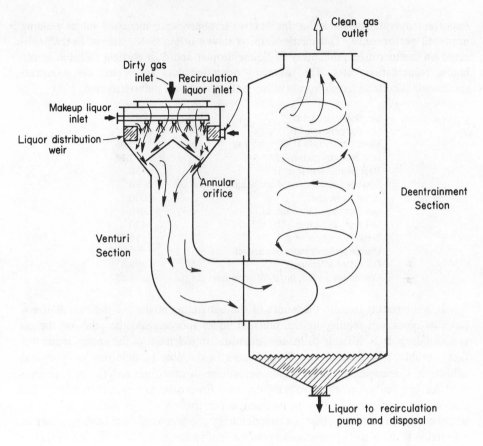

Figure 5-50. Venturi scrubber.

Referring to Fig. 5-50, the dust-laden gases enter the venturi section and impinge against the liquor stream in the wetted cone and throat sections. As the gases pass through the annular orifice formed by the cone and throat, at velocities of 12,000 to 20,000 fpm, the shearing action results in atomization of the liquor into innumerable fine droplets. The high differential velocity between the gas stream and atomized droplets promotes impaction of the dust particles on the fine droplets. As the gas leaves the venturi section and decelerates, further impaction and agglomeration of the liquor-dust particles occur. These droplets are then removed by centrifugal action in the cyclone separator section. In the venturi scrubber the pressure drop can be increased by increasing the gas velocity and/or by introducing recirculated liquor to the throat. In some designs, the cross-sectional area of the throat can be adjusted to maintain a constant pressure drop at varying gas flow conditions. The venturi scrubber is applied almost exclusively to the collection of submicron dust particles at pressure drops most commonly in the range of 10 to 60 in. w.g.

Performance characteristics as a function of the dust particle size and operating pressure drops are shown in Fig. 5-41. The liquid/gas ratio (L/G) is an extremely important operating variable for the venturi scrubber, with increased values yielding improved performance. This relationship, as shown in Fig. 5-43, is most likely predicated on the increased probability of liquor droplet and dust particle collision at the higher recirculation rates. The following operating data are typical for a venturi scrubber that collects flyash from a utility boiler fired with pulverized coal.

Gas flow rate, acfm	100,000
Gas temperature, °F	300
Water content, lb H_2O/lb dry gas	0.06
Dust loading, grains/acf	2.18
Dust size: % less than 10 μ	30
Scrubber dimensions, diam × hgt., ft.	16 × 30
Water flow, gpm	1,000
Gas discharge flow rate, acfm	85,000
Gas discharge temp., °F	130
Pressure drop, in. w.g.	10.5
Discharge dust loading, grain/acf	0.018
Collection efficiency, %	99.5
Contacting power, hp/1000 acfm inlet gas	1.65

It is interesting to note the values of the contacting power for the two different scrubber types, one relying on high-pressure liquor spraying and the other on the gas pressure drop. It is difficult to make a quantitative judgment of the energy inputs for these scrubbers because of the discrepancies in the dust particle size analyses and collection efficiencies. Both the higher percentage of dust fines and the performance level for the venturi scrubber explain the need for greater energy requirements. The one major conclusion to be drawn from a comparison of the various and many scrubber types is not only that the overall energy requirements must be determined as a function of dust particle size and expected performance, but also that an optimum selection of scrubber types must be made to suit the specific process conditions. The required collection efficiency, liquor effluent characteristics, energy availability, equipment arrangement, and economic burden are some of the factors to be considered in the final selection of the optimum scrubber type.

Proper selection of materials of construction is one of the most important considerations in the design of a wet scrubber. Both erosion and corrosion are problems peculiar to the wet scrubber and are not ordinarily encountered in the design of the cyclonic, electrostatic, or fabric filter collectors. For handling ventilation air, simple mild steel fabrication is suitable. For flue gas, a low-grade corrosion resistant steel such as Corten is advisable. Such gases as SO_2, SO_3, HF, HCl, Cl_2, and HNO_3 are most difficult to handle in a wet scrubber. Usually, plastic-coated steel has an economic advantage over such exotic alloys as the stainless steels, Hastelloy, titanium, and lead lining. However, most plastics do have a temperature limitation in the neighborhood of 180 to 220° F, which eliminates their use for many applications.

To overcome the temperature limitations of the various plastic materials, advantage is sometimes taken of the fact that in saturating moist hot gases, their temperature is reduced to a level within the operating limits of most plastic materials.

Aside from their temperature limitations, such lining materials as polyvinyl chloride, epoxy-phenolic, fiberglass polyester, polyethylene, and the furfurans are very resistant to most corrosive gases under wetted conditions. Either natural or neoprene rubber linings are used for chemical services, particularly in the fertilizer industry. One materials combination commonly employed to handle hot SO_2-laden metallurgical gases is mild steel with a protective rubber liner against which temperature and acid resistant brick is laid. Aluminum, wood, concrete, fiberglass-reinforced polyester, glass-lined steel, and glazed tile are some other materials used in the fabrication of wet scrubbers. There follows a very approximate index of fabrication costs for the more common materials of construction, based on mild steel with a value of 1.0.

Mild steel	1.0
Plastic-lined steel	1.4
Rubber-lined steel	1.6
Fiberglass-reinforced polyester	2.0
Type 304 stainless steel	2.5
Type 316 stainless steel	3.2

Installation Features

Wet scrubbers are usually limited in their capacity by gas and liquor distribution problems. Spray towers and packed scrubbers are commonly fabricated in the neighborhood of 10 to 15 ft diam with maximum equivalent capacities of 25,000 to 50,000 cfm. For venturi type scrubbers, gas flows up to 100,000 cfm are accommodated in a single throat or venturi section. In some installations, multiple throats can be furnished with a single separator section. Most wet scrubber manufacturers limit capacities to about 100,000 cfm, based on the saturated gas flow rate. Shipping limitations pose a further restriction so that shop-fabricated vessels are limited to about 12 ft diam. According to Fig. 5-46 the maximum sized prefabricated scrubber would have a capacity of 65,000 cfm. For capacities in excess of this value, field fabrication would be necessary.

Figure 5-51 shows a partial view of a wet scrubber installation designed to handle the gaseous effluent from an aluminum reduction plant. A total of twelve 18 ft diam scrubbers, each having a capacity of 140,000 cfm, was required. The service involved the collection of carbon, alumina, and fluoride particulates plus hydrogen fluoride gaseous absorption. In this application the gases are withdrawn from the process and pass through a fan that discharges them through tubular type cyclonic separators and then through the wet scrubbers. The cleaned gases are discharged from the top of these packed scrubbers into the rectangular duct supported above the scrubbers, and then continue to a stack that discharges them to the atmosphere. The ductwork, fan, and tubular collector ahead of the scrubber are of mild steel construction. Starting at the scrubber inlet and continuing to the stack, plastic linings and fiberglass-reinforced polyester (FRP) materials were used. The scrubbers were field-constructed, their FRP fabrication and erection involving novel and sophisticated techniques for this size of vessel.

The location of the fans upstream of the scrubbers was made possible by the relatively low particulate loadings in the gas stream passing through the fans, on the

Figure 5-51. Aluminum reduction plant wet scrubber. (Courtesy of Chemical Proof Corp.)

order of 0.2 to 0.8 grain/acf. In wet scrubber installation practice, wherever possible, the fans should be located on the dry side ahead of the scrubber, thereby reducing corrosion effects. Fans that handle wet saturated gases on the discharge side of a wet scrubber must be of exotic material construction, which adds to both capital investment and maintenance costs. In locating the fan upstream from the scrubber, larger gas volumes must be handled, and this factor together with the maintenance problems associated with erosion and dust buildup must be recognized in considering this arrangement. One interesting practice with regard to fan location is the dilution of the wet saturated gas stream leaving a scrubber with sufficient hot gases to bring its temperature above the dewpoint. With this system, mild-steel fan construction can be utilized, thereby eliminating the maintenance difficulties associated with exotic fabrication materials.

Wet scrubbers can handle an effluent gas stream at practically any temperature level. Saturation of the gases takes place in accordance with the psychrometric relationships illustrated in Fig. 5-40. For gases in excess of about 600° F, it is recommended that a presaturation chamber be located ahead of the scrubber for initial temperature reduction. Final temperature adjustment to the saturation point is then effected in the wet scrubber. For example, in one installation, the effluent gases from a municipal waste incinerator leave the furnace at about 2600° F and are cooled in a baffle type waterwash section to about 1200° F. The gases then pass through a spray chamber, where the temperature is further reduced to 600° F, and then enter the wet scrubber where final cooling and dust collection is accomplished. The final clean gaseous effluent is then discharged to the atmosphere, saturated with water vapor, at a temperature of 160° F. Performance of the total gas-cooling duty in the wet scrubber sometimes adversely affects the gas and liquor flow patterns in the scrubber, thereby reducing the collection efficiency. The amounts of water required at each cooling station can be accurately determined from the psychrometric chart. Temperature control instrumentation regulates the water flow for each temperature reduction.

In the application of the wet scrubber to a specific dust collection problem, serious consideration must be given to disposal of the waste liquor. In the chemical, paper, and primary aluminum industries, the liquor is often treated for recovery of the chemical values so that equipment costs can be charged off against by-product income. However, in the majority of applications, the treatment of the scrubber liquor effluent is of prime importance, particularly in the current climate of more stringent water quality control regulations. In dry collection, disposal of the collected dust usually requires conveying equipment and dry-bulk storage facilities. In some systems, however, the dry collected material is sluice-conveyed, and therefore liquor effluent handling facilities must be provided. The most simple system comprises a settling pond and recirculation pumps. This system, favored by the aggregates processing industry, involves precipitation of the collected dust from the liquor effluent in a pond or settling tank, with recirculation of the clarified liquor to the scrubber. The precipitated mud is dredged from the pond periodically, the frequency depending on the dust loading.

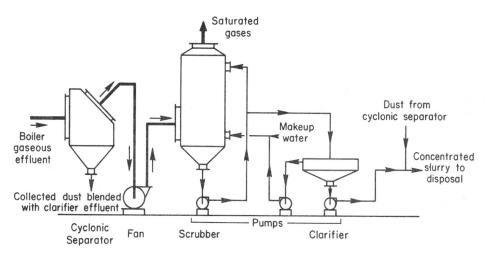

Figure 5-52.   Typical scrubber liquor disposal system.

A typical liquor effluent disposal system for a wood-fired boiler flyash wet scrubber application is shown in Fig. 5-52. Because of the continuous nature of this process, the "mud" is continuously discharged from the clarifier as a 30% solids slurry. It is blended with the dry material from the cyclonic separator and conveyed by truck to a land-fill area. In both the aggregates and flyash disposal systems described, the effluent liquor is neutral or slightly alkaline; this means that relatively simple equipment of mild-steel construction can be utilized. However, for those wet scrubber applications where acidic particulates or gases are processed, more sophisticated liquor disposal facilities are necessary. For example, in the control of flyash and gaseous sulfur oxides from power-plant boilers, the scrubbing liquor must be sufficiently alkaline to neutralize the SO_2/SO_3 contents of the flue gas stream. Usually, $CaCO_3$ or $Ca(OH)_2$ liquors are used and therefore storage, transfer, and clarification equipment must be provided to handle the insoluble $CaSO_3$-$CaSO_4$ liquor. Because acidic values are involved, stainless steel or rubber-lined equipment is necessary. For many wet-scrubbing systems, the capital investment requirements for the liquor-handling facilities can easily be equivalent to the wet scrubber costs. It becomes apparent that in the application of the wet scrubber, serious consideration must be given to the total system.

Applications

The wet scrubber can collect dust particles over a large size range by the expenditure of varying energy inputs. The overall energy requirements, whether on the gas side or scrubbing liquor side, varies inversely as the dust particle size. Single scrubbers are most commonly designed to handle gas flows in the range of 2,000 to 100,000 cfm. Some special designs have had capacities up to 250,000 cfm.

Some of the advantages to be found in the use of the wet scrubber are:

1. Particulate collection and gas absorption can be accomplished simultaneously.
2. High collection efficiencies can be obtained for any particle size range at a specified energy input.
3. Viscid materials can be collected without plugging.
4. High-temperature gaseous effluent streams can be handled.
5. The effluent gas moisture content and/or dewpoint is not critical to the scrubber operation.
6. Flammable dusts and gases can be processed without hazard.
7. Other process duties such as heat transfer, chemical reaction and evaporation broadens the application range for the wet scrubber.
8. Initial costs are relatively low.

Some of the major disadvantages of the wet scrubber are:

1. It requires high energy for the collection of the finer dust particles.
2. It is accompanied by corrosion and erosion problems characteristic of wet processing.
3. An effluent liquor disposal system is needed.
4. It causes discharge of a water-saturated gas stream, which in most applications produces a steam plume.

Generally, wet scrubbers are utilized in those process areas where wet collection appears desirable—after consideration of such factors as capital investment, power requirements, slurry disposal, by-product recovery, and heat transfer or absorption duties. Descriptions of some typical wet scrubber installations in various process industries are discussed below.

Pulp and Paper. In the kraft paper process, the recovery of sodium hydroxide involves the calcination of a lime slurry to obtain calcium oxide. A 65% $CaCO_3$ slurry containing some Na_2CO_3 is calcined in a rotary gas-fired kiln to yield CaO. The kiln is maintained at a temperature of 2000° F with the flue gases being discharged at about 600° F.

A venturi type scrubber was installed for this service and the following operating results were obtained:

CaO production rate, tons/hr	20
Gas flow, acfm	60,000
Gas temperature, °F	580
Gas moisture, vol. %	24.0
Inlet loading, grains/scf	8.5
Dust particle size, % less than 10 μ	50
Scrubber dimension, dia. × hgt., ft	12 × 36
Recirculation liquor, gpm	400
Pressure drop, in. w.g.	12.0
Collection efficiency, %	99.2

Fertilizer. In the production of diammonium phosphate, phosphoric acid and anhydrous ammonia are reacted. Emissions from the reactor are free ammonia, diammonium phosphate, and fluoride dusts. Because both phosphate particulates and ammonia gas have value, their collection and recovery become necessary.

In one installation, a low-energy venturi scrubber was utilized for this service. The scrubbing liquor was phosphoric acid, which reacted with the free ammonia to produce additional product. At a pressure drop of 6 in. w.g., approximately 80% of the particulates and 96% of the ammonia pollutants were removed. The use of a venturi was based on the precipitation of additional product inside the scrubber, which could cause plugging problems in other scrubber types. Materials of construction were neoprene-lined mild steel.

Steel Industry. In the basic oxygen-steel process, the effluent produced contains iron oxide fumes. The gas volumes to be processed are usually in the neighborhood of 500,000 to 1,000,000 cfm at temperatures of 800 and 1000° F. The dust particles are in the size range of 0.1 to 0.5 μ and because of the surface area of these fine, red colored particles, they are highly visible at very low concentrations. Thus, the allowable concentration of these particulates permitted in the effluent gas stream being discharged to atmosphere is usually set at about 0.02 grain/scf.

The venturi scrubber, shown in Fig. 5-50, is used for this collection duty. It is operated at pressure drops in the range of 60 to 80 in. w.g. The gases are precooled before entering the scrubber, and are discharged to the atmosphere at a final temperature of 160° F. The collection efficiency is usually on the order of 99.5$^+$%. The power requirements for the operation of these scrubbers are prodigious.

Aggregates. In one type of aggregates plant, various size fractions of crushed stone are classified, blended, and dried in preparation for the addition of hot petroleum stock to make asphalt. The blended aggregates are usually dried in a gas-fired kiln that produces an effluent containing silica dust at relatively heavy loadings.

Up to the present time, medium-energy wet scrubbers in series with a cyclonic separator represented the standard solution for this emissions problem. However, continuously upgraded regulations have made it necessary to consider either a venturi high-energy scrubber or a fabric filter for this service. There are still a number of impingement type scrubbers operating in this area. In one plant processing aggregates at the rate of 30 tons/hr, an impingement scrubber, as shown in Fig. 5-45, handles a gas flow of 60,000 acfm at a temperature of 450° F and a dust loading of 15 grains/acf. Actually, the effluent leaving the drier contains dust at a concentration of about 50 grains/acf with a 70% collection efficiency being accomplished in the cyclonic separator. The scrubber is operated at a pressure drop of 10 in. w.g. and the total collection system yields an efficiency of 99.4%. At the present time this system is barely meeting the current codes, which require an emissions of 0.10 grain/acf. To meet the anticipated tightened codes, plans are being considered to convert this design to a venturi configuration and increase the pressure drop to about 18 in. w.g.

Power Plants. With the regulations demanding the control of both particulates and SO_x from power-plant flue gases, wet scrubbers are being considered for the solution to this problem. This area of application for the wet scrubbers has been developing since 1965, and at the present time the total full-scale installations number about 20.

In the combustion of pulverized coal, flyash concentrations in the range of 2 to 4 grains/acf and SO_x concentrations of 1500 to 3000 ppm can be present in the discharged flue gas. For central steam generation plants the flue gas flow rate can be in the order of 500,000 to 2,000,000 acfm at 300 to 350° F. Venturi, mobile packing, liquor impingement, wetted baffle impingement, and even high-pressure spray types of scrubbers have been installed in either prototype or commercial scale for the collection of these pollutants. Particulates collection data for both high-pressure liquor spray and venturi scrubber types are shown on pages 198 and 200. Relatively good performance for both particulate collection and gas absorption can be obtained at moderate energy-input levels. The major problem associated with the wet scrubber in this application is disposal of the calcium sulfite liquor effluent.

Primary Aluminum. The wet scrubber installation illustrated in Fig. 5-51 handles an effluent gas stream from a primary aluminum reduction plant. Each of the electrolytic cells or "pots" is individually vented, the contaminated gases being discharged through ductwork to cyclonic separators, fans, wet scrubbers, and finally through a stack to the atmosphere. The gas stream leaving the pots contains air, CO, CO_2, and HF plus carbon, alumina, and fluoride particulates. Fluoride compounds, both gaseous and particulate, are the major pollutants to be removed.

In the installation shown in Fig. 5-51, the total gas flow is about 1,700,000 cfm at a scrubber inlet temperature of 200° F. Each of the 12 scrubbers is 18 ft in diameter by 32 ft overall height and are of FRP (fiberglass reinforced polyester) construction. Each scrubber contains a two-stage contacting section using 1¼ in. diam lightweight polypropylene spheres as the mobile packing. The superficial gas velocity is 550 fpm at a liquor circulation flow of about 1400 gpm per scrubber. The liquor is a weak sodium carbonate solution that reacts with the hydrogen fluoride and alumina to form cryolite, which is recovered for re-use in the electrolytic cells. The particulates collection and absorption efficiencies are about 75 and 98% respectively at an overall pressure drop across the scrubbers of about 5 in. w.g.

In addition to these specific applications, wet scrubbers are used for the removal of particulates and/or gases in a variety of industrial processes. The food, chemical processing, metallurgical, metal fabrication, mining, and petroleum industries are well represented in their use of the wet scrubber. In each of the industries, wet collection is utilized to control the pollutants from such divergent process equipment as kilns, furnaces, screens, crushers, mixers, reactors, grinders, conveyors, and ventilation systems.

Sample Design Calculations

The problem solved in this section will be for the same type of application, process conditions, and performance requirements stated for the cyclonic separator, electrostatic precipitator, and fabric filter. A statement of the problem and its solution follows.

Problem Statement

Gas flow, acfm (avg)	168,600
Gas flow, acfm (peak)	210,000
Gas temperature, °F	400
Gas moisture, vol. %	8.5
Gas analysis, vol. %	
CO_2	14.2
H_2O	8.5
SO_2	0.2
Dust type	Flyash
Dust sp gr	2.5
Dust bulk density, lb/ft^3	30.0
Dust loading, grains/acf (avg)	1.1
Dust loading, grains/acf (peak)	1.8
Dust size analysis, % less than 10 μ (assumed)	17.0
Barometric pressure, psia	14.71

Required Performance. The scrubber discharge emissions must not be greater than 0.03 grain/scf at the average gas flow and dust-loading values. Although compliance to the emission limits cited is not required at peak conditions, the expected performance should be specified at this production level.

Consideration for the near-future control of SO_2 should be incorporated into the wet scrubber design.

Problem Solution

Gas Flow Rate. The gas flow rate entering the scrubber is 168,600 acfm at 400° F and contains 8.5 vol % moisture:

$$\frac{8.5 \text{ mols } H_2O}{100 \text{ mols gas}} \times \frac{18 \text{ mol wt.}}{29 \text{ mol wt.}} = 5.1 \text{ wt \%}$$

$$\frac{5.1}{94.9} = 0.054 \text{ lb } H_2O/\text{lb dry gas}$$

From Fig. 5-40, gases entering the scrubber at 400° F, containing 0.054 lb H_2O/lb dry gas, would be discharged at 134° F with a moisture content of 0.128 lb H_2O/lb dry gas. The specific volume of entering gas is:

$$0.054 \text{ lb } H_2O \times \frac{1}{18} = 0.003 \text{ mol } H_2O$$

$$1.000 \text{ lb gas} \times \frac{1}{30} = \frac{0.033 \text{ mol gas}}{0.036 \text{ mol wet gas}}$$

$$0.036 \times \frac{379}{520} \times (400 + 460) = 22.6 \text{ ft}^3 \text{ wet gas}$$

$$\text{Spec. vol.} = \frac{22.6 \text{ ft}^3}{1.00 \text{ lb gas}} = 22.6 \text{ ft}^3/\text{lb dry gas}$$

Specific volume of discharged gas is

$$0.128 \text{ lb } H_2O \times \frac{1}{18} = 0.007 \text{ mol } H_2O$$

$$1.000 \text{ lb gas} \times \frac{1}{30} = \frac{0.033 \text{ mol gas}}{0.040 \text{ mol wet gas}}$$

$$0.040 \times \frac{379}{520} \times (134 + 460) = 17.3 \text{ ft}^3 \text{ wet gas}$$

$$\text{Spec. vol.} = \frac{17.3 \text{ ft}^3}{1.00 \text{ lb gas}} = 17.3 \text{ ft}^3/\text{lb dry gas}$$

The gas flow discharged from the scrubber is

$$168,600 \times \frac{17.3}{22.6} = 129,000 \text{ acfm}$$

Therefore, the wet scrubber will be designed to process a gas flow of 129,000 acfm saturated at 134° F and containing 0.128 lb H_2O/lb dry gas (equivalent to 17.1 vol. %)

Collection Efficiency. Inlet dust load is

$$1.1 \times 168,600 \times \frac{1}{7000} \times 60 = 1590 \text{ lb/hr}$$

Discharge dust load is

> Allowable emission = 0.03 grain/scf
> Gas flow leaving scrubber = 129,000 acfm

$$129,000 \times \frac{(60 + 460)}{(134 + 460)} = 113,000 \text{ scfm}$$

$$0.03 \times 113,000 \times \frac{1}{7000} \times 60 = 29.1 \text{ lb/hr}$$

Required collection efficiency is

$$\text{Eff} = \left(1 - \frac{29.1}{1590}\right) 100 = 98.3\%$$

Water Requirements. Saturation water:

Inlet gas flow = 168,600 acfm
Inlet gas specific volume = 22.6 ft^3/lb dry gas

$$\text{Dry gas inlet flow} = 168,600 \times \frac{1}{22.6} = 7460 \text{ lb/min}$$

Inlet moisture = 7460 × 0.054	= 403 lb/min
Discharge moisture = 7460 × 0.128	= 955 lb/min
Required makeup moisture	= 552 lb/min

Slurry water:

> Inlet dust load   = 1590 lb/hr
> Discharge dust load =   29 lb/hr
>   Collected dust   = 1561 lb/hr

Assuming 5% flyash slurry,

$$\text{Water required} = 1561 \times \frac{0.95}{0.05} \times \frac{1}{60} = 494 \text{ lb/min}$$

$$\text{Total water required} = \frac{552 + 494}{8.33} = 126 \text{ gpm}$$

Scrubber Selection

Assuming that prior operating experience is available for the collection of flyash, and keeping in mind the requirements for gas absorption capabilities, the scrubber illustrated in Fig. 5-45 will be selected.

Scrubber Sizing. At a saturated gas flow of 129,000 acfm, a Model 1300 scrubber with a capacity of 130,000 cfm will be chosen. The overall dimensions, as shown in Fig. 5-46, will be 18'-0" diam × 32-0" overall height.

As an alternate, two parallel scrubbers might be chosen. Two Model 650 scrubbers, each having a capacity of 65,000 cfm, could accommodate the saturated gas flow. With such an arrangement, some flexibility could be obtained at reduced load operation. For example, at 30% load, one scrubber could be shut down and the remaining unit operated at 60% of capacity. This arrangement could also permit maintenance attention for the inactive scrubber. The dimension of each scrubber would be 12'-0" diam × 23'-3" height, so that it might be possible to consider plant rather than field fabrication, with considerable capital economies.

Collection Efficiency. Based on the requirement of 98.3% collection efficiency and a dust size of 17% < 10 μ, a pressure drop of about 3 in. w.g. would be required. In view of the constantly upgraded regulations, an efficiency of 99.5% will be promised, equivalent to an emissions of 0.008 grain/scf. The required pressure drop for this dust size would be 6.0 in. w.g.; see Fig. 5-41.

At the peak gas flow rate of 210,000 acfm, an increased pressure drop can be expected across the scrubber. Based on the standard orifice equation, a pressure-drop increase of about 50% would be expected at the higher gas flow rate. At this increased value of 9 in. w.g., the collection efficiency would be 99.7%. At this performance value, the peak inlet loading of 1.8 grains/acf would be brought to an emissions level of less than 0.01 grain/scf.

Gas Absorption. Normally, the contact mechanism for particulate collection in a wet scrubber does not coincide with effective absorption requirements. Counter-current flow of gas and liquor over a suitable packing with adequate retention represents the optimum approach for good gas absorption. Because the emphasis in this problem is particulate collection, the gas-removal potential of the selected scrubber will be only approximately indicated.

It is estimated that by utilizing a calcium hydroxide slurry as the scrubbing liquor, a gas absorption of 70% can be obtained [24]. At an inlet concentration of 0.2 vol %, or 2000 ppm SO_2, the discharge level would be 600 ppm. By the use of the more soluble sodium hydroxide, the efficiency could be increased to about 80% [24].

Materials of Construction. At the relatively high level of SO_2 gas in the effluent stream, corrosion resistant construction is necessary. Either Type 316 stainless steel or phenolic-coated mild steel fabrication can be used. The upper temperature limit of the latter is about 200° F; this means that the gases must be precooled before entering the scrubber. The most severe corrosion conditions will be encountered when water is used for collection of the particulates. The liquor effluent under these conditions will have a pH on the order of 1.5 to 2.5. Some after-treatment of this discharge liquor will be essential before it is discharged to any settling area.

Weights and Pricing. Assuming that the added operating flexibility justifies the application of two scrubbers, weights and pricing are as follows:

Weight of each Model 650 scrubber of
    Type 316 stainless steel construction   = 16,000 lb
Cost of each scrubber, delivered basis   = $43,000

Therefore, the total cost of both scrubbers unerected would be $86,000, equivalent to a unit cost of $0.510/acfm. This price is for the scrubber only, essentially as shown in Fig. 5-45. It is estimated that a liquor treatment system common to both scrubbers, similar to that shown in Fig. 5-52, would add at least 80% to the capital investment value.

5.8. PERFORMANCE AND COST SUMMARY

In describing the four major types of particulate collection equipment, an effort has been made to assess their suitability for a particular application. The problem involved the collection of flyash from a coal-fired industrial size boiler.

The problem has been originally stated in Section 5.4 concerned with the cyclonic separator and then restated in subsequent sections concerned with the electrostatic precipitator, fabric filter and wet scrubber, in that order. The solution to the problem, under each section, has been predicated on the performance characteristics of the equipment, rather than on a fixed and current emissions level. Thus, for the cyclonic separator, with its limited capabilities, there was imposed a collection efficiency slightly under 90%. The other three collectors were designed to perform at an efficiency in excess of 98%.

As discussed in the foregoing sections, there are innumerable factors involved in the selection of the optimal equipment type. It is not the purpose of this summary to specify "the" equipment for the solution of this particular problem, but rather to review the various factors to be considered in the evaluation of particulate emissions control methods.

Table 5-20 summarizes the process conditions, space requirements, materials of construction, performance characteristics, capital costs, and operating costs for each of the four collector types. This is followed by a discussion of the various considerations to be evaluated for selection optimization.

Performance

The cyclonic separator is incapable of providing a sufficiently high collection efficiency to satisfy the current regulations. The emission requirements specified in the problem definition for this collector have long since been upgraded.

The electrostatic precipitator has been designed to *just meet* the required performance because any increase above the specified 98.3% efficiency level would require a considerably higher capital-cost expenditure. The fabric filter is inherently capable of high collection efficiencies and therefore the 99.5% value does not add to either the

Table 5-20. Particulate Collection Equipment Performance and Cost Summary

Gas Flow, acfm	Gas Temp., °F	Dust Load, grain/acf	Equipment Type	Plan Space, ft × ft	Fabrication Material	Collection Eff., %	Equipment Costs $1000	$/acfm	Operating Cost Index, hp/1000 cfm
168,600	400	1.1	Cyclonic separator	20 × 30	Mild steel	93.9	31	0.184	0.394
168,600	400	1.1	Electrostatic precipitator	20 × 30	Mild steel	98.3	121	0.718	0.340
168,600	400	1.1	Fabric filter	50 × 70	Mild steel	99.5	180	1.068	0.504
168,600	400	1.1	Wet scrubber	20 × 30	Stainless steel	98.3	86	0.510	0.473
168,600	400	1.1	Wet scrubber	20 × 30	Stainless steel	99.5	86	0.510	0.945

capital or operating costs. The wet scrubber can be operated at either a collection efficiency of 98.3 or 99.5% by increasing the operating pressure drop to obtain the higher performance level. As indicated in Table 5-20, there would be a greater operating cost incurred at the 99.5% value, but the capital investment for the basic scrubber would remain the same.

The peak gas flow of 210,000 acfm would influence the performance of each control equipment type differently. The cyclonic separator and wet scrubber, both depending on the gas velocity for their collection mechanism, would yield improved collection efficiencies at the peak gas flow rates. In the wet scrubber the entrainment separator capacity must be evaluated to prevent liquor droplet carryover at the increased velocity. As determined in the Sample Design Calculations, the performance of the fabric filter could be maintained at the peak gas-flow rate, but the bags would have to be cleaned more frequently. The electrostatic precipitator performance would suffer at the increased gas flow rate, dropping off to 96.3%.

Costs

The equipment-delivered capital costs were determined from actual current costs quoted by various equipment manufacturers and cross-checked by referring to Figs. 5-3 through 5-6. The spread in costs is characteristic for the four equipment types. The relatively high value for the wet scrubbers reflects the stainless steel fabrication costs. If mild steel construction could have been used, the cost would have been reduced by a factor of 3.2. Installation costs which were not included in Table 5-20 encompass site preparation; equipment erection; accessory equipment such as fans, ductwork, stack, motors, and control instrumentation; and the connection of all services such as electrical, water, slurry disposal, and gas discharge. Approximate installation cost values, expressed as percentages of equipment purchase costs, are shown in Table 5-5.

The operating costs favor the selection of the electrostatic precipitator for this particular duty. Actually, the precipitator, usually in series with the cyclonic separator, has been the standard choice of control equipment for the collection of flyash. The requirement for gaseous emissions control, which has only been recently demanded by APC regulations, has diluted the application of the precipitator for this service.

Process Considerations

In the past, the major process consideration to be evaluated for the collection of flyash from the effluent gas stream of a coal-fired power plant was that of particulates disposal. At one time, some by-product recovery was thought possible, with the flyash being used as an aggregates base for cinder-block manufacture. At the present time, this material has negligible value and is usually truck-conveyed in the dry state to land-fill areas or is sluice-conveyed to a settling pond on the plant property. In either case, disposal is not a serious problem.

With the advent of gaseous emissions control regulations, the straightforward concept of particulates dry collection has been replaced by the wet scrubber approach. Some consideration is being given to the combustion of low-sulfur coals and the use of fabric filters for flyash collection, but this solution has not received serious acceptance at present. Although the application of wet scrubbers for the simultaneous control of flyash and SO_x is gaining favor, this practice is still hampered by the problems attending the disposal of the flyash-calcium sulfite slurry.

Usually, the flue gases discharged from a coal-fired boiler are at a temperature well above the dewpoint. The gases are maintained at this temperature level as they pass through the cyclonic separator, electrostatic precipitator, or fabric filter. Therefore, the finally discharged effluent gas passes through the stack to the atmosphere in a relatively buoyant state at temperatures in the range of 300 to 350° F. Gases treated in a wet scrubber, however, are at their dewpoint, and condensation of sulfurous and sulfuric acids continue as the gases pass through fans, ductwork, and stack to the atmosphere. This precipitation of acid compounds in the accessory equipment, following the wet scrubber, demands the use of expensive, exotic materials of construction for these items. The plume delivered by the stack to the atmosphere poses an additional problem because of its appearance and the fact that more often it settles rather than disperses. Such plumes have sufficient residual acid values to become a source of corrosion and contamination in the plant vicinity. In some of the more recent power-plant wet scrubber installations, provision has been made to heat the scrubbed effluent gases above their dewpoint before discharging them to the atmosphere.

Despite some of the disadvantages inherent in wet collection, there is one very important fact to keep in mind. When both particulate and gaseous emissions controls are demanded, the wet scrubber is the only equipment type available in today's marketplace which can undertake this service.

Maintenance

Maintenance costs represent expenditures required to sustain the operation of control equipment at its designed efficiency. Maintenance problems with the cyclonic separator and electrostatic precipitator are relatively minor. Based on an arbitrary scale of 10, the cyclonic separator would have a rating of 2, the precipitator would be set at 3, the fabric filter at 7, and the wet scrubber at 8. Erosion and plugging are the two factors associated with the cyclonic separator. The high-tension discharge electrodes and electrical system are the major areas requiring attention for the precipitator. With the fabric filter, bag replacement is involved, and care must be taken to prevent fabric binding. In the wet scrubber, corrosion, erosion, and plugging problems are encountered in that order. Maintenance costs are usually assumed to be proportional to the capacity of the equipment.

Equipment Arrangement

The equipment arrangement is reflected in the installation costs. The elevation of the various types of equipment and their location in plan will determine the ductwork configuration and accessory equipment arrangement. The wet scrubber is usually located at ground level because of the weight of the liquor contained in the sump. This often requires some interesting ductwork arrangement to minimize installation costs.

The location of liquor disposal facilities is another important factor influencing the wet scrubber arrangement.

The space requirement and weight of the various collectors decides their location in the plant. Accessibility is also an important factor to facilitate maintenance. In view of the space requirements for the fabric filter, ground level installation is favored, as illustrated in Fig. 5-36.

Accessory Equipment

The ductwork, fans, and dust-handling equipment for the three dry particulate collectors contribute nominally to the overall cost of the total system. However, for the wet scrubber, the accessory equipment is a significant cost factor in the evaluation of "total installed" costs. For a venturi scrubber, the fan is sometimes equal in cost to the scrubber, and to this value there must be added the cost of settling tanks, pumps, and piping. The use of exotic fabrication materials is frequently necessary for the wet scrubber, and therefore the economic impact of these materials on the fans, ductwork, and liquor disposal system must be taken into consideration.

It would not be too unreasonable to consider that wet scrubber accessory equipment demands, particularly for corrosive service, could double the capital costs. As indicated in Table 5-20, this added element could easily place the total investment value for the system at about the same level as that required for the fabric filter.

Background and Experience

When the evaluation of the various particulate collection equipment types has been completed, the decision eventually made commits the operator to a selection that best fulfills his particular situation. In Table 5-20, and in the foregoing discussion, an attempt has been made to encompass most of the factors relevant to an optimal selection. However, there is one most important additional element yet to be discussed. This can be summed up in two very important questions that every operator should address to the equipment supplier: "How many similar or identical applications exist where your equipment has performed satisfactorily?" and "May I contact the plants where this equipment is installed and discuss its operation with the personnel involved?" Because the design of air pollution control equipment is based on an empirical approach, proven performance can best be evaluated by the examination of an existing similar facility. If there is no prior experience available, then a pilot plant program should certainly be undertaken.

In the sample problem evaluated in this chapter, there is a very strong background in the use of cyclonic separators and electrostatic precipitators so that almost 100% assurance can be offered by most equipment suppliers as to the validity of the promised performance for this particular application. However, limited experience is available for the wet scrubber and negligible background has been developed for the fabric filter. Pilot plant and prototype equipment installations are currently being undertaken to determine reliable design criteria for these latter two equipment types.

REFERENCES

1. Air Quality Control Law, Georgia Laws of 1967, pp. 581 et seq., Air Quality Control Board, Georgia Dept. of Public Health. Approved April 14, 1967.
2. *Industrial Ventilation—A Manual of Recommended Practice,* Committee on Industrial Ventilation, 13th Ed., 1974 American Conference of Governmental Industrial Hygienists, P.O. Box 453, Lansing, Mich.
3. J. P. Tomany (private communications), Air Correction Division, Darien, Conn., 1961–1968.
4. J. P. Tomany, *A Guide to the Selection of Air Pollution Control Equipment,* Air Correction Division, Darien, Conn., 1970.
5. J. M. Kane, "Guideposts Tell How to Select Dust Collecting Equipment." *Plant Engineering* (November 1954).
6. *Criteria for Performance Guarantee Determinations,* Electrostatic Precipitators Division, Publication No. 3, Industrial Gas Cleaning Institute, Stamford, Conn., June 1965.
7. R. D. Ross, *Industrial Waste Disposal.* Reinhold Book Corp., New York, 1968, chap. 2.
8. N. G. Edmisten and F. L. Bunyard, "A Systematic Procedure for Determining the Cost of Controlling Particulate Emissions from Industrial Sources," *J-APCA* Vol. 20, No. 7 (July 1970), p. 446.
9. J. A. Danielson, ed., *Air Pollution Engineering Manual,* 2d ed., Air Pollution Control District, County of Los Angeles. Environmental Protection Agency, Research Triangle Park, N.C. May 1973.
10. "Control of Particulate Emissions," Air Pollution Training Course, Robert A. Taft Sanitary Engineering Center, Cincinnati, Ohio; U.S. Dept. of Health, Education, and Welfare, November 1959.
11. W. Deutsch, Ann. der Physik, Vol. 68, 1922, p. 335.
12. R. E. Kirk and D. F. Othmer, *Encyclopedia of Chemical Technology.* Interscience Encyclopedia Inc., New York, 1947.
13. "Information Required for the Preparation of Bidding," *Specifications for Electrostatic Precipitators,* Publ. EP-5, Industrial Gas Cleaning Institute, Stamford, Conn., November 1968.
14. Cottrell Electrical Precipitators, Western Precipitation Division, Los Angeles, Calif., 1952.
15. P. L. Magill, F. R. Holden, and C. Ackley (eds.), *Air Pollution Handbook.* McGraw-Hill Book Co. Inc., New York, 1956.
16. R. T. Pring, "Bag Type Cloth Dust and Fume Collectors," *Proceedings of the U.S. Technical Conference on Air Pollution*—L. C. McCabe, (Ed.), McGraw-Hill Publishing Co., Inc., Vol. 35 (1952), p. 280.
17. W. C. L. Hemeon, *Plant and Process Ventilation.* The Industrial Press, New York, 2d ed., 1963.
18. C. E. Williams, T. Hatch, and L. Greenburg. "Determination of Cloth Area for Industrial Air Filter," *Heating, Piping, Air Conditioning,* Vol. 12 (April 1940), pp. 259–263.
19. American Air Filter Co. Inc., Louisville, Ky., Bulletin 310B, May 1973.
20. R. C. French, "Filter Media," *Chemical Engineering,* Vol. 70, No. 21 (October 1963), pp. 171–192.
21. *Air Pollution Manual, Part II, Control Equipment,* American Industrial Hygiene Association, 1968.
22. K. T. Semrau, "Dust Scrubber Design—A Critique on the State of the Art," 56th Annual APCA Meeting Presentation, June 1963.

23. Zimmerman and Lavine, *Psychrometric Tables and Charts.* Industrial Research Service Inc., Dover, N.H., 2d ed., 1964.
24. J. P. Tomany (private communications), Western Precipitation Division, Los Angeles, Calif., 1971.
25. A. W. Kielback, "The Development of Floating Bed Scrubber and Application in the Aluminum Company of Canada's Smelters," Amer. Inst. of Chem. Engr., from *Chemical Engineering Progress Symposium Series*, Vol. 57, No. 35, December 1961, pp. 51–54.

Chapter 6

GASEOUS POLLUTANTS
CONTROL EQUIPMENT

6.1. INTRODUCTION

In Chapter 2, gaseous pollutants were classified as organic and inorganic gases. The organic group consists entirely of compounds of hydrogen and carbon, and their derivatives. Such hydrocarbons as paraffins, olefins, and aromatics are included in this classification. Derivatives formed by the replacement of the hydrogen atom(s) by oxygen, halogens, nitrogen, and other substitutes are likewise classified as organic gases. Petroleum refining processes and the automobile are the two major emission sources of organic gases. The hydrocarbon derivatives are most commonly discharged to the atmosphere by industrial cleaning processes. Aldehydes, ketones, alcohols, organic acids, and chlorinated hydrocarbons are the atmospheric contaminants requiring the application of control equipment for this industry.

The common inorganic gases include oxides of sulfur, oxides of nitrogen, carbon monoxide, ammonia, hydrogen sulfide, and chlorine. The major source of these contaminants is the combustion of fossil fuels. Various industrial processes contribute minor quantities of ammonia, hydrogen sulfide, and chlorine to the atmosphere. In Los Angeles County, automobile emissions are currently responsible for approximately 70% of the total nitrogen oxides and 95% of the carbon monoxide emissions [1]. Although carbon monoxide does not enter into the reactions producing photochemical smog (PAN), the oxides of nitrogen and hydrocarbons do. Both pollutants are emitted by the automobile, and the technology for their control will be discussed in Chapter 8.

A wide range of these pollutants can be destroyed by incineration. Most of the organic gases, plus such inorganic gaseous pollutants as ammonia, hydrogen sulfide, nitrogen oxides, and cyanide gases, can be reacted to form innocuous end products. Such combustible gases are usually exhausted from an industrial process below the flammable range and cannot be directly ignited. Therefore, in incineration type emissions-control equipment, a fuel source must be utilized to heat the effluent stream to the contaminant oxidation temperatures.

6.2. PROBLEM DEFINITION

As in the case of particulate pollutants control, the various gaseous contaminant concentrations must be first defined in the effluent gas stream. When the applicable

Table 6-1. Gaseous Effluent Definition

Plant type	Paint drying oven
Effluent source	Oven exhaust
Duty	Gaseous contaminant control
Gas flow, scfm	10,000
Gas temperature, °F	150
Solvent contaminants	Toluene, xylol, methylethyl ketone (MEK), benzene
Solvent load, gph	60
Heating fuel	Natural gas at 1000 Btu/ft³

control regulations have been evaluated, only then can the degree of emissions control be determined and suitable equipment selected and designed. This procedure is best illustrated by consideration of a typical gaseous emissions condition as defined in Table 6-1.

Evaluation of these operating data should generally follow the procedure discussed in Chapter 5 for particulate emissions control. In the process being considered in Table 6-1, the problem of particulate emissions might be present. Under a variety of general "nuisance" regulations, the odor level of the contaminants being discharged to the atmosphere must also be considered. Usually, there is required a reduction of odor level from a maximum of several thousand times above the detectable threshold to about a hundred times above it. A well-designed stack of sufficient height is usually sufficiently effective to meet those demands for most applications.

For gas-fired ovens, the nitrogen oxide emissions level should be considered. Within the past year, regulations enacted for the control of this gaseous contaminant have become more restrictive. Normally, close stoichiometric combustion control is sufficient to meet the current demands. However, should the regulations continue to be upgraded, emissions control equipment may be required.

The gas flow rate determines the control equipment sizing. For the gaseous effluent described in Table 6-1, where the solvent contaminants may be oxidized for their control, stoichiometric calculations are required to determine the additional combustion gas volume to be handled by the control equipment. Depending on the contaminant loadings, the waste gas temperature, and the required oxidation temperature, the total gas volume flowing through the equipment may be increased by as much as 50% over the waste gas stream, because of the combustion of the fuel gas flow necessary for oxidation of the solvents load. Heat balances, discussed later in the text, will illustrate the determination of the total volumetic gas flow.

Accurate data that fully define the various contaminants to be removed are most difficult to obtain. The design of gaseous emissions control equipment, unlike that for particulates control, is predicated on the thermodynamic interrelationships of the various constituents of the effluent gas stream. In incineration control processes, the reaction conditions must be closely specified, for efficient conversion of the contaminants to nontoxic gases. Some of the most important data required for the design of effective gaseous emissions control equipment are:

1. Definition of contaminants.
2. Physical state: liquid, solid, gas.
3. Ultimate analysis: C, H_2, O_2, N_2, H_2O.
4. Metals: calcium, lead, sodium, copper, etc.
5. Halogens: bromides, chlorides, fluorides.
6. Heating values: Btu/lb.
7. Special characteristics of contaminants: toxicity, corrosiveness, flammability limits, solubility, etc.
8. Effluent flow rates: average, peak, minimum, and future.

6.3. THE REGULATIONS

The basic approach to the enactment of control measures for gaseous emissions is directed at the reduction of atmospheric discharge quantities as they relate to their contribution to the total contamination level. Thus, although a particular industrial process may not be creating a local nuisance, its emissions may be limited by the authorities because of their influence on the overall air quality in the community. In poorly ventilated geographic areas such as the Los Angeles Basin, where the air pollution levels are relatively high, legislation was first enacted to accomplish control of combustible gases and vapors [2]. The identification of such combustibles as contributors to photochemical smog was the major factor in causing them to be included in control regulations.

There are two Los Angeles County Air Pollution Control District (LAAPCD) regulations governing the plant effluent defined in Table 6-1. These are Rule 66a, which is concerned with the control of solvent emissions from processes where heat is applied; and Rule 51, which generally pertains to the allowable nuisance level or, in this particular case, the odor characteristics of the effluent. In Rule 66, if it can be proved that the discharged solvents are not likely to react in the atmosphere to form noxious photochemical by-products, then they are exempt from the rule. If the solvent contains reactive compounds, then the process effluent is limited to 15 lb of organic material per day, if no controls are applied, or to 15% of the solvent that is normally emitted after control equipment is installed.

The solvent emissions in Table 6-1 are stated as 60 gal/hr, equivalent to 10,320 lb per day. It will be assumed that these solvents are subject to the demands made by Rule 66, so that the removal of 85% of the solvents is required. Therefore, the allowable atmospheric emissions would be

$$60 \text{ gal/hr} \times (8.33 \times 0.86) \text{ lb/gal} \times \frac{(15)\%}{100} = 64.5 \text{ lb/hr}$$

If incineration equipment is to be used, then (60 × 8.33 × 0.86 × 0.85), or 365.4 lb/hr, of the hydrocarbon solvents must be converted to carbon dioxide and water vapor. The conversion efficiency must be based on the combustible carbon content of the solvents.

6.4. EQUIPMENT SELECTION

The control of gaseous emissions is usually an expensive process, in terms of both capital and operating costs. Before the selection of control equipment is considered, the process must be carefully examined to ensure the validity of the process data. Because fuel costs represent the most predominant operating burden for most gaseous emissions control equipment, process heat and material balances should be evaluated before final equipment selection is decided.

The total control equipment available for the abatement of both particulate and gaseous emissions is shown in Fig. 5-2. For gaseous emissions, thermal incineration, catalytic combustion, adsorption and wet scrubbing control equipment can be applied. In thermal incineration, the gaseous contaminants are converted by combustion to such harmless waste products as carbon dioxide, water vapor, and nitrogen. Catalytic combustion can accomplish the decomposition of pollutant materials at reduced temperatures and at corresponding lower fuel consumption costs. Both thermal incineration and catalytic combustion involve both oxidation and reduction reactions. Thus, in the destruction of a combustible hydrocarbon, the carbon and hydrogen atom are oxidized to CO_2 and H_2O, respectively. In the control of nitric oxide, this compound is converted to N_2 and H_2O by catalytic reduction with CH_4.

Adsorption is particularly favored where the pollutants are noncombustible and are present in very dilute concentrations in the effluent gas stream. This control method relies on the attraction of certain gases to such materials as activated carbon, silica gel, and alumina, where they are retained on the surface of such adsorbents by intermolecular forces. Control of gaseous emissions by wet scrubbing is applicable when the gases are highly soluble and/or reactive in the scrubbing liquor. Regeneration and recovery of gaseous contaminants may be practiced in the adsorption and wet scrubbing processes.

A combination of these equipment types is sometimes applied in series. For example, in handling an effluent containing a halogen-substituted hydrocarbon, such as vinyl chloride, thermal incineration is utilized as the primary control device, followed by a wet scrubber. When vinyl chloride is thermally decomposed, CO_2, H_2O, HCl, and some free chlorine are produced. By utilizing an alkaline liquor in the wet scrubber, the gaseous chlorine compounds can be removed from the effluent stream. A combination adsorption-incineration system has also been proposed wherein a primary adsorption-regeneration cycle effectively concentrates the solvent loading in the waste gas stream being directed to the incinerator, thereby reducing both the equipment and fuel costs in the combustion step.

Three general approaches can be used to reduce solvents emissions [3]. These are: (1) process modifications, (2) recovery, and (3) incineration. One process modification might require the substitution of "nonreactive" solvents so that the LAAPCD Rule 66 would not be applicable. For example, a formulation of acetone and a paraffin hydrocarbon mix can be combined with the mixture defined in Table 6-1 so as to exempt the emissions from the current regulations. In this approach, a total systems perspective must be adopted, involving all three paint-processing areas: the manufacture, application, and drying/baking. Any modification of the solvent system to reduce air pollution control costs must be evaluated in terms of the corresponding capital and operating cost variations for all three process operations.

By-product recovery would involve separating the gaseous pollutants from the effluent gas stream and subsequently concentrating them. Adsorption and wet scrubbing can be utilized in such a recovery system. Actually, the current operating practice is to reduce the process pollutant emissions to a minimum so that very rarely is by-product recovery economically justified in air pollution control.

The final and most common solution is that of incineration, both catalytic and noncatalytic. Although either method is applicable to the emissions described in Table 6-1, there are some limitations to be considered. The presence of particulates generally, and some inhibitor particulates specifically, interferes seriously with the performance of catalytic combustion equipment. Heavy particulate loadings will also eliminate consideration of thermal incineration and adsorption for emissions control.

Theoretically, the four gaseous emissions-control equipment types shown in Fig. 5-2 may be considered as candidates for the solution of the particular atmospheric pollution problem in Table 6-1. However, as was shown in Chapter 5 for particulate pollutants control equipment, there are many variables to be evaluated before a final selection can be made. A discussion of some of the major variables follows.

Collection Efficiency

The currently demanded gaseous pollutant collection or removal efficiencies are relatively easy to obtain in thermal incineration and catalytic combustion equipment. The 90 to 95% efficiencies usually required to meet LAAPCD Rule 66 can be easily attained by thermal incineration. Although there are some marked economic advantages favoring catalytic combustion, a history of operating difficulties, involving the fouling and deterioration of the catalyst elements, has favored thermal incineration.

Adsorption of solvent vapors can be accomplished in the range of 97 to 98%. With a single pollutant compound, acceptable adsorption levels can be realized. However, with a mixture of solvents, such as that shown in Table 6-1, there is preferential adsorption of the least volatile components so that the total adsorption potential of the system is limited. However, adsorption and subsequent recovery can be accomplished, although the capital investment requirements for recovery are usually prohibitive.

Wet scrubbing is limited by the solubility of the various solvents in the mixture. For example, toluene is completely insoluble in water; therefore, for a typical solvents mixture composition, overall wet scrubber removal efficiencies on the order of 40 to 60% might be expected. As a result, the wet scrubber cannot be considered as a viable solution to many solvent control problems.

Gas Flow

Both capital and operating costs for all four types of gaseous emission control equipment are very sensitive to gas flow rate and contaminant concentrations. This is particularly true for incineration equipment, both thermal and catalytic. In the conditions presented in Table 6-1, the gas flow and solvent load are actually fixed by insurance regulations, which demand that the solvent-to-air ratio lies well outside the lower explosive limit. These regulations prescribe a ventilation rate equivalent to 10,000 scf per gallon of solvents in the effluent gas stream, or 10,000 scfm per gal/min. This ratio is based on maintaining the gas stream mixture at 25% of the lower explosive limit for most solvents.

Thermal incineration or catalytic combustion control equipment is extremely costly except for applications involving small gas flows. In coatings application, the volume of air handled is predicated on a ventilation rate adequate to carry off the overspray. For example, a spray booth for the application of 100 gph of solvent would require a ventilation rate on the order of 200,000 scfm. In addition to high capital costs for coating applications, operating difficulties contribute to increased operating costs.

Both thermal incineration and catalytic combustion equipment are designed to provide a specified dwell time or retention period in the combustion area. For thermal incinerators, the dwell time in the combustion chamber is in the range of 0.10 to 1.0 sec. In a catalytic system, the gas dwell time spent in the catalyst bed can be extremely short, a few hundredths of a second or less. The desired retention period in the combustion chamber or catalyst bed is determined by the chamber length or catalyst depth and the design velocity. For reasonable chamber lengths and catalyst bed depths, gas velocities in the range of 20 to 200 fps are used. Because the completeness of combustion is directly dependent on the retention period, an increased gas flow will reduce this period and thus adversely affect performance.

The critical design parameter for adsorption processes is the allowable superficial velocity through the adsorbent beds. For atmospheric pressure systems, such as might be applied for the conditions in Table 6-1, superficial velocities in the neighborhood of 50 to 100 fpm are recommended. For a high-pressure system, such as one operated at about 600 psig, the velocity might be reduced to as low as 15 fpm.

In the design of wet scrubbers, both the retention period and the liquor recirculation rate are important design parameters affecting the absorption rate. Velocities in the range of 200 to 800 fpm are utilized for the various scrubber types. The gas volume to be treated must first be converted to the saturated volumetric flow rate, particularly when high temperatures are involved.

Gas Temperature

The three most important design parameters influencing effective combustion are time, temperature, and turbulence. Of these, temperature is the most important, as it relates to equipment operating costs. Thermal incinerators operate at temperatures of 1200 to 1500° F. With process exhaust gases entering a thermal incinerator at about 150° F, considerable fuel is required to elevate the gas stream temperature to that level required for reaction. Thus, for a thermal system handling 20,000 scfm of effluent gases, the fuel consumption would be equivalent to about 30×10^6 Btu/hr. Catalytic combustion requires reaction temperatures on the order of 500 to 1000° F, thereby demanding roughly one-half the fuel rate as thermal incineration.

Adsorption depends on the vapor-pressure/temperature characteristics of the pollutant being removed from the effluent gas stream. Because most organic compounds have negligible vapor pressures at ambient temperature, the adsorption process is well suited for their control in the low temperature range of 100 to 200° F. Higher temperature exhaust gas streams may be cooled by dilution with air to reduce the volatility of the organic pollutants and thereby increase their adsorption potential. Adsorption is an exothermic process that causes a temperature rise in the adsorbent bed. For paint-spraying pollutant control applications, similar to that defined in Table 6-1, a temperature rise of about 15° F can be expected. This relatively small temperature increment does not affect the capacity of the adsorbent.

The performance of wet scrubbers for gas adsorption is influenced by the temperature of the waste gas stream that is being treated. Improved absorption can be expected at lower temperatures, where solubility controls the absorption process. In some absorption/reaction systems such as the removal of SO_2 in a $Mg(OH)_2$ scrubbing liquor, an elevated temperature in the range of 120 to 130° F yields the best performance. In a power-plant service that might adopt this wet scrubbing system, the exhaust gas stream would enter the wet scrubber at about 300° F, but would be quickly saturated by the scrubbing liquor and its temperature would be reduced to about 125° F. This mechanism was explained in Chapter 5. In other systems where the saturated gas temperature is too high for effective absorption, increased makeup water can be introduced into the scrubber for further reduction of the gas temperature.

Pollutant Properties

Some of the more important gaseous pollutant properties to be evaluated for optimum application of the most effective type of control equipment are: composition, flammability, reactivity, vapor pressure, and solubility.

Composition of the pollutant stream is important to the design of thermal and catalytic combustion equipment. A reaction retention period must be chosen so as to satisfy the least reactant component in the solvent mixture. As the reaction rate of a particular solvent decreases, the necessary retention period increases. Therefore, for any system of solvents, the composition must be known to determine the maximum dwell time for complete combustion.

In adsorption there is a selective removal action of the least volatile pollutant, which must be considered in the design of the equipment. Thus, the physical properties of the least adsorptive of the solvents contained in the effluent mixture should provide the system design basis. In coatings applications the recovered solvent will not have the same composition as the originally applied formulation because of the relative volatilities of the individual components in the spray carryover and the adsorbent's variable affinity for each component.

The gas composition is important to the wet scrubber selection because the concentration of the various components, with their different vapor pressures and solubility characteristics, will determine the design. In some solvent mixtures, a definition of the composition might indicate the presence of a slightly soluble compound. However, its concentration in the mixture may be sufficiently small so as to have a negligible effect on the overall removal efficiency.

Flammability is an important property of those pollutants whose emissions are to be controlled by combustion equipment, either thermal or catalytic. Flammable or combustible materials are usually hydrocarbons that are oxidized to yield carbon dioxide and water vapor. Such noncombustible compounds as ammonia, nitric oxide, and sulfur dioxide may be processed in catalytic combustion equipment. For these gases, the reaction is one of reduction, with the pollutants being converted to nitrogen, sulfur, carbon dioxide, and water vapor. Processes are presently being investigated for the simultaneous catalytic reduction of NO_2 and SO_2.

Adsorption can be utilized for the control of nonflammable compounds. Carbon disulfide, nitric oxide, and chlorinated hydrocarbons are some of the nonflammable pollutants that may be collected by adsorption. In many industrial processes, where emission control is a secondary function, flammable gases and vapors removed by adsorption are usually recovered.

The performance of wet scrubbers is unaffected by the flammability characteristics of the pollutants. One advantage favoring the application of the wet scrubber is that it can process flammable gases at any concentration without the hazard of explosion.

The reactivity of organic compounds need be considered only for the thermal and catalytic combustion processes. When the reaction is one of oxidation, it is usually 90 to 95% completed in the equipment. These reactions have been established as first-order types with respect to the combustible component. Thus, the rate of combustion is doubled if the concentration is doubled. With specific rate values, which have been developed for the most common solvents, the required retention period can be determined at a specific reaction temperature to ensure maximum combustion.

The vapor pressure of pollutants has a negligible influence on the design or the performance of thermal or catalytic combustion equipment. However, adsorption efficiences are often expressed in vapor-pressure terms. An adsorbent will preferentially attract and hold those gases and vapors having the smallest vapor-pressure values.

When air containing a mixture of organic vapors is passed through an adsorbent, the vapors are equally adsorbed initially. As the high-boiling constituent is retained on the adsorbent in increasing amounts, the more volatile compounds are displaced; thus, when the adsorbent becomes saturated, the exit vapors will consist largely of the more volatile components.

In absorption processes, as carried out in the wet scrubber, mass transfer of the gaseous pollutants from the gas phase to the liquor phase depends on the difference between the partial vapor pressure of the gas in the gas stream and the vapor pressure of the solute gas above the scrubbing solution. Thus, for an SO_2-alkaline liquor scrubbing system, this differential vapor pressure for a 1500 ppm SO_2 gas stream would measure about 1 mm Hg.

The solubility of gaseous pollutants does not affect the incineration control processes. However, the application of adsorption and wet scrubbing equipment does depend on pollutant miscibility. For the recovery of water immiscible solvents from an adsorption process, simple decantation can yield a very pure by-product. For miscible or highly soluble pollutants, a more complex distillation recovery system becomes necessary.

It is desirable in the application of wet scrubbing control methods that pollutants be water-soluble. Although alternate scrubbing liquors can be employed to take advantage of their affinity to absorb components of a specific pollutant mixture, this approach is seldom used because of the cost of such special liquors. One of the major industrial exceptions is the use of amine compounds for the absorption of CO_2 and H_2S. Water-insoluble gaseous pollutant emissions are not usually controlled by wet scrubbing equipment. For the absorption of such acidic gases as SO_2 and HF, the addition of alkaline salts such as NaOH and $Ca(OH)_2$ to aqueous scrubbing liquors is commonly practiced.

Pollutant Concentration

In handling a waste gas stream with combustible gases and vapors, flammability hazards must be avoided. As illustrated in the sample problem described in Table 6-1, the gas flow rates and solvent loadings from industrial equipment that is handling flammable compounds are usually fixed by insurance regulations so as to maintain the pollutant concentration at a value of 25% of the lower explosive limit (LEL).

For such applications as large paint-spray processes, the operating costs for an incineration system could be prohibitive. However, by the use of a combination adsorption-incineration control system, the solvents concentration leaving the adsorbent during the regeneration cycle can be increased. Thus, it is possible to treat smaller volumes of exhaust gas in the incineration equipment with a corresponding reduction in fuel requirements.

In the wet scrubber the various pollutant concentrations influence the mass transfer of the gases from the gas phase to the liquid phase. The concentration of each component is a function of the partial pressure-driving forces that control the degree of absorption.

Contaminants

The presence of contaminants such as lead, arsenic, or phosphate compounds, or of other types of noncombustible particulates in the exhaust gas stream, interferes with the catalytic combustion process. The heavy metals and phosphates act as poisons to most industrial catalysts, rendering them completely ineffective. Such particulates as alumina, silica, and metallic oxides tend to foul the catalyst surface, thereby reducing its activity. Thermal incineration equipment is similarly adversely affected by noncombustible particulates. However, carbon particles, a common pollutant emitted from malfunctioning combustion processes, can be controlled by thermal incineration through their conversion to carbon dioxide.

Adsorption equipment cannot tolerate particulate loadings. In paint-spray applications, a regenerable dry filter must protect carbon adsorbents from resinous solids carried in the solvents effluent stream. Wet scrubbers are ideally suited for the collection of particulates, either alone or as a precleaner for incineration, catalytic combustion, or adsorption equipment.

Disposal Methods

With incineration and catalytic combustion control methods, the pollutants are usually converted to CO_2, H_2O, and N_2, which are discharged to the atmosphere. With both adsorption and wet scrubbing, disposal of the collected effluents must be effected. If recovery is economically feasible, then the additional capital and operating costs for the recovery facilities must be traded off against the value of the recovered solvents.

In considering recovery of pollutants, it must be realized that the performance of both collection and recovery systems must be evaluated to ensure that the total solvents discharged to atmosphere fall within the APCD allowances.

Economic Factors

The economic evaluation of a particular emissions control system must certainly involve the equipment performance. The pollutant emissions defined in Table 6-1 obviously require control. The wet scrubber would appear to be eliminated because of the presence of the insoluble toluene as one of the pollutants in the effluent gas stream. Toluene is normally present in industrial solvents in the range of 45 to 50 wt %. Assuming a removal efficiency of only about 5% for this constituent would yield an overall collection efficiency of 45%.

Adsorption equipment can easily provide the 90% removal efficiency required to satisfy the regulations. However, the adsorption method is most economically applied to conditions involving very low pollutant concentration. The pollutant loadings shown in Table 6-1 are at the allowable maximum value as set by plant safety limits. Recovery of the collected solvents might be considered, but in addition to increasing the equipment costs, this practice is not too favorably received by the paint application industries because of the need for reconstitution of the formulation following the selective adsorption of the individual solvents.

Thermal incineration or catalytic combustion is the preferred control system to be considered for this particular problem. Either system can easily provide a 95% removal efficiency. Although this method of control is costly, it does eliminate the problems of disposal or recovery usually associated with wet scrubbing and adsorption. Thermal incineration provides the most flexible system with the least maintenance demands. However, it is the most costly to operate because of the high fuel demands, although the use of heat exchangers can reduce these costs. Catalytic combustion requires considerably less fuel because of its lower reaction temperature, but, catalyst maintenance may make it difficult to attain a continuous high performance operation. A rigid catalyst service and replacement program is necessary to maintain high-level performance.

The major operating costs for the various control equipment types are utilities. The fuel gas requirements for incineration equipment are a direct function of the gas volume to be treated, the specific heat of the gas, and the outlet gas temperature. The inlet gas temperature at the source may vary from 100 to 400° F. For thermal incineration the outlet gas temperature is usually in the range of 1200 to 1500° F. In catalytic combustion, temperatures of 500 to 1000° F are required for complete reaction. The use of heat exchangers may reduce the fuel requirements by as much as 65%. In Table 6-2 there are listed fuel costs based on natural gas costing $0.60/10^6$ Btu. Catalyst reactivation and replacement is another source of operating costs associated with catalytic combustion.

Adsorption and wet scrubbing equipment require electric power to overcome the gas pressure drop through the equipment. Thus, for adsorption, pressure drops in the range of 2 to 8 in. w.g. might be expected. The actual value is a function of the gas velocity, adsorbent bed depth, and number of beds. In wet scrubbers, the pressure drop is usually somewhat greater than that for adsorption. Thus, for a packed tower designed to remove 95% of the ammonia from a gas stream containing 10% of this gas, approximately 15 ft of packing would be required. At the gas and water flow rates necessary for the desired absorption duty, a pressure drop of about 12 in. w.g. would be required.

In addition to power requirements, the adsorption process has need of a hot gas stream to remove the pollutants from the adsorbent. Low-pressure saturated steam is the usual source of heat and has the ability to remove most adsorbates. In addition, superheated steam up to 600° F in temperature may be required, on an intermittent basis, to reactivate the adsorbent. Desorption steam rates of 2 to 10 lb/lb of adsorbate may be necessary. Adsorbent life is an important variable influencing the operating

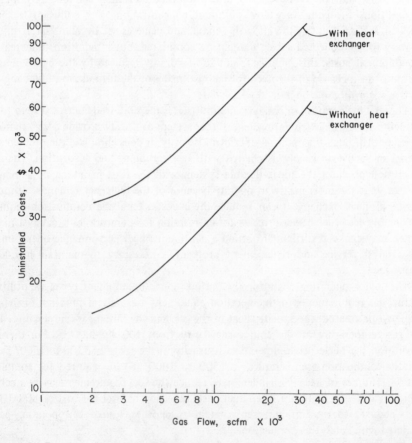

Figure 6-1. Purchase costs for thermal incineration equipment with and without heat exchangers. (Adapted from U.S. HEW publication, Ref. 5.)

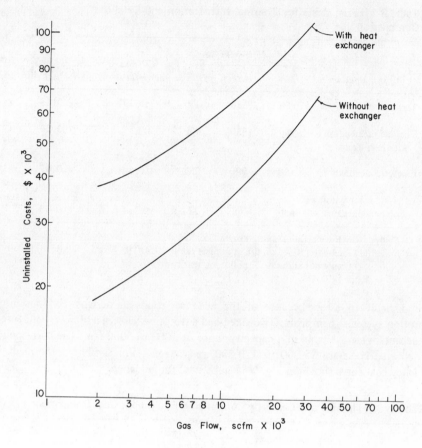

Figure 6-2. Purchase costs for catalytic combustion equipment with and without heat exchangers. (Adapted from U.S. HEW publication, Ref. 5.)

costs for the adsorption process. The use of relatively inexpensive activated carbon and the absence of particulates in the waste gas stream represent optimized minimum cost conditions.

In both the adsorption and wet scrubbing processes, additional steam is required if recovery is practiced. Distillation is the usual method of solvent recovery for both control processes. One other operating cost source is that of alkaline additives for the absorption of acidic gases.

Average purchase costs [5,6] for thermal incineration equipment is shown in Fig. 6-1 and for catalytic combustion equipment in Fig. 6-2. In both curves, the additional purchase costs are shown for accessory heat-exchange equipment. Fuel costs with and without heat exchange are listed in Table 6-2. Installation costs may vary from 50 to 100% of the equipment purchase costs.

Purchase costs for both adsorption and wet scrubbing equipment is somewhat

229

Table 6-2. Fuel Costs for Thermal Incineration and Catalytic Combustion

Control Method	Air Temperatures, °F			Cfh Gas per scfm Air	Fuel Costs, $/10³ scf Air
	Inlet	Outlet	Diff.		
Thermal incineration	150	1350	1200	1.630	0.0163
Thermal incineration with heat exchange	950	1350	400	0.535	0.0054
Catalytic combustion	150	900	750	0.969	0.0097
Catalytic combustion with heat exchange	650	900	250	0.334	0.0033

Source: North American Manufacturing Co., Ref. 4.
Notes: (1) Fuel costs based on use of natural gas at $0.60/10⁶ Btu.
(2) Credit not allowed for pollutant heat content.

more difficult to assess because of the wide variations in designs. Skid-mounted adsorption systems—comprising two fixed-bed adsorption vessels, a blower, condenser, and decanter—are available in a capacity range of 1000 to 5000 cfm. The corresponding delivered costs are $5,000 to $20,000, respectively. For scrubbers, the medium efficiency cost curve shown in Fig. 5-6 can be used for guidance.

Table 6-3. Thermal and Catalytic Combustion Equipment Cost Comparison

Equipment Type	Thermal Incineration	Thermal Incineration with Heat Exchange	Catalytic Combustion	Catalytic Combustion with Heat Exchange
Gas flow, scfm	10,000	10,000	10,000	10,000
Gas inlet temp., °F	150	150	150	150
Gas discharge temp., °F	1,350	700	900	550
Solvents loading, gph	60	60	60	60
Removal efficiency, %	95	95	95	95
Heat input, 10⁶ Btu/hr	13.0	4.3	8.1	2.7
Gas costs, 10³ $/yr	67.3	22.3	42.0	14.0
Equipment purchase cost, $	30,500	57,000	34,600	62,000
Installation cost, $	15,300	40,000	17,300	43,000
Annual costs, 10³ $	3.1	6.5	3.5	7.0
Total Annual Costs, 10³ $	70.4	28.8	45.5	21.0

Note: Estimating bases are 360(24 hr operating days)/yr and a 15-year amortization period.

A review of the various economic and performance factors informs us that the gaseous emissions defined in Table 6-1 can be best controlled by the application of either thermal incineration or catalytic combustion equipment. Estimated capital and operating costs for both thermal and catalytic systems, with and without heat exchange accessory equipment, are shown in Table 6-3. Costs are based on Table 6-2 and Figs. 6-1 and 6-2.

From the data presented in Table 6-3, two facts are apparent: first, that the operating costs for incineration equipment to handle the waste gas stream, as defined in Table 6-1, are extremely high; second, that the savings to be realized by selection of the catalytic combustion system with heat exchange are most attractive. The cost of fuel so far outweighs the annual equipment depreciation costs that investment in the catalytic unit is economically justified.

One additional operating cost associated with catalytic combustion equipment is maintenance. For the particular size of unit chosen for this duty, an annual charge of $1200 might be estimated for the catalytic unit as compared with $400 for thermal incineration. These additive costs do not detract from the recommendation that the catalytic combustion equipment, complete with heat exchanger, is the optimum equipment selection for the control of the gaseous emissions described in Table 6-1.

6.5. EQUIPMENT CHARACTERISTICS—THERMAL INCINERATION

Each of the four gaseous pollutant control equipment types will be described under the following headings: operating principles, design parameters, mechanical design, installation features, applications, and sample design calculations. The first to be discussed is the thermal incinerator.

Operating Principles

Combustible gases are released from many different industrial processes. When they are exhausted to the atmosphere, these streams are usually below the flammable range and cannot be directly ignited. To oxidize these pollutants, it is necessary to heat the fume to high temperatures in the range of 1200° to 1500° F, to effect thermal incineration. Oxygen, usually in air, must be present in slightly greater than stoichiometric quantities for total combustion. The oxidation of gaseous pollutants at these temperature levels is accomplished in a thermal incinerator.

In thermal incineration or catalytic combustion, both oxidation and reduction of the gaseous pollutants can be provided. Thus, the oxidation of combustible pollutants will cause them to be converted to CO_2 and H_2O, as typified by the following reactions for toluene and benzene.

$$\text{Toluene: } C_7H_8 + 9O_2 \rightarrow 7CO_2 + 4H_2O \tag{6-1}$$

$$\text{Benzene: } C_6H_6 + \frac{15}{2}O_2 \rightarrow 6CO_2 + 3H_2O \tag{6-2}$$

For these reactions to occur, requirements are the proper reaction temperature, an optimum retention period, and sufficient turbulence of the reactants to ensure complete mixing. The reaction temperature determines the fuel gas requirements and reaction rate, while the latter decides the necessary retention period for a particular degree of conversion. Thus, temperature, time, and turbulence are important parameters influencing the incinerator mechanical design.

A schematic diagram of a thermal incinerator is shown in Fig. 6-3, where the approximate indicated temperatures correspond to the conditions shown in Table 6-3. The exhaust gases containing the gaseous pollutants enter the fan at 150° F and are then discharged through an annular space surrounding the firing chamber, where some preheating takes place. The heated gases enter the firing chamber and are intensely mixed with the fuel combustion gases. The gas mixture, at a reaction temperature of 1350° F, travels through the firing chamber where the oxidation of the pollutant gases takes place. The chamber is of sufficient length to provide the retention period necessary for the required degree of conversion. The combustion gases plus the oxidized pollutants, now converted to CO_2 and H_2O, are discharged to the atmosphere.

In the determination of a heat-and-material balance, the major process value is the heat demand. To determine the heat loads listed in Table 6-3, the following equation was applied:

$$Q = v \times m \times Cp \times \Delta t \times 60 \qquad (6\text{-}3)$$

where

Q = heat required, Btu/hr
v = gas flow, scfm
m = gas density, lb/ft^3
Cp = specific heat, Btu/lb-°F

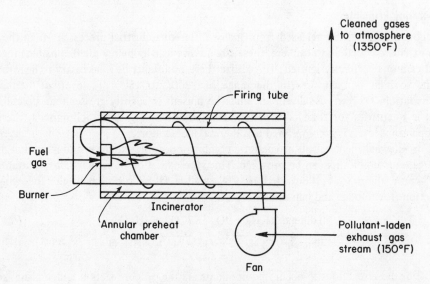

Figure 6-3. Thermal incinerator process diagram.

Δt = temperature rise, °F

60 = constant, min/hr

For the process conditions defined in Table 6-1,

$$Q = 10,000 \times 0.076 \times 0.24 \, (1350 - 150) \times 60 = 13 \times 10^6 \text{ Btu/hr}$$

To obtain this heat release, a certain amount of fuel gas and combustion air are required. At the elevated discharge temperature of 1350° F, the available heat of combustion for natural gas, which is the fuel source, would be approximately 700 Btu/ft^3. Therefore, the fuel requirement would be

$$13 \times 10^6 \text{ Btu/hr} \times \frac{1 \text{ ft}^3}{700 \text{ Btu}} \times \frac{1 \text{ hr}}{60 \text{ min}} = 300 \text{ scfm} \qquad (6\text{-}4)$$

For the combustion of natural gas, 2 mols of oxygen are required per mol of gas, approximately equivalent to 10 volumes of air per volume of fuel gas. Therefore, the combustion air required is

$$\text{Air} = 300 \text{ scfm} \times 10 = 3000 \text{ scfm} \qquad (6\text{-}5)$$

The total gas flow to be accommodated in the firing tube is (10,000 + 300 + 3,000), or 13,300 scfm. The volumetric equivalent of the solvents loading of 60 gph is about 30 scfm, and therefore can be considered negligible. Combustion air for the conversion of the solvents to CO_2 and H_2O in accordance with reactions 6-1 and 6-2 is available in the waste gas stream. Some fuel credit can be allowed for the oxidation of the solvents in the firing chamber. For this particular problem, the equivalent combustion temperature rise would be about 250° F. Therefore, the temperature of the mixed gases at the inlet end of the firing chamber could be maintained at (1350 − 250), or 1100° F. This heat gain was not considered in the fuel values determined in the sample problem.

As illustrated in Table 6-3, considerable savings in fuel consumption are achieved by the addition of a heat exchanger to the thermal incinerator. This system is shown in Fig. 6-4. The hot gases leaving the firing tube at 1350° F are passed countercurrently, in the heat exchanger, to the incoming waste gas stream at 150° F. The temperature of the waste gas stream with its solvents load is increased to 950° F, while the total combustion gases undergo a temperature drop of 1350° F to 700° F. The fuel savings is directly proportional to the temperature rise sustained by the waste gas stream in the incinerator. Thus, without the heat exchanger, a temperature increase of 150° F to 1350° F was necessary, whereas with the exchanger, the rise was reduced to the difference of 1350° F and 950° F. In this way, consumption was reduced by a factor of 3.

The increase of chemical reaction rate with rising temperatures can best be expressed by a logarithmic plot derived from the empirical Arrhenius equation. Figure 6-5 shows a series of such plots, relating values of k, the reaction rate constant, with the temperature expression $10^5/T°R$ for a number of common solvents. Thus, for toluene, the reaction rate increases from 3.3 sec^{-1} to 29.0 sec^{-1} as the temperature is increased from 1200° F to 1350° F. From Eq. 6-1, the required retention period for

233

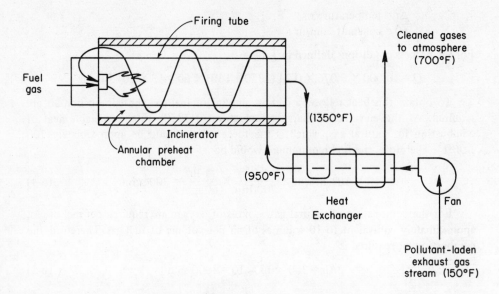

Figure 6-4. Thermal incinerator with heat exchanger.

toluene at a combustion temperature of 1350° F and a conversion of 95% would be calculated as follows:

$$\frac{100 - C}{100} = e^{-kt} \tag{6-6}$$

where

C = reaction conversion, %
k = reaction rate constant, sec^{-1}
t = retention period, sec
e = constant, 2.718

Therefore,

$$\frac{100 - 95}{100} = e^{-29t}$$

$$\frac{1}{1.301} = \frac{1}{29t \times \log e} \ ;$$

$$t = 0.104 \ sec$$

At the lower temperature of 1200° F, the required time would be inversely proportional to the reaction rate, so that (29.0/3.3 × 0.104), or 0.92 sec, would be required for the reaction to proceed to 95% of completion.

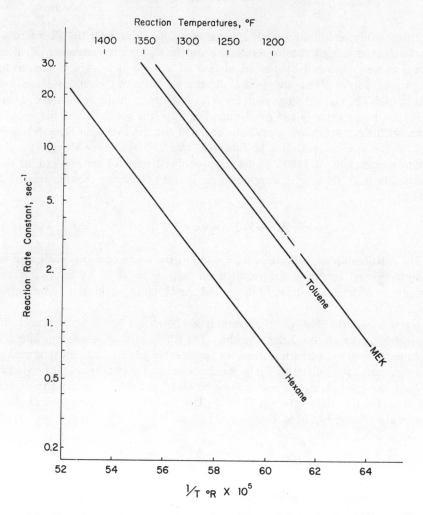

Figure 6-5. Reaction rate constants for thermal incineration of various solvents. (From Tomany, Ref. 6.)

Design Parameters

The major parameters that dictate the design of thermal incineration equipment are conversion efficiency, gas flow rate, gas temperature, and pollutant type and concentration.

Conversion Efficiency. Reference to Eq. 6-6 demonstrates the relationship of conversion efficiency to the pollutant reactivity and retention period. Reactivity values for various hydrocarbon pollutants were shown as a function of temperature in Fig. 6-5. Thus, for any desired conversion of a pollutant to CO_2 and H_2O, the incineration temperature determines the reactivity rate and Eq. 6-6 predicts the required retention period, or dwell time.

Theoretically, an infinite retention period would be required for 100% conversion. Fortunately, the current control regulations can be equated to reasonable retention periods, as was illustrated by the calculation of a dwell time at 95% conversion efficiency by Eq. 6-6. Thus, the physical dimensions for most incinerator designs can be made viable. Equipment dimensions are actually those of the firing chamber and are decided by the gas velocity and dwell time. For a specific gas flow rate, the velocity determines the cross-sectional area of the chamber and the dwell time fixes its length. Thus, for the problem described in Table 6-1, at a 95% conversion efficiency and a reaction temperature of 1350° F, the required dwell time for toluene and air was calculated to be 0.104 sec. At an assumed gas velocity of 50 fps, the chamber length would be

$$50 \, \frac{ft}{sec} \times 0.104 \, sec = 5.2 \, ft \qquad (6\text{-}7)$$

The relationship of conversion efficiency to the firing chamber length can be demonstrated by increasing the required conversion to 98%. At this value, the necessary dwell time would be 0.135 sec and the chamber would have to be lengthened to 6.8 ft.

Gas Flow Rate. The gas flow rate at a chosen gas velocity determines the cross-sectional area of the firing chamber. The total gas flow rate, comprising the pollutants, process exhaust stream, and the fuel combustion products, must be used as the design basis to calculate the firing chamber dimensions. For the sample previously discussed, the total gas flow was calculated to be 13,300 scfm, equivalent to 46,600 acfm at the reaction temperature of 1350° F. Therefore, the cross-sectional area of the chamber at a velocity of 50 fps would be

$$46,600 \, \frac{ft^3}{min} \times \frac{1 \, sec}{50 \, ft} \times \frac{1 \, min}{60 \, sec} = 15.5 \, ft^2 \qquad (6\text{-}8)$$

The diameter of the chamber would be 54 in., so that the inside dimensions of the incinerator firing chamber would be 4'-6'' diam × 5'-3'' length.

Gas Temperature. The gas temperature determines the required heat input into the system as well as controls the reaction rate and equipment size. Because of the relationship between the exhaust gas temperature and the fuel requirements, the application of both thermal incineration and catalytic combustion equipment favors processing high-temperature gas streams. The increasing heat input requirements are illustrated as being a function of the exhaust temperature in Table 6-3.

The influence of temperature on pollutant reactivity has been previously discussed. In calculations based on Eq. 6-6, it was shown that for a toluene conversion of 95%, reaction retention periods of 0.104 sec and 0.92 sec would be required at 1350° F and 1200° F, respectively. Thus, for a reaction temperature increase of 150° F, the combustion chamber length could be reduced by a factor of approximately 9. This reduction in equipment capital costs must be traded off against higher fuel costs, more expensive high-temperature-resistant materials, and increased equipment maintenance.

Pollutant Type and Concentration. Both the pollutant type and concentration in the exhaust gas are important variables. Most industrial exhaust gases contain mixtures of pollutants which vary considerably from one process to the next. In considering regulations concerned with odor removal, the problem becomes more complex than merely reducing the total pollutant concentration to some arbitrary value. There is considerable variation in the odor threshold for different gases and vapors. Thus, the detectable odor level for ammonia is 53 ppm, whereas certain aldehydes require discharge concentrations below 0.25 ppm to escape detection. Therefore, a complete definition of all gaseous, liquid, and solid pollutants in the exhaust gas stream must be available. This information is sometimes difficult to ascertain, and the overall problem may be complicated by the possibility that incomplete combustion in the emissions control equipment may convert relatively odorless pollutants to those having a highly objectionable odor. The problem of meeting regulations demanding odor-level compliance must often be solved by the use of pilot or laboratory control tests.

Thermal incineration can be used to promote both oxidation and reduction reactions. For the reduction of nitrogen oxides to nitrogen and water vapor, thermal reduction has been successfully applied for an exhaust gas stream containing a high concentration of nitrogen oxides. For lower concentrations of this pollutant type, in the range of 2000 to 4000 ppm, catalytic combustion equipment has been used.

When the pollutants are combustible, the concentration of the various components in the exhaust gas stream becomes critical. Lower explosive limits (LEL) for some of the common industrial solvents are shown in Table 6-4. As mentioned previously, a concentration of 25% of these LEL values is specified for many industrial operations, to reduce the hazard of explosions.

Mechanical Design

A thermal incinerator comprises a combustion chamber, gas burner, and burner controls. A heat exchanger and exhaust fan can be furnished as optional equipment.

Table 6-4. Lower Explosive Limits for Some Industrial Solvents

Solvent	Lower Explosive Limit (LEL) vol % in air at 20° C
Acetone	2.55
Methylethyl ketone (MEK)	1.81
Isopropyl alcohol	2.0
Benzene	1.4
Toluene	1.27
Ethyl acetate	2.25
Isopropyl acetate	2.0
Diethyl ether	1.85
Methyl alcohol	7.2
Ethyl alcohol	4.3
Hexane	1.2

Source: Handbook of Industrial Loss Prevention, Ref. 7.

The combustion chamber must provide complete mixing of the exhaust gas stream with the fuel combustion gases and achieve conversion of the pollutants to harmless end products. The combustion chamber is usually cylindrical and may be fabricated of firebrick or castable refractory with a mild-steel shell. An alternate fabrication material is lightweight stainless steel sheet with external structural steel supports. In one such design, an annular space is provided between the combustion or firing chamber and outside shell, as shown in Fig. 6-3. Heat is thus transferred to the exhaust gas stream as it flows countercurrent to the high-temperature combustion gases in the firing chamber. Besides accomplishing heat transfer, the exhaust gas stream acts as an insulator between the combustion chamber and the outside shell. Thus, minimal insulation can be applied to the outside shell. A mixing chamber is provided at the inlet end of the combustion chamber. Its function is to cause intimate mixing of the products of combustion from the gas burner with the exhaust gas stream and to impart a rotary motion to the gas mixture to ensure turbulence in the combustion chamber.

Various gas burner designs can be used with thermal incineration equipment. Atmospheric mixing, pressure mixing, and multijet burners are some of the types available for this service. One burner arrangement with a vertical refractory-lined thermal incinerator is shown in Fig. 6-6. In this design, multiple gas burners are located around the circumference of the combustion chamber. The combustion gases from the burners enter the chamber tangentially at the same elevation as the waste gas stream. Complete mixing of both gas streams is accomplished by the circular motion of the

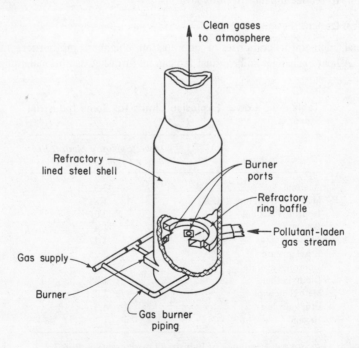

Figure 6-6. Direct-fired thermal incinerator.

gases as they enter the chamber. This turbulent action continues as the mixed gases flow upward vertically through the combustion chamber, ensuring complete incineration of the pollutants. In some incinerator designs the combustion air supplied to the burners is furnished by the waste gas stream. This burner design is used in the horizontally arranged incinerator shown in Fig. 6-3. The system eliminates the need for a large combustion air blower, simplifies controls, and achieves the most economic auxiliary fuel utilization.

The minimum requirement in combustion controls is a flame-failure device, which shuts off the gas supply on the loss of flame. In addition to this precautionary device are a vast number of partially automated control assemblies that can provide dependable operation. A recommended piping diagram for the almost completely automatic control of a gas-fired incinerator is shown in Fig. 6-7. In this system, the exhaust gases are brought to the incinerator and maintained at the proper temperature by controlling the burner firing rate. The burner controls must permit considerable flexibility in the automatic adjustment of the burner input to accommodate variations in the temperature, flow, and concentration of the pollutants. The function of the various control devices is explained in the following operational sequence (refer to Fig. 6-7):

1. The system exhaust fan is started, thereby closing the air flow switch (AFS).
2. The auxiliary combustion air blower is started and the air pressure switch (APS) closes.
3. The manual gas cock (MGC) is opened; the gas high-pressure switch (GHPS) and gas low-pressure switch (GLPS) are both closed. The gas pressure is regulated by (GR) and indicated by (PI).
4. The ignition push button (IPB) is depressed, thereby opening the pilot valve (PV), the main gas valve (MGV), and the block valve (BV).
5. The main safety valve (MSV) is manually opened, which causes the block valve (BV) to open and the vent valve (VV) to close.

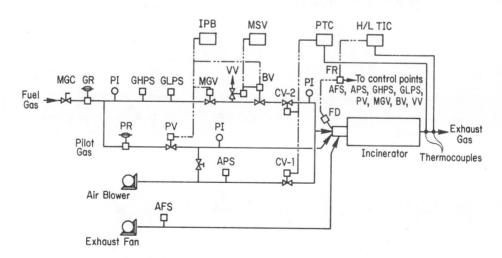

Figure 6-7. Thermal incinerator control instrumentation.

239

6. The main burner is now "on," with its operation controlled by the indicating proportioning temperature controller (PTC) through control valves (CV-1) and (CV-2). The burner flame is supervised by the flame detector (FD) and flame relay (FR).

7. The temperature limits in the incinerator are controlled by the high/low temperature indicating controller (TIC).

In this control system, each of the functional switches is so connected in a series circuit that if any single function is interrupted, the total system is shut down. For example, if the air blower fails, the discontinued flow of combustion air will be sensed by the air pressure switch, the switch will open and the exhaust gas, air, and fuel gas streams will shut down and/or be diverted. All control elements must meet insurance and local code requirements.

Heat exchangers are usually furnished as optional equipment, but in many industrial designs they are integral units inside the combustion chamber shell. The exchanger is a tube-and-shell type, with one tube sheet fixed and the other free to float to allow for thermal expansion. The tubes are of welded steel or stainless steel, with the latter material usually recommended for temperatures in excess of 1200° F. Some heat exchanger configurations are shown in Fig. 6-8. With the system illustrated in scheme C, the overall process operating costs can be reduced almost to the cost of operation without incineration control equipment. Scheme A was applied to the problem summarized in Table 6-3.

A typical thermal incinerator with integral heat exchanger is shown in Fig. 6-9. The combustion instrumentation and manual controls are contained in the cabinet located at the firing end of the incinerator. The capacity of this thermal incinerator is 32,000 scfm and its approximate overall dimensions are 12 ft diam by 30 ft long.

The exhaust fan is necessary if the process gas stream has insufficient pressure to force it through the incinerator. The fan is a single-blade centrifugal type. Because the pressure drop through the incinerator is quite low, frequently less than 1.0 in. w.g., a

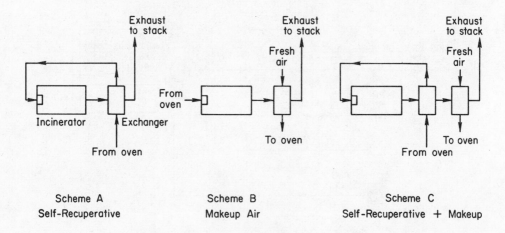

Figure 6-8.   Incinerator heat exchanger configurations.

Figure 6-9. Thermal incinerator with heat exchanger arrangement. (Courtesy of Air Correction Division.)

fan may be provided to replace the process exhaust fan and perform both functions. The fan is usually located between the process and the incinerator. With heat exchange, it can still be placed in this position or located between the heat exchanger and discharge stack.

Because of the almost infinite variation in gaseous emission problems, thermal incinerators are usually custom designed. However, the recurrence of specific and typical problems has encouraged several manufacturers to fix certain design parameters and offer a series of standard incinerator designs based on these fixed values. As discussed previously, the three major parameters affecting incinerator physical design are gas flow, temperature, and dwell time. Some of the more knowledgeable and experienced emissions regulatory bodies have concurred with manufacturers of control equipment in specifying the temperature and dwell time for certain industrial process emissions. For example, it has been agreed that the design of incineration equipment applied for the control of varnish-cooking emissions will be based on a temperature of 1200° F and a retention period of 0.3 sec. For solvents emissions from lithographing ovens, a temperature and dwell time of 1400° F and 0.6 sec are specified, respectively.

Capacities, dimensions, and weights for such a series of standard incinerators are shown in Fig. 6-10. These incinerators will provide a retention period of 0.6 sec and can be operated in the temperature range of 1200° F to 1500° F. The combustion chamber, annular preheat space, burner, support steel, instrumentation, and control cabinet comprise the total incineration system. The cylindrical combustion chamber is fabricated of lightweight stainless steel and the outer shell is of mild steel. The burner is designed to utilize the oxygen in the exhaust gas stream. The burner equipment, control valves, pressure switches, safety cutoffs, etc., are completely prepiped and wired. The purchase costs for these thermal incinerators are approximately as shown in Fig. 6-1.

Installation Features

Thermal incineration equipment, similar to that shown in Fig. 6-9, is usually shipped preassembled. The control cabinet must be connected at the plant site to utility connections and the individual equipment items comprising the system. The connecting ductwork and insulation are field-installed.

The horizontal thermal incinerators shown in Fig. 6-10 are relatively light and therefore can be located on the roof of the building housing the process. In this arrangement, the exhaust gas ductwork is minimized and the discharge stack has the advantage of the added elevation. Refractory-lined incinerators are usually installed at ground level because of their mass. Refractory type incinerators are usually positioned vertically so that the incinerated pollutants can be discharged upward through a stub stack to atmosphere. Whether the incinerator is installed at ground or roof level, the control instrumentation panel should be located inside the plant, close to the process control center. Actually, in gaseous emissions control the incineration equipment must be considered as a continuation of the process train. In some installations, the combustion air and fuel source, instrumentation, and heat exchangers may be common to both "ends" of the complete process, so a total systems approach becomes mandatory.

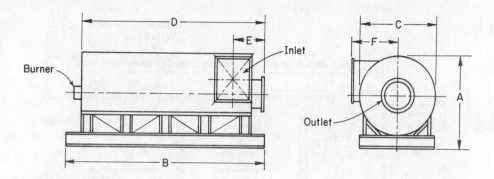

Figure 6—10   Thermal Incinerator Capacities and Dimensions
(Courtesy of Air Correction Division)

Capacity, scfm	Dimensions, ft-in.								Weight lb
	A	B	C	D	E	F	Inlet	Outlet	
2,000	4-5	21-7	3-9	18-7	2-8	2-2	3-9 × 2-0	1-3	5,100
4,000	5-5	22-4	4-9	19-4	2-8	2-8	3-9 × 2-8	1-9	6,700
6,000	6-3	22-10	5-7	19-10	2-8	3-1	3-9 × 3-2	2-2	7,900
8,000	6-11	22-10	6-3	19-10	2-8	3-5	3-9× 3-7	2-6	8,800
10,000	7-6	23-2	6-10	20-2	2-8	3-8	3-9 × 4-0	2-10	9,700
12,000	8-0	23-4	7-4	20-4	2-8	3-11	3-9 × 4-4	3-1	10,600
14,000	8-6	23-7	7-10	20-7	2-8	4-2	3-9 × 4-7	3-4	11,400

A thermal incineration system for the control of gaseous emissions from a varnish-cooking operation is shown in Fig. 6-11. Preheating of the exhaust gas from the cooking kettles is accomplished in the collecting duct, which directs the pollutant-laden gas stream to the incinerator. The installation in Fig. 6-11 was custom-designed with sufficient flexibility to handle simultaneously the gaseous effluent from two transportable cooking kettles. The major portion of the combustion air required for fuel combustion is furnished by the kettle off-gases. The difference is introduced into the system as secondary air at the burner. For operation of a single kettle or during the startup period, heat and material balance conditions are maintained in the incinerator by recirculation of the total combustion products through the recycling damper. Some dilution air is introduced into the system to protect the manifold ductwork against temperatures in excess of 1500° F. The ductwork is well insulated and made as short as possible to minimize the precipitation of condensates. Allowances are made for ductwork and equipment expansion during cold-to-hot operating changeovers.

With both thermal and catalytic incineration equipment, it is necessary to establish a maintenance schedule, particularly for the control instrumentation. Therefore, complete accessibility of all elements comprising the system should be allowed.

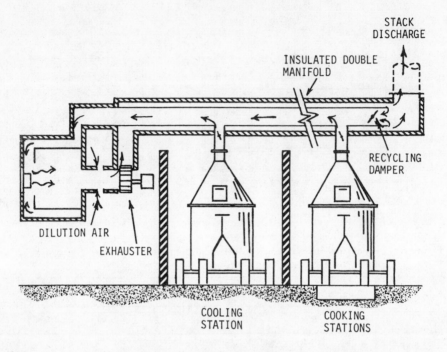

Figure 6-11. Thermal incinerator for varnish-cooking kettles. (From Brewer, Ref. 3.)

Because a combustion process is involved, the installation and maintenance of incineration equipment will be influenced by the strict requirements of the Factory Insurance Association or Factory Mutual.

Applications

Thermal incineration equipment can be applied for the control of gaseous emissions from a large variety of industrial processes. Emissions of both gas and vapor types of pollutants and some particulates can be reduced to a sufficient degree to permit compliance with air pollution control regulations. Pollutant concentrations, total emissions, opacity level, and odor nuisances can be controlled by incineration means.

In the foregoing discussion, thermal incineration was described as that process wherein an exhaust gas stream, carrying various pollutants, is mixed with and exposed to the flame and flue gases resulting from the combustion of the fuel gas. There is an alternative to this system, known as "direct flame" incineration, in which the contaminated gases are oxidized (or reduced) directly in the flame along with the fuel gas. In thermal incineration, the contaminated gases have a combustible content below the LEL and the oxidation of the pollutants is accomplished by exposure to a fuel/air-fired flame and by a residence of sufficient duration in the firing chamber. In direct flame incineration, the contaminated gases contain sufficient combustibles to develop a flame in the presence of a proper amount of combustion air. Some auxiliary fuel may be necessary to sustain the combustion.

244

Direct flame incineration is not too often applied to the control of pollutant emissions because of the unusual condition that the exhaust gas stream must contain sufficient contaminants so as to be self-combustible. One example of a direct flame incineration application is that of the "open flare" oxidation of hydrocarbon waste gases as practiced by the petroleum refineries. One other interesting application of direct flame incineration involves the control of nitrogen oxides from a nitration process in which the mixed oxides (NO, NO_2) concentration is almost 100%. By direct flame combustion with methane, the nitrogen oxides are reduced to nitrogen, while the methane, which acts as a reducing agent, is oxidized to carbon dioxide and water vapor.

Thermal incinerators are often referred to as afterburners because they are so often located downstream from a primary combustion process. Their major disadvantage is that of operating costs associated with high fuel consumption rates. However, as shown in Table 6-3, this operating burden can be reduced considerably by the use of heat exchangers. First costs for thermal incineration equipment are much less than those for the catalytic combustion type. Although the use of catalysts ensures lower fuel costs, sustained performance sometimes poses a problem, and therefore catalytic incineration is not fully accepted as an emission control method by some APCD groups. Even though it is more costly in operation, thermal incineration offers the most dependable solution.

Some of the industrial applications for thermal incineration control equipment are resin manufacturing, paint and varnish cooking, metal decorating, coil and strip coating, paint application and baking, carbon baking ovens, tar and asphalt blowing

Table 6-5. Thermal Incinerator Operating Data

Case No.:	C-725	C-566	C-318
Process equipment:	Three varnish cook kettles	Five meat smokehouses	Paint bake oven
Burner type:	One inspirator	One multijet	Three nozzle mixing
Incinerator combustion temp., °F	1200	850	1520
Inlet gas flow, scfm	200	1600	1400
Discharge gas flow, scfm	920	2000	1800
Inlet particulate loading, lb/hr	5.70	1.66	0.40
Discharge particulate loading, lb/hr	0.20	0.56	0.09
Removal efficiency, %	96.5	66.3	77.5
Inlet organic acids, lb/hr	0.24	1.88	—
Discharge organic acids, lb/hr	0.00	0.27	—
Inlet aldehydes, lb/hr	0.29	0.49	0.19
Discharge aldehydes, lb/hr	0.02	0.22	0.03

Source: J.A. Danielson, Ref. 8.

and coating, meat smokehouses, fat-rendering plants, and various chemical processes. Of these various emission sources, organic solvents, either as a vapor or in aerosol particulate form, constitute the most significant portion of the total air pollution problem. In many of these applications, the removal of solid carbon particles, or soot, is also required to satisfy the emissions standards.

Descriptions of three typical industrial applications [8] of thermal incinerators are given below. Operating and emissions data for these three cases are summarized in Table 6-5.

Case C-725, Varnish-Cooking Kettles. The air pollution problem involved the reduction of total particulate emissions, both liquid and solid, and the elimination of opacity and odor to satisfy current nuisance regulations. The thermal incinerator consisted of a horizontal combustion chamber fired by an inspirator type gas burner. The contaminated gas stream entered the incinerator chamber tangentially, adjacent to the burner. The exhaust fan was located at the incinerator outlet.

Emission testing indicated that the incinerator achieved high removal efficiencies for both particulates and organic acids.

Case C-566, Meat Smokehouses. The major problem in this application was the emissions of solid particulates (soot), characterized by excessive opacity. The smoke and gases from the five smokehouses were vented directly to the thermal incinerator, which was a vertical, cylindrical type. The exhaust gas stream entered the incinerator axially at the base where it was mixed with burner flue gases entering tangentially at this same location. All the oxygen for combustion of the fuel gas was obtained from the contaminated gas stream.

The incinerator performed at a reasonably efficient level at a relatively low temperature. No visible emissions were observed during the test.

Case C-318, Paint Bake Oven. The operation involved spray painting of drums with epon-phenolic and oleoresinous coatings followed by baking at 420° F in a conveyorized, gas-fired, recirculating type bake oven. The major pollution problem was the emissions of particulates beyond the allowable regulation limits.

A portion of the hot recirculated gases containing solvents and particulates was directed to the incinerator. The incinerator consisted of a vertical cylinder with three nozzle-mixing gas burners located around the circumference. Both the contaminated gases and burner flue gases entered the base of the incinerator tangentially.

The incinerator was operated at temperatures of 1410° F and 1520° F. Data for the 1520° F operating level is shown in Table 6-5. Actually, satisfactory emissions control was obtained at both temperatures.

Sample Design Calculations

To illustrate the design of a typical thermal incinerator, a sample problem will be stated and solved.

Problem Statement

A metal-decorating oven discharges a contaminated gas stream to the atmosphere at the rate of 6000 scfm. The major portion of the pollutants are solvent vapors.

These, plus some particulates, must be converted at an efficiency of 98% to satisfy local regulations with regard to odor and opacity requirements.

The solvent system comprises hexane, representing the aliphatic hydrocarbons; toluene, for the aromatics; and methylethyl ketone (MEK), for the ketone group. The alcohol component is to be considered negligible. The particulates are mostly carbon. Operating data for the operation are as follows:

Process	Metal decorating
Equipment	Baking oven
Exhaust volume, scfm	6000
Exhaust temperature, °F	400
Solvents loading, lb/min	1.4
Solvents composition, wt %	
Hexane	35
Toluene	45
MEK	20
Operating period, 24-hr days/yr	360
Fuel type	Natural gas
Fuel costs, $/1000 scf	0.60
Performance, conversion of organics, %	98

A basic thermal incinerator is to be designed to process this gas stream and satisfy the performance requirements.

Problem Solution

1. *Explosive Limits Check.* Contaminated air per gallon of solvents is calculated as

$$\frac{6000 \text{ scfm}}{1.4 \text{ lb/min} \times 1 \text{ gal/7 lb}} = 30,000 \text{ ft}^3/\text{gal}$$

The individual solvent having the most critical LEL value is hexane; see Table 6-4. Therefore, the allowable air/solvent ratio, at a value of 1.2 for the hexane LEL, would be

$$\frac{(100 - 1.2) \text{ ft}^3 \text{ air}}{1.2 \text{ ft}^3 \text{ solvent}} \times \frac{1 \text{ ft}^3 \text{ solvent}}{(0.075 \times 3.0) \text{ lb solvent}} \times \frac{7 \text{ lb solvent}}{1 \text{ gal solvent}} = 2560 \text{ ft}^3/\text{gal at } 70° \text{ F}$$

At 400° F, the allowable ratio is

$$2560 \times \frac{(460 + 400)}{(460 + 70)} = 4150 \text{ ft}^3 \text{ air/gal solvent}$$

$$\text{Safety factor} = \frac{30,000}{4,150} = 7.2$$

Therefore, the exhaust volume is satisfactory, being well below the required 25% LEL value.

247

2. *Particulates Loading.* The total evaporated solvents load exhausted from the oven is 1.4 lb/min. At an oven temperature of 400° F, it is estimated that 5% of the solvents is emitted as particulates, mostly carbon.

Grain loading = 1.4 lb/min × 5/100 × 7000 grains/lb × 1 min/6000 scf
= 0.08 grain/scf

3. *Heat Duty.* An incineration temperature of 1450° F will be chosen for this duty. Using Eq. 6-3,

$$Q = 6000 \times 0.076 \times 0.24 \, (1450 - 400) \, 60 = 6.9 \times 10^6 \text{ Btu/hr}$$

Radiation and convection losses from the thermal incinerator will be assumed at 10%; therefore,

Total duty = 7.6×10^6 Btu/hr

4. *System Gas Flows.* Because of the large air dilution in the contaminated exhaust gas stream, sufficient oxygen is present to furnish all the combustion air required for the incinerator fuel gas.

The gross heating value of natural gas is 1000 Btu/ft^3. The net heat value at 1450° F, with combustion air being furnished 100% by the contaminated gas stream, is 650 Btu/ft^3. Therefore, fuel gas required is

$$\text{Flow} = \frac{7.6 \times 10^6}{650 \times 60} = 195 \text{ scfm}$$

The products of combustion from natural gas are in an approximate volumetric ratio of 11:1. Therefore, the total gas flow at 1450° F, representing the products of combustion from the fuel gas, is

$$195 \times 11.0 \times \left(\frac{1450 + 460}{60 + 460}\right) = 7850 \text{ acfm}$$

The contaminated exhaust gas flow rate is 6000 scfm, equivalent to 22,000 acfm at 1450° F. Of this flow, there is required

$$195 \times 10.0 \times \left(\frac{1450 + 460}{60 + 460}\right) = 7170 \text{ acfm}$$

for the combustion of 195 scfm of natural gas. Therefore, the net volume of contaminated gases being treated in the incinerator is (22,000 − 7170), or 14,830 acfm.

The total volume of gases to be processed in the incinerator, calculated at 1450° F, is = 7850 + 14,830 = 22,680 acfm.

A schematic flow diagram indicating gas flow distribution and temperatures for the system is shown in Fig. 6-12.

5. *Retention Period.* The retention period in the incinerator must be based on the least reactive of the solvents. According to Fig. 6-5, normal hexane would be the least reactive, having a reaction rate constant of about 27.0 sec^{-1} at 1450° F. Actually, based on heats of reaction for the solvents mixture, a temperature rise of about 200° F

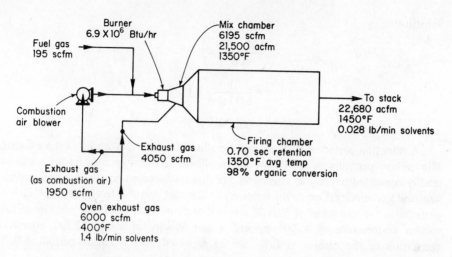

Figure 6-12. Thermal incinerator design flowsheet.

would be expected from their combustion, which would influence the reaction rate constant.

The most rigorous approach to determining the retention period in the combustion chamber would involve a stepwise integration procedure. A combustion chamber inlet temperature of 1250° F would be chosen, on the assumption that 200° F temperature rise would occur across the combustion chamber. Then, based on the inlet temperature of 1250° F, a retention period increment would be chosen, the expected temperature rise estimated, an average temperature for that time increment calculated, a *k* value established for each component, the conversion for each component computed, and the actual temperature rise determined. The actual temperature rise would then be compared with the estimated value, and if sufficiently close agreement was obtained, an added increment would be chosen and the computation procedure repeated. This integration process would be continued until the desired conversion value had been exceeded.

The accuracy required in the solution of this problem does not justify such a strict approach. Instead, the reaction rate constant *k* for hexane will be determined from Fig. 6-5 for the average combustion chamber temperature of (1250 + 1450)/2, or 1350° F, and this value together with the required conversion efficiency *C* will be substituted in Eq. 6-6.

$$\frac{100 - C}{100} = e^{-kt}$$

where

$k = 6.0 \text{ sec}^{-1}$
$C = 98.0$
t = retention period, sec

249

Substituting,

$$\frac{100 - 98}{100} = e^{-6.0}$$

$$t = \frac{1}{6.0 \log e} \times 1.70$$

$$= 0.65 \text{ sec}$$

A retention period of 0.70 sec will be allowed for the conversion of the solvents. The carbon particulates must now be considered. Carbon or soot particles can be readily converted to CO_2 at a reaction rate that is dependent on temperature, particle size, and gas velocity. Due to the tendency of the soot particles to agglomerate, carbon particles in the size range of 5 to 20 μ would be expected. For carbon particles of this size, a temperature of 1350° F, and a gas velocity of about 30 fps, complete conversion of the carbon to CO_2 can be expected at a retention period of 0.70 sec [6].

6. *Equipment Sizing*. Based on a velocity through the combustion chamber of 30 fps, the cross-sectional area of the chamber would be

$$\frac{22,680}{30 \times 60} = 12.6 \text{ ft}^2$$

equivalent to a diameter of 4'-0''. At a retention period of 0.70 sec and a velocity of 30 fps, the combustion chamber length would be 21'-0''.

7. *Temperature Corrections*. Because of the strong influence of the operating temperature on the reaction rate (27.0 sec^{-1} at 1450° F and 6.0 sec^{-1} at 1350° F), the temperature rise caused by combustion of the solvents was considered in the determination of the retention period.

There are two other design factors where temperature corrections can be applied: the combustion chamber diameter and the heat duty. The former is negligible, for instead of a diameter of 4'-0'' at the gas temperature value of 1450° F, this dimension could be decreased to 3'-10'' for the corrected average temperature of 1350° F. The heat duty value should be corrected because it represents the major operating cost. Therefore, the revised heat duty is

$$7.6 \times 10^6 \times \left(\frac{1350 - 400}{1450 - 400}\right) = 6.9 \times 10^6 \text{ Btu/hr}$$

8. *Summary*. Pertinent design, operating and cost data for the thermal incinerators designed in this problem are summarized in Table 6-6.

6.6. EQUIPMENT CHARACTERISTICS—CATALYTIC COMBUSTION

Operating Principles

Catalytic combustion equipment has much in common with thermal incinerators. Both types control gaseous emissions by chemical conversion of the pollutants to harmless

Table 6-6. Thermal Incinerator Design Data

Application	Metal decorating oven
Exhaust volume, scfm	6000
Exhaust temperature, °F	400
Solvents loading, lb/min	1.4
Solvents composition, wt %	Hexane, 35; toluene, 45; MEK, 20
Particulate loading, grain/scf	0.08
Operating period, 24-hr days/yr	360
Fuel type	Natural gas
Fuel costs, $/$10^6$ Btu	0.60
Performance, % conversion	98.0 for organics
Incinerator type	Direct thermal without heat exchange
Dimensions, ft diam × ft long	4.0 × 21.0
Equipment furnished	Combustion chamber, burner, instrumentation, and support steel
Equipment purchase cost, $	24,000
Installation cost, $	12,000
Annual depreciation costs, 10^3 $	2.4
Heat input, 10^6 Btu/hr	6.9
Fuel costs, 10^3 $/yr	35.8
Total annual operating costs, 10^3 $	38.2

end products. By the addition of catalyst elements, the retention period and operating temperatures for catalytic equipment can be reduced considerably below that required for thermal incineration. Both benefits are derived from the fact that catalysts increase the rate of chemical reaction without being consumed in the reaction.

The greatest single reason for considering catalytic incineration is that of fuel economy, as illustrated in Table 6-3. This advantage reflects the operating temperature range of 500 to 1000° F for catalytic conversion as compared with the values of 1200 to 1500° F required for thermal incineration. Although there has been a history of operating difficulties with catalysts in certain application areas, the considerable reduction in operating costs favoring their use still makes catalytic combustion a prime candidate for the control of air pollution problems. Usually, thermal incineration is more reliable in its performance, but is considerably more expensive to operate than catalytic combustion.

There are two catalysis mechanisms involved in industrial practice. One employs acidic catalysts that rely on the transfer of an electron or proton from the reacting material to the catalyst, or vice versa. Alumina is such a catalyst and has the capability of contributing a hydrogen ion to other chemical compounds in the vicinity. Acid type catalysts like activated alumina, silicates, phosphates, sulfates, and anhydrous chlorides are effective in promoting various industrial reactions such as hydration, polymerization, and esterification.

The second mechanism is an adsorption process in which solid catalysts are

251

generally used. In gas purification processes utilizing such catalysts, contact between the catalyst and gaseous reactants is most important. This contact is established by adsorption of the gaseous reactants on the catalyst surface, causing an increase in reactant concentration. The complete mechanism by which gaseous pollutants are converted to harmless end products may be described as a stepwise process in which:

1. The gaseous pollutants are transferred from the carrier gas stream to the catalyst surface.
2. The pollutants diffuse into the catalyst pores.
3. Surface reactions of adsorbed pollutants take place to form end products.
4. End products diffuse out of the catalyst pores.
5. The gaseous end products are transferred from the catalyst surface to the carrier gas stream.

The catalysts used for gaseous pollutants control are usually metals or metal salts. These can be supported on inert carriers such as alumina or porcelain, or they can be used directly in the unsupported state. For oxidation reactions, platinum, palladium, nickel, cobalt, iron, silver, copper, tungsten chromium, molybdenum, and similar metals may be used.

Metals serve as adsorption catalysts when the surface layer is unstable or incomplete. Thus, within the body of a platinum metal crystal, all bonds are linked with and satisfied by other platinum bonds. However, at the surface of the crystal the unattached bonds must be satisfied by adsorption of atoms from the gas or liquid surrounding the catalyst. Metals such as aluminum and iron readily attract and hold oxygen atoms by this adsorption mechanism so that a strong oxide layer is formed at the surface, thereby disqualifying these metals as catalysts. Platinum and members of the platinum family function well as catalysts because they do not form strong attachments with oxygen, but do attract and adsorb other atoms on their surfaces.

The two most important factors influencing the effectiveness of a catalyst to accelerate a particular reaction are catalyst material and its surface area. Platinum and palladium are the most common metals used with various inert support materials. For example, one commercial catalyst consists of a combination of platinum and alumina deposited on porcelain carrier rods. Another catalyst system is made up of metals of the platinum-palladium group supported on a nickel alloy ribbon. This latter catalyst type is most commonly used for oxidation reactions. It has also been successfully applied for the control of nitrogen oxides by their reduction to nitrogen. A catalyst consisting of cobalt, copper, and chromium oxides impregnated in a ceramic support media has also been used. Although this catalyst is not as effective for oxidation reactions as the platinum-palladium type, it does have a considerable price advantage.

The same reactions that occur in thermal incineration, illustrated by Eqs. 6-1 and 6-2, can be accomplished in catalytic combustion equipment. The function of a catalyst is to reduce the activation energy required to cause these reactions to take place. The net result is to bring about these reactions at a reasonable rate and at temperatures lower than those that would be required without a catalyst. Thus when the reaction rates k, defined in Eq. 6-6, are plotted against temperature for hexane, toluene, and MEK in a catalytic system, the values shown in Fig. 6-13 are obtained. It can be seen that at any given temperature, the catalytic reaction rate constant is

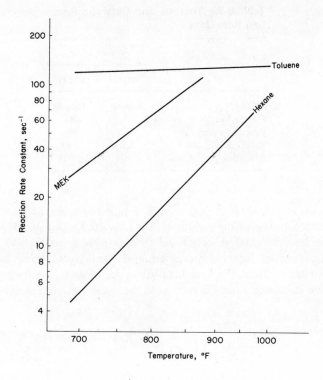

Figure 6-13. Reaction rate constants for catalytic combustion of various solvents. (From Tomany, Ref. 6.)

considerably greater than for the thermal process. Thus, for toluene, it was found earlier, by reference to Fig. 6-5, that at a temperature of 1200° F, the reaction rate for thermal incineration was 3.3 sec^{-1}. As indicated by Fig. 6-13, the catalytic reaction rate constant at 900° F would be 128 sec^{-1}. This latter value can be substituted directly in the reaction rate expressed by Eq. 6-6, so that for a reaction conversion of 95%, a retention period of 0.024 sec would be required in the catalyst bed, rather than 0.104 sec demanded in the combustion chamber of the thermal incinerator. A summary of incinerator design conditions for the two processes is shown in Table 6-7.

A schematic diagram of a catalytic combustion system is shown in Fig. 6-14. The temperatures indicated conform to the catalytic combustion conditions shown in Table 6-3. The process exhaust gas stream passes through a preheat chamber where the temperature is increased from 150 to 900° F by the introduction of gas-fired flue gases. Some conversion of the pollutants is achieved in the preheat chamber. The gaseous components pass through the fan where they are thoroughly mixed and then uniformly discharged across the catalyst elements. Final destruction of the gaseous pollutants is accomplished in the catalyst section. Actually, combustion of the solvent hydrocarbons in the catalyst bed is a highly exothermic process. Consequently, a

Table 6-7. Thermal and Catalytic Reaction Rate Data

	Process Type	
	Thermal	*Catalytic*
Pollutant	Toluene	Toluene
Conversion, %	95	95
Reaction temp., °F	1350	900
Reaction rate, sec^{-1}	3.3	128
Retention period, sec	0.104	0.024

temperature rise on the order of 200 to 600° F might be expected, depending on the nature and concentration of the pollutants. The converted pollutants plus the combustion flue gases leave the catalyst section and are discharged to atmosphere.

Heat and material balances are determined, similarly to those calculated for thermal incineration. Thus, the heat load values for catalytic combustion shown in Table 6-3 were calculated from Eq. 6-3. As in the case for thermal incineration, credit

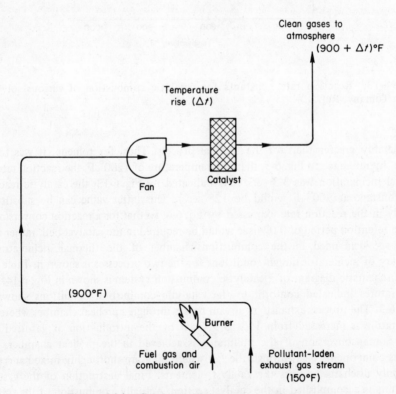

Figure 6-14. Catalytic incinerator process diagram.

was not allowed for the temperature rise caused by oxidation of the solvents in the catalyst section. The benefits in heat economy provided by the use of a heat exchanger were indicated in Table 6-3 for the catalytic combustion system. A flow sheet for such a system is shown in Fig. 6-15. The hot gases leave the catalyst section at a temperature of 900° F and flow countercurrently in the heat exchanger against the contaminated exhaust gas stream at 150° F; thereby the temperature of the cleaned gases being discharged into the atmosphere is reduced to a value of 550° F. The temperature of the contaminated gases is thus increased from 150 to 650° F, and as they leave the exchanger they are joined by the combustion gases, which further increase their temperature to 900° F. The combination gas stream passes through the fan where thorough mixing is achieved and the gases are uniformly discharged across the catalyst bed.

Design Parameters

The same major parameters affecting thermal incinerator design are applicable to catalytic combustion equipment. In addition, there is a need for carefully selecting those emission control applications that will provide the stringent process conditions demanded by catalytic combustion systems. The various parameters influencing catalytic combustion equipment design are discussed below.

Conversion Efficiency. The relationship defined by Eq. 6-6 also correlates catalytic conversion efficiency with the reactivity and retention period for a specific pollutant. Reactivity values, as a function of temperature, for the catalytic conversion of the more common industrial solvents is shown in Fig. 6-13. With these data and a

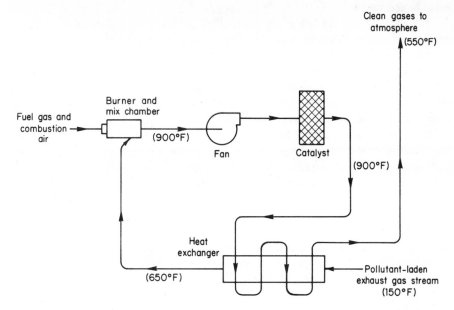

Figure 6-15. Catalytic incinerator with heat exchanger.

specified conversion efficiency, the required retention period in the catalyst bed for the oxidation of a specific solvent can be determined.

The retention period calculated from Eq. 6-6 is the actual dwell time occupied by the total gas flow, which comprises the contaminated gases plus the combustion gases in the catalyst bed. Gas velocities in the range of 10 to 20 fps are commonly used. Thus, for a specific dwell time and gas velocity, the catalyst bed depth can be computed. Unlike the thermal incinerator, the overall dimensions of catalytic combustion equipment are not wholly determined by the dimensions of the combustion section. By referring to Fig. 6-17 it can be seen that the major portion of the equipment is occupied by the preheater section and gas flow ducts.

Gas Flow Rate. The gas flow rate at a specified gas velocity determines the cross-sectional area of the catalyst bed. Therefore, knowing the total gas flow and retention period, and assuming a reasonable gas velocity, the dimensions of the catalyst bed can be determined. For example, consider the total gas flow of 13,300 scfm for the problem defined in Table 6-1. At a catalytic reaction temperature of 900° F, the volumetric flow of gas entering the catalyst bed would be 34,600 acfm. Assuming a gas velocity of 15 fps, the required cross-sectional area of the catalyst would be 38.5 ft^2. Industrial catalyst elements for this type of service could be made available in standard sizes measuring 30 in. wide by 30 in. long by 4.5 in. deep. Thus, by using a configuration comprising three elements wide by two elements long, a total cross-sectional area of $(2.5 \times 3.0 \times 6)$, or 45 ft^2 can be provided.

If the same 95% conversion efficiency for toluene is required, then the retention period of 0.024 sec, previously calculated, will be necessary. At a gas velocity of 15 fps and dwell time value of 0.024 sec, the required catalyst bed depth would be 15 ft/sec $\times$ 0.024 sec $\times$ 12 in./ft, or 4.3 in. Thus, the standard elements, which are 4.5 in. deep, would be acceptable.

The gas flow rate for any particular catalyst configuration influences the pressure drop across catalytic incinerators. Pressure drops less than 1 in. w.g. are normally experienced and at this level do not normally present a serious problem. In the catalyst configuration calculated above, it is estimated that the pressure drop across a single element, at the chosen design gas velocity, would be in the range of 0.2 to 0.4 in. w.g. The gas flow through the catalyst bed must be turbulent to promote intimate contact of the contaminants with the catalyst surface.

Gas Temperature. Similar to thermal incinerator design concepts, the gas temperature determines the required heat input for the system and controls the reaction rate. Heat balances can be determined from Eq. 6-3; the pollutant reactivity and retention period have been calculated for the sample problem in Table 6-1 from Fig. 6-13 and Eq. 6-6.

Because of the lower temperature levels at which catalytic incineration is accomplished, it is sometimes possible to eliminate preheating so that the contaminated gases can be directed through the catalyst bed without being accompanied by a preheat gas stream. Most importantly, the need for fuel may be eliminated, or at least very sharply reduced. For example, for a coatings oven operated at a high exhaust gas temperature of about 450° F, and with a solvents discharge rate of about 60 gph, the need for reheat fuel has been completely eliminated. In fact, in one particular design the overall fuel savings credited to the use of the more efficient catalytic combustion method has resulted in a very reasonable payout period for the incineration equipment [9].

Catalytic incineration is essentially a flameless combustion process that occurs at relatively low temperatures, usually in the range of 500 to 1000° F. There is considerable heat released in the catalyst, which is absorbed by the gases, thereby causing a temperature rise as they pass through the bed. Very approximately, for each gallon of solvent oxidized in the catalyst bed, the release of 120,000 Btu can be expected. For the sample problem (Table 6-1), 60 gph of solvents would therefore produce 7.2×10^6 Btu/hr. At a total gas flow rate through the catalyst bed of 13,300 scfm, the expected temperature rise would be 490° F. Theoretically, the gases could enter the catalyst at a temperature of 655° F and leave at 1145° F, thereby maintaining the average reaction temperature at the specified value of 900° F. Thus, it can be seen that the preheat load and fuel consumption is inversely proportional to the concentration of the combustible pollutants. In many applications involving very high solvents loadings, the preheat burner fuel flow is regulated as a function of the temperature of the gases being discharged from the catalyst section. With this arrangement, a constant incinerator discharge gas temperature and catalyst temperature can be maintained.

Different pollutant types require various operating temperatures for their catalytic conversion. Combustion temperatures for the more common solvents are shown in Fig. 6-13. For example, tall oil emissions from varnish-cook operations require a temperature of 650° F, while asphalt fumes are catalytically incinerated at 850° F. The relationship of conversion efficiency to temperature for a number of coatings operations is shown in Fig. 6-16. As previously pointed out, the conversion efficiency for catalytic incineration, although strongly dependent on temperature, is also influenced by turbulent gas mixing, uniform flow of the gases through the catalyst bed, and adequate contact time with the catalyst surface.

Pollutant Type and Concentration. Although the physical condition of pollutants processed by catalytic incinerators may be in the gaseous liquid or solid state, for effective oxidation the combustible materials should be converted to the vapor phase as they pass through the catalyst. The major portion of this duty is performed in the preheat section. Entrained particles and liquid droplets do not effectively contact the catalyst because of their greater mass and lack of diffusional activity.

The combustible materials may be present in any concentration below the flammability limit. Varying pollutant concentrations, even down to trace amounts, can be converted to harmless end products by catalytic oxidation. For example, in the control of dimethylamine odors, an inlet concentration of 0.2 vol %, equivalent to 2000 ppm, has been reduced to 5 ppm by catalytic incineration. Oxygen must be present in sufficient concentration to achieve complete combustion of the various pollutants in the contaminated gas stream.

In most industrial process emissions, there is usually a mixture of pollutants that must be destroyed. The required catalyst surface area must therefore be determined for the least reactive compound in the contaminated gas stream. Thus, by referring to Fig. 6-13, it can be seen that for a mixture of solvents containing toluene, MEK,

257

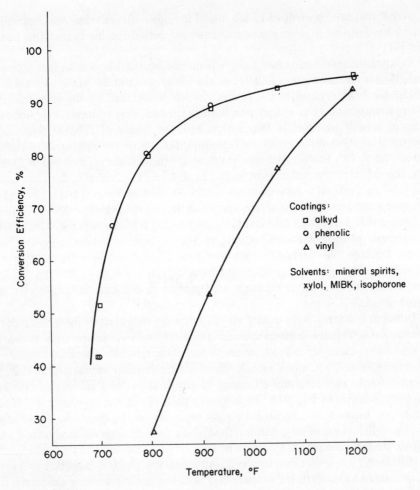

Figure 6-16. Hydrocarbon conversion efficiencies for catalytic incineration. (Adapted from Krenz et al., Ref. 10.)

and normal hexane, the design should be predicated on the hexane component, assuming that equal conversion efficiencies are necessary. If, because of disparate concentrations of the three compounds or some particular regulatory demand, identical conversion values are not required, then a series of temperature-reactivity-time calculations must be performed for the three components, to achieve the desired overall conversion efficiency.

The more stable compounds are generally the least reactive. Methane is an example of a stable, low molecular weight compound that requires a relatively high temperature for its catalytic conversion, about $1,000°$ F [11]. Hydrogen, however, is extremely reactive, having a conversion temperature of about $500°$ F. In some cases, hydrogen can be completely reacted at room temperature.

Contaminants. The one serious problem standing in the way of universal acceptance of catalytic incineration is that of maintenance and deterioration of performance. Catalysts undergo a gradual loss of activity through fouling and erosion of their surfaces. Two distinct classes of contaminants must be avoided when considering the application of catalytic incineration: poisons and fouling agents.

Some of the most damaging catalyst poisons are metallic vapors of lead, zinc, phosphorus, and arsenic. Iron oxide also has an adverse effect on catalyst activity, although the overall life is not too severely limited. Halogens and sulfur, the latter under certain process conditions, can also act as catalyst suppressants. However, recent improvements in catalyst formulations and application techniques have almost entirely eliminated the harmful influence of sulfur.

Fouling of catalytic surfaces is the more common problem detracting from the continuous activity required to maintain acceptable emissions control. Limited combustible particulate loadings can be processed in catalytic combustion equipment However, if the catalyst temperature is not maintained at the design level, a coating of organic material or carbon will be deposited on the surfaces and catalyst activity will suffer. Another problem involves the reaction of a contaminant in the catalyst bed to form solid oxides, which coat the surface area. Generally, if the problem is one of organic solid deposits, these can be burned off by a carefully controlled increase in the bed temperature. If the contaminants are inorganic materials, the catalyst must be removed from service and subjected to an acid and/or detergent wash.

Under ordinary operating conditions, it is recommended that the all-metal catalysts, previously described, be washed about twice a year to remove loose dust and other accumulations. Under ideal operating conditions, reactivation of this type of catalyst should be undertaken about every four years.

Mechanical Design

The mechanical design of an effective incinerator should incorporate those features that will provide adequate temperature, time, and turbulence to accomplish the desired level of combustion. In addition, catalytic incinerators must be equipped with catalysts suited to the required conversion duty.

Catalytic combustion equipment is available in almost an infinite number of configurations, and for most emissions problems a custom design approach is favored. For applications with existing processes, the basic components can be assembled inside a separate, insulated housing, which is usually installed in the vicinity of the process. For new plant installations, catalyst elements only are provided, being designed for incorporation into the process ductwork. Thus, the control equipment becomes an integral component of the process train. Some of the possible equipment arrangements for one type of design are shown in Fig. 6-17. For the package design concept, a common capacity range is on the order of 1,000 to 10,000 scfm. For a 10,000 scfm catalytic incinerator designed to process the gas stream defined in Table 6-1, an arrangement similar to scheme A of Fig. 6-17 would be chosen, with approximate overall dimensions of 10 ft wide by 12 ft long by 15 ft high.

The heart of catalytic combustion equipment is obviously the catalyst. The adsorption capacity, reaction rate, and desorption capacity of a specific catalyst for

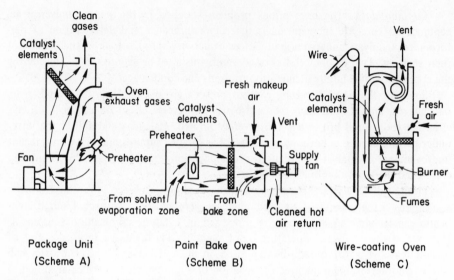

Figure 6-17. Catalytic incinerator equipment configurations. (Courtesy of Air Correction Division.)

the chemicals system being processed will characterize the overall reactivity and suitability of the catalyst for the specific control equipment being designed. Different concentrations of the various noble metals that comprise the catalyst family exhibit specific reactivity rates for each hydrocarbon. Because of the infinite variation in pollutant type and concentration, it becomes impractical to furnish a special catalyst for each control problem. Therefore, the manufacturers attempt to make available a general catalyst type that will provide optimum conversion of the maximum number of pollutants. There are a variety of commercial catalysts available. Two of the most commonly applied types are described below.

Air Correction Division manufactures an all-metallic catalyst based on a specially activated platinum-group metal. The catalyst is deposited on thin nickel-alloy ribbon, which is crimped and packed to provide the maximum available catalyst surface. The ribbon is uniformly distributed in mat form and is protected by alloy screens attached to alloy channel frames. Elements are available in a range of standard sizes and thicknesses, either in rectangular or circular shapes. Two typical elements, one measuring approximately 18 by 24 in. and the other with an 8 in. diam, are shown in Fig. 6-18. Each shape is rated according to the volumes of gases handled in standard cubic feet per minute. Pressure drops across these elements lie in the range of 0.2 to 0.4 in. w.g. Operating temperatures of 200 to 1400° F are permissible.

Oxy-Catalyst, Inc., produces a catalyst consisting of a combination of platinum and alumina supported on a porcelain carrier, mounted in a specially designed element. The Oxycat assembly is shown in Fig. 6-19. Multiple porcelain rods are held in place by two porcelain end plates. The assembly measures approximately 5½ in. long by 3 in. square, and weights 1.6 lb. The number of Oxycat elements required are based on a space velocity of 5 to 15 cfm per element for operation at atmospheric pressure. The

Figure 6-18. Typical UOP catalyst elements (Courtesy of Air Correction Division.)

Figure 6-19. The "Oxycat" catalyst element. (Courtesy of Oxycat Division of Research-Cottrell, Inc.)

elements are supported on suitable metal grids of either carbon steel or special alloys, depending on the operating temperature. The pressure drop through a multiple-element catalyst bed of this type is usually less than 1 in. w.g. The operating temperature range is 400 to 1800° F.

A completely assembled catalytic incinerator is shown in Fig. 6-20. This unit corresponds to scheme A in Fig. 6-17. It is furnished complete with catalyst elements, preheat burner, fan, and instrumentation. The combustion chamber may be fabricated of black iron, heat-resistant steel, stainless steel, or refractory materials. Heat-resistant steel is used for operating temperatures in the range of 750 to 1100° F, stainless steel for temperatures in excess of 1200° F, and refractories for temperatures greater than 1300° F. The exterior housing is usually fabricated of mild-steel, light-gage plate.

In the arrangement illustrated in Fig. 6-17A, where the fan is downstream from the preheat burner, an atmospheric burner can be used, since a negative pressure usually exists in the combustion chamber. The burner is usually rated for the total heat duty without reliance on the heat content of the solvents in the contaminated gas

Figure 6-20. Package catalytic incinerator. (Courtesy of Air Correction Division.)

stream. The burner may actually be operated at low air/fuel ratios ranging 7-8: 1, with the contaminated gas stream providing sufficient oxygen to complete the combustion. The preheat chamber volume is based on heat release rate values in the range of 40,000 to 60,000 Btu/hr-ft^3.

The process exhaust gas stream can be delivered to the incinerator by the process exhaust fan, or (see Fig. 6-17) the incinerator fan can be designed with sufficient capacity and pressure so as to eliminate the exhaust fan. The incinerator fan must also mix the fuel combustion gases with the contaminated gas stream and distribute them

263

evenly over the catalyst. Uniform distribution may be accomplished by splitting vanes or perforated baffles located ahead of the catalyst bed. In its location downstream from the preheat burner, the fan is not troubled with the deposition of condensates on the blades, since it is operated well above the condensation temperature. However, in this location a "hot" fan design must be utilized, with suitable fabrication materials and bearings.

Because of the high rate of heat release in the catalyst bed, control instrumentation is usually furnished to regulate the preheat burner fuel-injection rate as a function of the catalyst discharge temperature. Thus, if the temperature increase across the catalyst bed falls off because of a decrease in pollutant concentration, then the fuel gas input is increased to maintain the temperature of the catalyst at the proper level. This temperature control instrument may be furnished with a recorder that continuously indicates the temperature rise of the gas stream as it passes through the catalyst. This temperature record is a valuable indication of process upset conditions. For example, the maladjustment of paint-spray nozzles in a coatings oven operation will cause a variation in pollutant loadings. This, in turn, will be reflected in the catalyst temperature rise, as indicated on the recording control instrument. The basic fuel control and safety instrumentation scheme illustrated in Fig. 6-7 is generally applicable to catalytic incineration.

When a catalytic incineration system operates at temperatures in the range of 800 to 1000° F, the application of heat recovery should be considered. Thus, in Fig. 6-17B, a catalyst system for a new paint-bake oven is aimed at emissions control and heat recovery. In this design the catalyst elements are located in a recirculation heater section, with only a portion of the recirculation stream passing through the catalysts. This portion of the total flow supplies additional heat to the oven by the combustion of the solvent vapors in the catalytic bed. A balanced heat supply is maintained by the addition of fresh makeup air. This type of integral oven-incinerator design must be initiated at the time a new oven installation is being considered.

Figure 6-17C shows a wire-coating oven incorporating a catalyst reheat system. In this design the oven volume is small and operating temperatures are high so that the catalyst can provide almost all the heat required by the oven. Many such catalyst-equipped ovens for this process are completely self-sustaining in their heat duty requirements, even at solvent loadings of 3 to 5 gph.

In summary, when catalyst elements are designed into an oven gas stream circuit, the contaminated gas stream is heated to the catalytic operating temperature by the oven heater. When a separate catalytic incinerator is installed, a preheat burner must be furnished to provide this heat duty.

Installation Features

The "package" thermal incineration unit shown in Fig. 6-20 is shipped completely assembled. It is prewired and prepiped so that field installation involves placing it into position and providing ductwork between it and the process, and connecting power leads to the control cabinet. The exhaust gas ductwork should be as short as possible and/or be adequately insulated to prevent the precipitation of organic condensibles. In many applications a single incinerator can handle the exhaust gas stream from multiple processes. In such an arrangement the cost of interconnecting ductwork and the restricted flexibility in the process-operating schedule must be traded off against incinerator equipment savings. Operations must be staggered so as to ensure a reasonably constant pollutants loading and gas flow to the incinerator.

In those catalytic incinerators provided for new or proposed oven designs, the manufacturer usually provides the catalyst elements only, accompanied by design data for optimally locating the catalyst in the oven ductwork. In such designs the preheat burner, fan, and instrumentation are integral components of the oven, and thus minimal installation efforts are required.

An Oxycat system for both emissions control and heat recovery is shown in Fig. 6-21. In this installation the solvent vapors from the lithographing oven are converted to carbon dioxide and water vapor in the catalyst section located downstream from the preheat burner. The clean gases leaving the catalyst are at a sufficiently elevated temperature to justify economically their recirculation to the oven. The remainder of the stream is discharged to atmosphere through the stack. In some applications the clean hot gases are diverted for building space heating.

Ordinarily, processes utilizing catalytic incineration are operated well below the lower flammability limit. A catalyst "per se" cannot cause an explosion or create a fire, providing the gas stream being processed remains outside the flammability range. Process safeguards must be provided to ensure this nonflammable condition. One possible fire source is that of utilizing existing ductwork or stack for a proposed catalytic incinerator installation. It is absolutely essential that any organic deposits

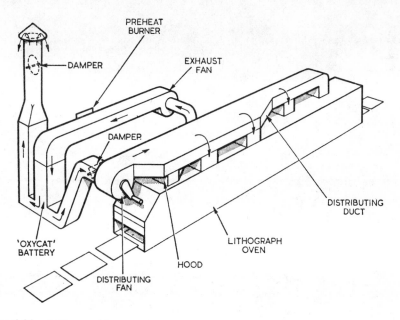

Figure 6-21. Lithographing oven catalytic system. (Courtesy of Oxycat Division of Research-Cottrell, Inc.)

that might be present in the existing equipment be completely removed, since the higher temperatures associated with catalytic combustion will create a fire hazard.

The need for a continuous maintenance program for the catalyst elements is mandatory if continuously acceptable performance is to be maintained. This require-ment demands easy access to the incinerator catalyst section. In Fig. 6-20, showing a catalytic incinerator, a removable panel is located just above the control panel, allowing convenient access to the catalyst. The individual rectangular elements, shown in Fig. 6-18, are clamped in place so that they can be readily removed for inspection and/or cleaning. The Oxycat catalyst elements are arranged in layers, and are sup-ported by a metal or ceramic gridwork, as shown in Fig. 6-22. Thus, in the incinerator arrangement shown in Fig. 6-21, these elements can be easily removed for servicing.

Applications

Air pollution control by catalytic incineration is generally applicable under the following conditions:

The fume stream to be processed contains vaporized or gaseous organic com-bustible material.

Figure 6-22. Catalyst arrangement. (Courtesy of Oxycat Division of Research-Cottrell, Inc.)

Catalyst poisons or suppressants are not present.
The contaminated gas stream contains no great amount of dust, flyash, or other solid inorganic material.

Even with these limiting conditions, there are a wide number of industrial processes to which catalytic combustion systems can be applied. Both heat-economy motives and air pollution control needs are well served by catalytic incineration.

As discussed previously, one of the most serious complaints directed at catalytic combustion is the requirement to maintain the catalysts at a continuous level of activity. Manufacturers of the equipment are very careful to point out the operating limitations, and continuously appeal to the process industries to operate their catalytic combustion equipment according to design specifications, to ensure acceptable results. However, catalysis is a sensitive mechanism and the reasons for a malfunctioning system are myriad.

Some APCD groups are especially skeptical of the application of catalytic systems in their community. One of the earliest regulations to be enacted for the control of solvents emissions was Rule 66 by the Los Angeles County APCD. In 1966 the advisability of recommending the use of catalytic systems for the control of solvents was questioned. LAAPCD comments on the situation were as follows [12] : "Any air pollution control equipment capable of reducing the organic materials to the required quantities is acceptable. The catalytic incineration devices that have been tested, *so far,* by the Air Pollution Control District are judged to be incapable of meeting the requirements of Rule 66." Since this statement was made, a number of catalytic combustion installations have performed successfully in the Los Angeles area.

It might be stated that the same classification of industrial processes served by thermal incineration could be similarly served by catalytic incineration except that poisons and noncombustible particulates must be absent. Thus, a more restricted group of industries might be represented as follows: paint baking, sewage treatment, chemical processing, wire enameling, paint and varnish manufacture, drying ovens, metal coating, food processing, and lithographing. Many of the specific pollutants controlled by catalytic incineration in the chemical process industries are characterized as nuisances because of their odorous and lachrymal properties. Acrylates, mineral oils, phthalic anhydrides, dimethylamines, hydrogenated fatty acid derivatives, and alkyd resins are some of the objectionable organic compounds that have been successfully controlled by the catalytic incinerator types illustrated in Fig. 6-17.

Catalytic combustion can accomplish gaseous emissions control by the promotion of both oxidation and reduction reactions. The organic compounds cited above have been successfully oxidized to carbon dioxide and water vapor. The nitrogen oxides, on the other hand, require a reduction technique for their destruction [13]. The nitrogen oxides NO and NO_2, and the higher oxides, are characterized by their pungent odor and reddish brown color. These gases are also toxic, and a continuous exposure safe level has been established at 5 ppm. In nitric acid manufacturing plants, nitrogen oxides are continuously released from the high-pressure absorber at concentrations of 3000 to 5000 ppm. To achieve complete reduction, the total oxygen must be removed from the discharged tail gas stream, both as free oxygen and as that present in NO and NO_2. By using methane as the reducing agent, these gases are reduced to nitrogen, carbon dioxide, and water vapor in a two-stage catalytic combustion system. This system comprises a two-stage reactor with startup burner, a heat exchanger, and a waste heat boiler. The reaction must be accomplished in two stages because of the temperature limitations of the catalyst and vessels. The burner in the system increases the temperature of the tail gases from 550 to 900° F. When this operating temperature is attained, the preheat duty is then continued in the heat exchanger. The gases pass through the first-stage catalyst, where partial reduction is achieved, and the gases are discharged at 1300° F. After exchanging heat with the inlet gases, the partially reacted gas stream enters the second stage at about 950° F, where reduction of the nitrogen oxides is completed and a final temperature of 1300° F is attained. The heat released by oxidation of the reducing fuel is recovered in the waste heat boiler, with the final effluent gases being discharged at 550° F.

The Oxycat system illustrated in Figs. 6-19 and 6-21 is capable of promoting both oxidation and reduction reactions. Heat recovery applications have also been undertaken. One of the earliest applications for the Oxycat system was for the control of automotive exhaust fumes emitted from heavy-duty trucks. A list of some of the more common applications is given in Table 6-8.

Sample Design Calculations

Heat balance calculations, fuel gas demands, and retention period requirements for catalytic incinerators follow the same basic principles as those demonstrated in the Sample Design Calculations at the end of Section 6-5. Therefore, that same problem, as defined for thermal incineration, will be presented in this section, followed by a solution that emphasizes fuel consumption economics.

Table 6-8. Oxycat Industrial Process Applications

Industry	Pollutants	Catalytic Oxidation Temp., °F
Asphalt oxidation	Aldehydes, anthracenes, oil vapors and hydrocarbons	600–700
Carbon black mfg.	H_2, CO, CH_4, carbon	1200–1800
Catalytic cracking units	CO, hydrocarbons	650–800
Coke ovens	Wax, oil vapors	600–700
Formaldehyde mfg.	H_2, CH_4, CO, HCOH	650
HNO_3 mfg.	NO, NO_2	500–1200
Metal-lithography ovens	Solvents, resins	500–750
Phthalic anhydride mfg.	Maleic acid, phthalic acid, naphthoquinones, CO, HCOH	600–650
Polyethylene mfg.	Hydrocarbons	500–1200
Printing presses	Solvents	600
Varnish cooking	Hydrocarbons	600–700
Wire coating and enameling ovens	Solvents, varnish	600–700
Octylphenol mfg.	C_6H_5OH (phenol)	600–800

Source: Data provided courtesy of Oxycat Division of Research-Cottrell, Inc.

Problem Statement. The same process and operating conditions prevail as for thermal incineration; see Section 6-5. The basic data and requirements are:

Process	Metal decorating
Equipment	Baking oven
Exhaust volume, scfm	6000
Exhaust temperature, °F	600
Solvents loading, lb/min	1.4
Solvents composition, wt %	
Hexane	35
Toluene	45
MEK	20
Operating period, 24-hr days/yr	360
Fuel type	Natural gas
Fuel costs, $/1000 scf	0.60
Performance, organics converted, %	98.0

The oven is to be redesigned for the incorporation of a catalytic incinerator system so as to satisfy the performance requirements and achieve maximum fuel economy.

Problem Solution. The oven is assumed to have a single compartment with flows and temperatures as shown in Fig. 6-23A. Besides the flow of 6000 scfm of fresh makeup air, there is a recirculation flow of 6000 scfm to ensure adequate gas flow turbulence in the oven, to achieve mass and heat transfer. The average operating temperature in the oven is 600° F at a temperature gradient of 50° F. This gradient value represents the heat absorption and radiation losses to the oven conveyor and walls as well as the work load being processed. Therefore, a total gas flow of 12,000 scfm enters the oven at 650° F with a recirculation load of 6000 scfm and an exhaust flow of 6000 scfm, both leaving at 600° F.

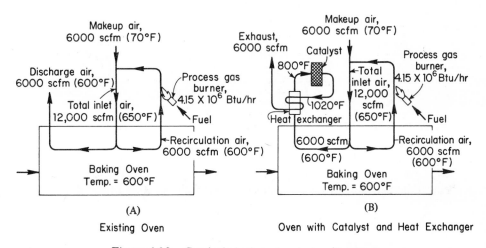

(A)

Existing Oven

(B)

Oven with Catalyst and Heat Exchanger

Figure 6-23. Catalytic incinerator design flowsheets.

1. Determination of Existing Oven Heat Load

Q_1 = fresh air heating
= 6000 × 0.076 × 0.24 (650 − 70)60 = 3.82 × 10^6 Btu/hr
Q_2 = recirculation load
= 6000 × 0.076 × 0.24 (650 − 600)60 = 0.33 × 10^6 Btu/hr
Total load = 4.15 × 10^6 Btu/hr

2. The flow arrangement for the catalyst installation, including a heat exchanger, is shown in Fig. 6-23B. The required catalyst bed temperature is 800° F; therefore the required preheat duty is

6000 × 0.076 × 0.24 (800 − 600)60 = 1.32 × 10^6 Btu/hr

3. Catalyst temperature rise is

Inlet solvent loading = 1.4 lb/min × 60 min/hr × 1 gal/7 lb = 12 gph

At a combustion heat value of 120,000 Btu/gal, the total heat release in the catalyst is

Total heat = 120,000 Btu/gal × 12 gal/hr = 1.44 × 10^6 Btu/hr

Temperature increase across the catalyst bed is

$$\text{Temp diff.} = \frac{1.44 \times 10^6}{6000 \times 0.076 \times 0.24 \times 60} = 220° \text{ F}$$

Average reaction temperature in catalyst bed is

$$\text{Avg temp} = (800 + 1020)/2 = 910° \text{ F}$$

4. Retention period: The required retention period will be based on the reactivity of hexane at 910° F. Refer to the catalytic reactivity rates in Fig. 6-13; the value of k for hexane at 910° F is 43 sec^{-1}.

Substituting in Eq. 6-6,

$$\frac{100 - 98}{100} = C^{-43t}$$

$$t = \frac{1}{43 \log e} \times 1.70 = 0.092 \text{ sec}$$

It will be assumed that the combustible particulates will be catalytically incinerated at these values of temperature and dwell time.

5. System gas flows: Air for combustion of the solvents is available in the discharge gas stream. The net heating value of natural gas at 910° F, with combustion air being furnished 100% by the contaminated gas stream, is 800 Btu/ft^3. Thus,

Catalyst preheat duty = 1.32 × 10^6 Btu/hr

$$\text{Fuel gas} = \frac{1.32 \times 10^6}{800 \times 60} = 28 \text{ scfm}$$

$$\text{Combustion products} = 28 \times 11.0 \times \left(\frac{910 + 460}{60 + 460}\right) = 810 \text{ acfm}$$

The contaminated discharge stream is 6000 scfm, equivalent to 15,800 acfm at 910° F. Of this flow there is required

$$28 \times 10.0 \times \left(\frac{910 + 460}{60 + 460}\right) = 740 \text{ acfm}$$

as a source of air supply for the combustion of 28 scfm of natural gas. The total gas stream flowing through the catalyst bed at an average temperature of 910° F is 810 + (15,800 − 740), or 15,870 acfm.

A schematic flow diagram indicating gas flow distribution and temperatures is shown in Fig. 6-31B.

5. Equipment sizing: Based on a velocity through the catalyst bed of 10 fps, the cross-sectional area of the catalyst elements would be

$$15,870 \div (10 \times 60) = 26.5 \text{ ft}^2$$

Any number of catalyst elements similar to those shown in Fig. 6-18 could be used. One arrangement might be six elements, each measuring 1.5 ft by 3.0 ft for a total of 27.0 ft^2 of catalyst surface area. By placing the mats askew, as shown in Fig. 6-17A, the cross-sectional area of the ductwork can be minimized.

The depth of the catalyst elements must be sufficient to allow a dwell time of 0.092 sec. Based on a velocity of 10 fps, there would be required a bed depth of 0.92 ft. This would demand at least two stacked elements, each approximately 6 in. deep. The use of the Oxycat element shown in Fig. 6-19 might also be considered. The capacity of each of these assemblies is rated at 5 to 15 scfm per unit. Therefore, the total elements required at a flow rate of 10 scfm per element would be

$$\frac{6000}{10} = 600$$

6. Overall heat balance and cost summary: The various heat balances may be interpreted as

(a) Existing oven duty, before installation of catalytic control device = 4.15 × 10^6 Btu/hr.
(b) Extra heat input required for contaminated gas stream preheat = 1.32 × 10^6 Btu/hr.
(c) Heat return from combustion of the solvents in the contaminated gas stream = 1.44 × 10^6 Btu/hr.

The heat balance in this system indicates that the extra heat input required for catalytic oxidation of the solvent pollutants is offset by the potentially recoverable heat derived from combustion of the solvents. The cost of a heat exchanger for this installation might be on the order of $7000, with installation charges at about $3000. If a relatively high thermal efficiency of 90% is allowed, then the recoverable heat is about equal to the catalytic oxidation preheat requirements.

At an amortization period of 15 yr, estimated depreciation costs would be (7000 + 3000)/15, or $670/yr. The savings accountable to heat recovery for a 24-hr, 360-day year, would be

$$(1.44 \times 10^6)\text{Btu/hr} \times 0.90 \times 360 \times 24 \times \$0.60/10^6 \text{ Btu} = \$6740 \text{ yr}$$

Therefore, the annual net fuel savings would be 6740 − 670, or $6070 up to completion of the 15 yr payout period for the heat exchanger, and $6740 thereafter—still a savings to be considered.

6.7. EQUIPMENT CHARACTERISTICS—ADSORPTION

Operating Principles

Although the principles governing adsorption have been known for some time, this process is not too often applied to air pollution control problems. Many manufacturing plants for the production of plastics, rayon fabrics, and rubber goods, which require solvents recovery, have relied on adsorption systems for this service. Unlike the incineration process that effects destruction of the pollutants, adsorption emphasizes product recovery.

In adsorption, the pollutant-contaminated exhaust gas stream is passed through the adsorbent bed where the gaseous contaminants are contacted and held fast on the adsorbent surface. In batch operations, this adsorption cycle is continued until saturation has been attained, as indicated by the presence of the pollutants in the gas stream being discharged from the adsorber. The contaminated gas flow is then directed to a parallel and identical vessel containing regenerated adsorbent, and the process continued. Concurrently, the adsorbent in the first vessel is regenerated by the passage of either hot inert gases or low-pressure steam through the bed. The pollutant-steam effluent is condensed and the pollutants recovered either by simple decantation, in the case of water-insoluble materials, or distillation. Thus, the operation of a batch adsorption system is intermittent, being governed by the adsorption and regeneration cycle periods. In some installations, the simultaneous switch of both contaminated gas and steam flows is accomplished automatically through a cycle controller. In a more advanced design, continuous adsorption is accomplished by passing the waste gas stream countercurrently through a number of stages, each comprising a fluidized bed of adsorbent. The pollutants are progressively adsorbed and the clean gas discharged to atmosphere from the top stage. The saturated adsorbent is continuously discharged from the vessel and transferred to a desorber, where it is stripped and then returned to the adsorption vessel.

An extensive variety of materials has adsorption properties. Activated carbon is the most common of those being applied for solvent recovery, gas purification, and the elimination of odors. Silica gel, alumina, bauxite, and molecular sieves also have the ability to adsorb various gaseous and vapor components from a gas stream. Activated carbon adsorbs organic gases and vapors, even in the presence of water vapor. Silica gel preferentially removes water vapor from a mixture of wet organic compounds present in a gas stream. Alumina, bauxite, and molecular sieves are applied for the dehydration of gases. Other adsorbents, such as Fuller's earth, bone char, and magnesia, are used to remove impurities from liquid solutions.

Activated carbon can be produced from such carbonaceous materials as wood charcoal, coconut char, coal, coke, bone char, and lignite. The adsorption properties of a particular carbon are a function of the porosity and capillary structure of the material. Physical adsorption is the result of an attraction between the gaseous phase and the solid surface effected by van der Waal forces. These forces cause condensation of the gas or vapor when the capillary pressures within the adsorbent pores become sufficiently large. On a smooth surface, adsorptive forces would produce a single condensate layer of a few molecules in thickness. However, in the pore structure of the adsorbent, the quantity of condensate is considerably increased by the capillary forces.

The most important characteristic of an effective adsorbent is a large surface-to-volume area and preferential attraction for the compound being adsorbed. The total surface area of commercial activated carbons lies in the range of 600 to 1200 m^2/g, with internal pore volumes measuring from 0.4 to 0.9 cm^3/g. Thus, a pound of adsorbent may contain a surface area as great as 4.5×10^6 ft^2. Most carbonaceous materials must be activated to develop additional adsorptive potential. High-temperature calcination in the presence of steam or carbon dioxide is one method of activation. The base material is usually pulverized and mixed with a suitable binder such as sugar or tar, and is then pelletized to yield a dense, compressed material. The pellets are then calcined at temperatures of 1300 to 1650° F. Various types of wood and coal are used as the base material. The prime objective of this treatment is to release carbon from its compounds, thus converting it to the active state. In some processes the carbonaceous materials being activated are treated with various salt compounds to prevent the formation of hydrocarbon complexes. The metallic oxides formed during the carbonization are subsequently leached out, thus yielding a porous structure with molecular-sized capillaries.

Activated carbons used for solvent removal differ widely in form, physical properties, and adsorptive capacity. They also exhibit varying affinity for different gases. Thus, an activated carbon, based on a coconut shell material, has a higher capacity, though not so great a selectivity for propane over ethane when compared to a coal-base material at identical operating conditions. For any specific problem, the activated carbon selected must have the ability to adsorb the maximum amount of the specific pollutant(s). The adsorbent must also possess sufficient strength to resist abrasion and be available in such a form as to offer low-pressure drop characteristics when exposed to the contaminated gas flow. The adsorptive capacity of any active carbon usually increases with the molecular weight of the organic compound being adsorbed. Additionally, unsaturated compounds are generally more completely adsorbed than saturated compounds, and cyclical compounds are more easily adsorbed than linearly structured materials.

Adsorption data are represented by isothermal plots that relate the amount of compound adsorbed (the adsorbate) to the equilibrium pressure at a constant temperature. Isothermal plots for the adsorption of several hydrocarbons are shown in Fig. 6-24. The carbonaceous base material for the activated carbon used to obtain these data is coconut shell. The increased adsorption activity for both higher molecular

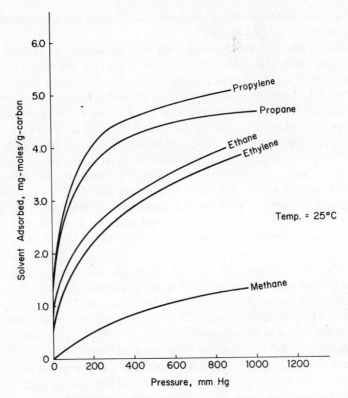

Figure 6-24. Adsorption isotherms for hydrocarbons on activated carbon. (Adapted from Kohl and Riesenfeld, Ref. 14.)

weight and unsaturated compounds are illustrated in these plots. Although propylene shows a greater adsorption potential over propane, the ethane-ethylene compounds do not follow the general rule.

The theoretical bases for isothermal plotting is developed as follows: In the adsorption process, energy is required to convert the gaseous compounds to the saturated liquid state and to accomplish the adsorption of this liquid. The energy requirements for the second step are negligible, so that the free energy change for the total process can be considered as that required for the change of state for the compound being adsorbed. This theory was advanced by Polanyi [15], who designated the free energy requirements for this change of state as the adsorption potential, expressed as

$$\Delta F = NRT \ln p_s/p \tag{6-9}$$

where

ΔF = free energy change
N = number of moles adsorbed

274

R = gas law constant

T = absolute temperature

p_s = saturated liquid vapor pressure

p = adsorption pressure

It was found by Polanyi that the adsorption potential is a function of the adsorbate volume and that the function is not seriously influenced by the temperature. Therefore, by plotting the volume of gas adsorbed versus the adsorption potential, isotherms can be predicted for the same gas at other temperatures. This equation was modified by Dubinin et al. [16], based on the assumption that similar types of compounds with a given adsorbent will have equal adsorption potentials when equal amounts are adsorbed. Dubinin suggested that the amounts adsorbed be expressed as the product of moles adsorbed (N) and the molal volume of the adsorbate (V'), measured as a saturated liquid. Therefore,

$$N_1 RT \ln (p_s/p)_1 = N_2 RT \ln (p_s/p)_2 \tag{6-10}$$

and

$$N_1 V_1' = N_2 V_2' \tag{6-11}$$

If Eq. 6-10 is divided by Eq. 6-11, then

$$\frac{RT}{V_1'} \ln \left(\frac{p_s}{p}\right)_1 = \frac{RT}{V_2'} \ln \left(\frac{p_s}{p}\right)_2 \tag{6-12}$$

where

V' = molal volume of saturated liquid at the adsorption isotherm temperature

Equation 6-12 was further modified by W.K. Lewis et al. [17] by the substitution of fugacities for vapor pressures. The final form of the equation for equal values of VN then becomes

$$\left[\frac{RT}{V} \ln \frac{f_s}{f}\right]_1 = \left[\frac{RT}{V} \ln \frac{f_s}{f}\right]_2 \tag{6-13}$$

where

V = molal volume of saturated liquid at a temperature where vapor pressure is equal to adsorption pressure

f_s = fugacity of the vapor corresponding to vapor pressure

f = fugacity of the vapor corresponding to adsorption pressure

The value of the group $(RT/V) \ln (f_s/f)$ should be the same for all gases at equal volumes adsorbed, in order to correlate the isotherm data. Lewis and his coworkers did succeed in correlating the value of $(RT/V) \ln (f_s/f)$ with the volume of adsorbed gas for a number of hydrocarbons. Such a plot is shown in Fig. 6-25 for ethylene and propane on silica gel at various adsorption temperatures. Both curves show a good

275

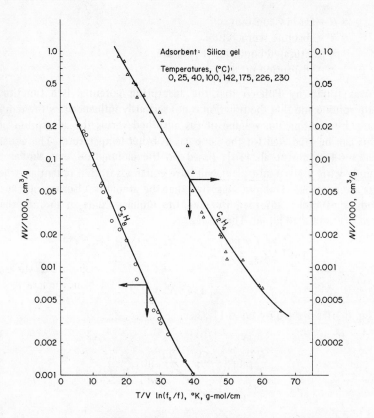

Figure 6-25. Correlation of ethylene and propane isotherms. (Adapted from Lewis et al., Ref. 17.)

degree of correlation over a temperature range of 0 to 230° C. The increased adsorption activity of the unsaturated ethylene is evident from this plot.

This method of correlation has been found most helpful in estimating isothermal data for different hydrocarbons on the same adsorbent, at various values of temperature and pressure. In addition, its development has lead to a definition of the important variables influencing adsorption phenomena. From many such isotherm plots it was possible to draw the following conclusions:

1. Both the molecular weight and structure of the hydrocarbon must be considered in the evaluation of its adsorption potential. For a carbon adsorbent, lower vapor pressure equivalent to high molecular weight can be the major factor favoring adsorption. Unsaturation does have an influence and can be the deciding factor when the volatilities are close. Silica gel seems to be more seriously influenced by hydrocarbon unsaturation.

2. Capacities of different adsorbents for hydrocarbons differ widely. One type of commercial activated carbon can adsorb 30 to 60% more than another. This difference

could be a function of both the base carbonaceous material and the degree of activation. In general, activated carbon showed a much greater adsorption capacity for the hydrocarbons than did the silica gel.

3. The influence of temperature on the adsorption isotherm, over a range of 100 to 200° C, was successfully correlated by this method.

Considerable isotherm data are available for the adsorption of different compounds on specific adsorbents. However, where additional adsorption data are required, experimental procedures have been established for obtaining them. Vapor-phase isotherms, similar to those shown in Fig. 6-24, can be developed experimentally in a static adsorption apparatus by measurement of the weight or volume of the gas adsorbed at a series of pressures. Isothermal data can be determined over a wide range of conditions from minimal experimental results by making use of the Lewis et al. [17] correlation developed in Eqs. 6-9 through 6-13.

Isothermal plots are usually determined under static equilibrium conditions, whereas adsorption proceeds dynamically in commercial operations so that considerable interpretation is required to convert isotherm data to design values. The isotherms indicate the capacity of the adsorbent for the solvent, but cannot relate the contact time or amount of adsorbent required to reduce the solvents concentration to a desired level. The major factors that influence the necessary contact times are the adsorbent bed depth, adsorbent mesh size, gas velocity, gas temperature, solvent concentration, moisture content of the gas and adsorbent, and the adsorptive capacity of the absorbent over the desired temperature and concentration range. The effects of these various factors are illustrated in Fig. 6-26, which demonstrates the adsorption of a particular solvent as it travels through an adsorbent bed at a specific velocity. Curve A shows the theoretical saturation under equilibrium conditions for the specific operating conditions; say, 20% by weight of the carbon charge. This limiting value could be obtained only by slowly passing the gas until the entire bed was saturated. In practice, the gas is passed at a high rate through the carbon bed and therefore the adsorbate content of the carbon varies from 100% at the inlet to 0% at the outlet, as indicated by curve B.

As the adsorption progresses, the carbon temperature is raised by the exothermic heat of the process and this heat is carried ahead of the adsorption zone by the gas. The further reduction in adsorption caused by this temperature rise is shown by curve C. The moisture content of the gas and adsorbent further reduces the adsorption capacity, so that curves D and E can be drawn. Curve E represents the adsorption profile through the bed when all these factors have been considered. The area enclosed by this curve, divided by the height, yields the mean capacity of the carbon at the end of the adsorption period. Actually, the adsorbate may only be 8% of the carbon at the end of the cycle as compared to the equilibrium value of 20% defined by the isotherm. Other factors that make it difficult to predict the adsorbent capacity are the presence of residual adsorbate at the end of the regeneration cycle and the processing of multiple solvents, which increases the difficulty of interpreting the isothermal equilibrium curves. There are no general formulas or relationships available for the prediction of adsorption performance, which are based on the physical and chemical properties of the solvent and adsorbent.

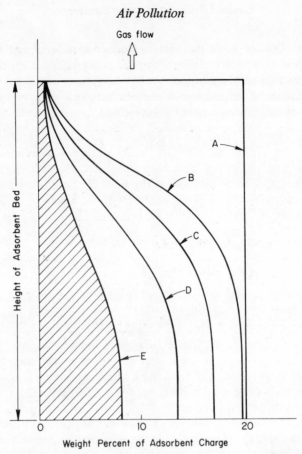

Figure 6-26. Adsorption profile through an adsorbent bed. (Adapted from Edeleanu, Ref. 18.)

Hougen and Marshall [19] developed methods for calculating the time, temperature, and concentration conditions in both gas and solid phases during the adsorption of a dilute gas flowing through granular adsorbent beds. "Height of transfer unit" formulas, based on both heat and mass transfer, were modified for the adsorption of moisture on solid desiccants. Problems were considered involving isothermal adsorption with both linear and complex equilibrium relationships and nonisothermal adsorption. Activated carbon adsorption equilibrium is seldom linear, which complicates the theoretical approach to a solution.

In actual practice, the adsorptive capacity, bed height, and regeneration conditions for a specific application are experimentally determined in the laboratory.

Design Parameters

The major adsorber design parameters are adsorption efficiency, regeneration, gas flow rate, gas temperature, pollutant type and concentration, contaminants, and adsorbent selection. A description of each parameter follows.

278

Adsorption Efficiency. In the removal of a vapor by activated carbon, the adsorption efficiency is initially 100%. However, as the process is continued, the rate of adsorption falls off until the vapor concentration leaving the adsorbent is equal to that entering. Since the carbon is saturated at this point, the adsorption efficiency beyond this operating period is rated at zero percent. The time at which the adsorption initially begins to decline is termed the "breakpoint" of the adsorbent. The curve in Fig. 6-27 illustrates this decline in adsorption effectiveness with time for a single solvent on activated carbon.

Theoretically, a 100% adsorption efficiency can be obtained for the system illustrated in Fig. 6-27 if adsorption is continued up to 8 hr, followed by regeneration, followed in turn by adsorption, and so on in the process. However, practical adsorption efficiencies in the range of 95 to 99% are usually attained, based on allowances for residual solvent in the adsorbent bed following regeneration and loss of activity of the adsorbent due to contamination. Where a very high efficiency is required, in the range of 99.5⁺%, a series operation of adsorbers is practical. For example, with four adsorbers in the system, one is on a regeneration cycle while the remaining three are removing solvents. Two of the three are operated in parallel as precleaners; the third, located in series with the most nearly saturated adsorber, accomplishes final cleanup.

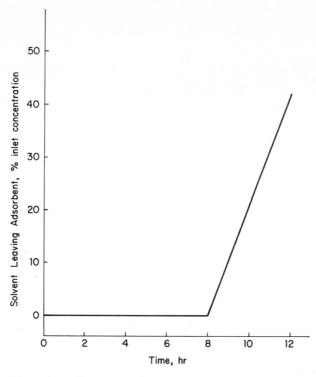

Figure 6-27. Adsorption effectiveness for a single solvent. (Adapted from Danielson, Ref. 8.)

Regeneration. The adsorbate is removed from the activated carbon by heating the bed to a temperature above that at which the solvents were adsorbed. Thus, for most processes, regeneration is accomplished by passing saturated steam through the adsorbent countercurrently to the direction taken by the waste gas stream. Adsorption and regeneration are usually performed cyclically, with adsorption taking place in one vessel and regeneration in the other. A typical system is illustrated in Fig. 6-28. During the regeneration cycle, the steam vaporizes the adsorbed solvent and the steam-solvent mixture is directed to a water-cooled condenser and then to storage via the decanter.

Steam flow rates can represent up to about 10% of the waste gas throughput. The steam consumption during regeneration is expended to heat the adsorber vessel and its contents, and to provide the latent heat of vaporization for the adsorbate. The major portion of the steam flow, about 60 to 70%, acts as a carrier gas for the adsorbed solvents. Steam consumption in the range of 2 to 10 lb per pound of adsorbate is common. The steam demand varies with time and the type of solvent being adsorbed. In Fig. 6-29, there is shown the steam requirements per pound of solvent recovered for toluene and perchlorethylene. The increased steam consumption for toluene is a function of its greater heat of vaporization.

The total regeneration cycle involves both steam and air flows. Following steam stripping, the adsorbent is hot and saturated with water vapor. Drying and cooling, in that order, are achieved by blowing solvent-free air through the beds. During the

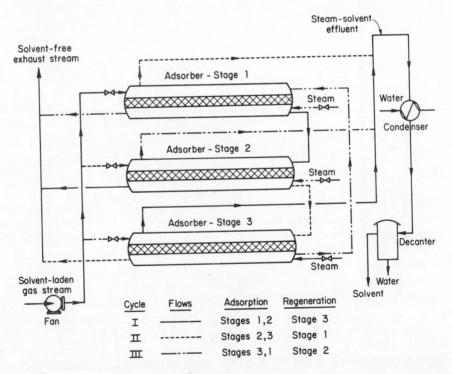

Cycle	Flows	Adsorption	Regeneration
I	———	Stages 1,2	Stage 3
II	- - - - -	Stages 2,3	Stage 1
III	—··—	Stages 3,1	Stage 2

Figure 6-28. Three-stage adsorption system.

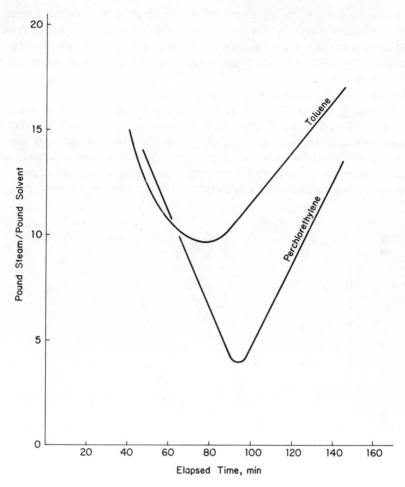

Figure 6-29. Regenerative steam requirements. (Adapted from Danielson, Ref. 8.)

drying portion of the cycle, moisture evaporation assists in cooling the adsorbent. In a typical adsorption process for the removal of a solvents mixture from a printing plant effluent stream, a steam/solvent ratio of 3.5:1 is used. The cycle for each adsorber vessel is 2 hr adsorption and 1 hr regeneration. A breakdown of the regeneration cycle is: 30 min steam, 15 min hot air at 220° F, and 15 min cooling air.

Drying the carbon bed with clean air is usually necessary to obtain maximum recovery of water-miscible solvents. In some cases it is desirable to allow residual moisture in the carbon so that the heat of adsorption may be used for its evaporation, thus minimizing the temperature rise.

Gas Flow Rate. The gas flow rate determines the area of the adsorbent bed as a function of the pressure drop and desired cycle time. Thus, the gas flow defines the equipment size and capital costs. The gas velocity must be sufficiently low to prevent upset or severe motion of the adsorbent. In systems operated at atmospheric pressure, superficial velocities in the neighborhood of 20 to 100 fpm are generally used. Pressure drop is a function of the gas velocity, bed depth, and adsorbent particle size. As illustrated in Fig. 6-30, increased pressure drop can be expected at decreasing adsorbent particle size.

Downflow of the waste gas stream is practiced to minimize movement of the adsorbent. Usually, sufficient adsorbent bed area is provided to yield a pressure drop in the range of 2 to 8 in. w.g. The bed depth at a given adsorbent area is then dictated by the solvents concentration, operating cycle, and allowable superficial gas velocity. Shallow bed depths at 1 to 3 ft, and cycle periods of 30 to 60 min are commonly used.

Gas Temperature. The retentive capacity of one type of activated carbon for a specific solvent is a function of the temperature. Thus, adsorption isotherms represent

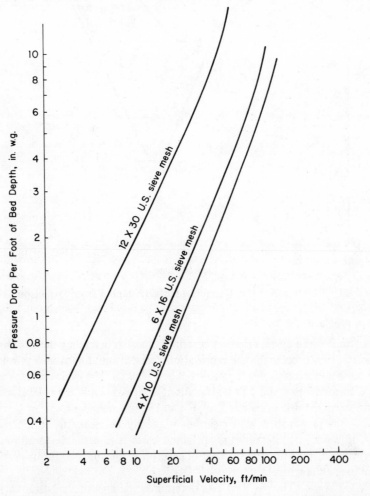

Figure 6-30. Gas flow rate versus pressure drop for various adsorbent sizes. (Courtesy of Pittsburgh Activated Carbon Co.)

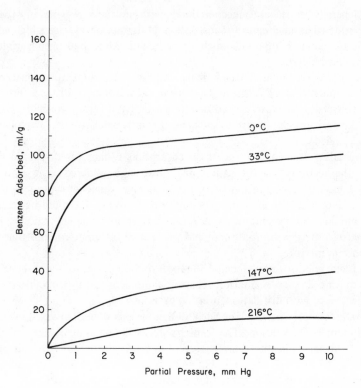

Figure 6-31. Benzene isotherms at varying temperatures with coconut-shell charcoal adsorbent. (Adapted from Kohl and Riesenfeld, Ref. 14.)

the solvent capacity of a particular adsorbent at a specific temperature. This capacity is increased at the lower temperatures as a function of the solvent volatility. This relationship of temperature and adsorption capacity is illustrated in Fig. 6-31, where isotherms for benzene at different temperatures are represented. Thus, a fivefold adsorption capacity is obtained as the temperature is reduced from about 150 to 0° C at a partial pressure of 2 mm Hg.

The heat released during adsorption increases the temperature of the carbon bed. For example, the solvent concentrations treated in paint-spraying or coating processes cause a temperature rise of about 15° F. This temperature increment has little influence on the adsorber performance. However, in other processes that handle high pollutant concentrations, dangerous temperature levels can be reached and some consideration must be given to providing temperature control in the adsorbent bed, either by cooling coils or by tempering the inlet gas flow.

Pollutant Type. The type of pollutant does not seriously influence the design of an adsorption system. In general, the solvents should have a boiling point less than 350° F so that they may be readily stripped from the adsorbent by the low-pressure steam. Hydrocarbons, chlorinated hydrocarbons, esters, ketones, and aromatic hydro-carbons can all be recovered from waste gas streams by adsorption on activated carbon. Generally, those organic compounds with molecular weights in excess of 45 can be considered as candidates for adsorption. Molecular sieves have a high adsorptive selectivity for polar compounds such as carbon dioxide, hydrogen sulfide, sulfur dioxide, and mercaptans.

Some organic compounds tend to decompose on active carbon surfaces. Thus, hydrogen sulfide will break down to yield elemental sulfur and sulfur compounds. This chemical conversion increases the adsorption potential of carbon for hydrogen sulfide. Solvents that are easily oxidized, such as the nitroparaffins and cyclohexanones, should be avoided.

In evaluating the application of an adsorption system for a specific emissions problem, the recovery of the pollutants must be considered. The solvent type, emissions concentration, and market value of the recovered end product should be defined, to determine the economic feasibility of recovery. It should be realized that when a number of solvents with differing adsorption characteristics is involved, adsorption selectivity will produce an end product of different composition than the original solvent mixture.

The fixed and operating cost differences between a water-immiscible solvents system requiring simple decantation recovery equipment and the miscible type, which must be recovered with distillation, must be examined.

Pollutant Concentration. The ideal design basis is minimum gas flow and maximum pollutant concentration. The former reduces the equipment size and the latter leads to a higher concentration of solvent being collected by the adsorbent. As indicated on the isotherm curves in Fig. 6-24, the degree of adsorption is proportional to the concentration of the solvent as expressed by its partial pressure.

The limiting factor on the pollutant concentration that can be processed is that imposed by the lower explosive limit (LEL) of the solvent in an air stream. As mentioned earlier, the use of combustion devices demands concentrations at 25% LEL. In adsorption, however, concentrations up to 50% LEL have been used. For example, benzene has an LEL of 1.4% by volume in air so that the ventilation rate may be adjusted to maintain the benzene vapor concentration below 0.7%.

Contaminants. The activity of an adsorbent progressively declines. The life of activated carbon in a solvents emission control system depends on the composition of the gas stream, and in some applications the carbon replacement has been found to be zero. When the gas stream contains traces of heavy organic compounds, such as those found in surface-coating operations, the carbon may become fouled and require replacement on a periodic schedule. With some solvent vapor streams it is necessary to use a scrubber ahead of the adsorption system to remove such particulates. A positive, regenerable filter is another device utilized to remove entrained dust or other particulates that might foul the adsorbent.

The life span of activated carbon depends on the nature of the pollutants being controlled. For clean solvents, a carbon life of 10 to 20 yr can be expected; for a stream containing trace amounts of high-boiling materials, 5 to 10 yr is reasonable; but the presence of polymerized solvents may require carbon reactivation every 1 to 3 yr.

The adsorbent charge may be regenerated in place or be removed and returned to the supplier for complete reactivation. In some surface-coating applications, superheated steam at about 650° F is used intermittently to strip out the high-boiling contaminants.

Before reactivation is necessary, an approximation of the expected adsorbent life for a known contaminant carryover can be calculated as follows:

Adsorbent requirement	120 lb/1000 scfm
Maximum contaminant loading	0.20 lb/lb adsorbent
Contaminant carryover	0.005 grain/scf

$$\text{Adsorbent life} = \frac{\text{max contaminent loading}}{\text{contaminant carryover rate}}$$

$$= \frac{0.20}{(0.005/7000) \times (1000/120) \times 60}$$

$$= 560 \text{ hr}$$

A life of only 560 hr between reactivations is not considered sufficient. Therefore, an efficient regenerable filter will be necessary to remove the particulate contaminants from the waste gas stream.

Adsorbent Selection. The quantity of solvent adsorbed per unit weight of adsorbent is dependent on the adsorbent/solvent equilibrium relationship and the operating conditions. These data are available from the various suppliers of adsorbent materials in the form of isotherms; see Fig. 6-24. Industrial activated carbons have solvent capacities in the range of 20 to 60% of adsorbent weight.

Activated carbons are most commonly furnished in U.S. Sieve mesh sizes of 4 by 10, 6 by 16, and 12 by 30. Additional intermediate mesh sizes are also available. The carbon granules must be sufficiently hard and abrasion-resistant to withstand attrition and deterioration. Disintegration of the adsorbent causes high-pressure drops and mechanical losses due to material carryover in the waste gas stream. The adsorbent should have maximum capacity to adsorb the pollutant and must be easily regenerated. Adsorbent costs must be traded off against capacity and physical properties for a specific duty. Low capacity at high solvent loadings must be balanced against adsorbent requirements and optimization of the cycle period.

Standard tests have been devised to permit evaluation of activated carbon adsorption activity, physical properties, and other characteristics. The more important of these tests are:

Adsorption activity: The quantity of carbon tetrachloride adsorbed is measured at 25° C and 760 mm Hg from air that has been saturated with carbon tetrachloride at 0° C. The value, expressed as a percentage of the original carbon weight, indicates the adsorptive capacity of the carbon for concentrated organic vapors.

Adsorptive retentitivity: the quantity of carbon tetrachloride remaining in the carbon after passing dry air through the bed, saturated from the activity test, for 6 hr at 25° C. The value, expressed as a percentage of the carbon, defines the ability of the carbon to retain a previously adsorbed vapor.

Hardness: A sample of 6–8 mesh carbon adsorbent is placed on a 14-mesh screen and agitated with steel balls for 30 min under specified conditions. The percentage of the carbon remaining on the screen is a measure of its resistance to attrition.

Mechanical Design

A fixed-bed adsorber is essentially comprised of a vessel containing a bed of adsorbent. For vapor and gaseous emissions control applications, the adsorbent is usually activated carbon. In fixed-bed adsorption, the process is usually semicontinuous, which imposes varying loads on the subsequent recovery system. Continuous adsorption can be accomplished by the use of a fluidized bed system.

The most common design for a fixed-bed adsorber comprises a vertical, cylindrical vessel provided with a grid type of support screen to contain the adsorbent. This type of design is shown in Fig. 6-34. The waste gas stream containing the solvents enters the vessel at the top and flows downward through the carbon bed. This flow arrangement permits the use of maximum gas velocities without causing serious bed movement. If the process being served is intermittent in its operation, a single vessel may be used, assuming the availability of downtime for regeneration.

Either vertical or horizontal vessels can be used to contain the adsorbent. Vertical units are more suited to high-pressure applications or where deeper or multiple beds are required. Horizontal cylindrical vessels, with the adsorbent bed parallel to the axis, is favored for the treatment of large gas flows and low-pressure drop requirements. Bed depths of 1 to 3 ft are usually employed. Where large gas flows are to be handled, baffles and distribution grids are necessary to ensure uniform gas flow through the beds. The regeneration fluids, which may be steam or hot air, flow upward through the adsorbent. Because the regeneration process flows are usually at a maximum of 10% of the waste gas throughputs, the gas velocity upward through the bed during regeneration is too low to disturb the adsorbent.

Materials of construction for the adsorber vessel, associated ductwork, and piping are selected as a function of the pollutants being processed. Mild steel may be used with alcohols, ethers, and hydrocarbons; stainless steels are used for esters and ketones because organic acids are produced from them through hydrolysis. For chlorinated hydrocarbons and fluorocarbons the use of highly corrosive-resistant nickel and cobalt alloys or graphite is recommended because of the formation of halogen acids during the adsorption of these compounds.

For the control of pollutants from a continuous process, a minimum of two adsorber vessels is required—one unit adsorbing and the other being stripped of solvent and regenerated. The cycle period for a specific configuration is determined by the stripping and regeneration requirements, with sufficient time being allowed for the adsorbent bed to be cooled before it is switched to adsorption service. For maximum adsorption efficiency, a three-stage system like that shown in Fig. 6-28 can be used. In this configuration two of the vessels are adsorbing in series. The gas stream from the first bed, after being stripped of solvent, is passed through the second bed, which has been regenerated but is still hot and wet. Thus, the solvent-free gas from the first unit removes the moisture from the second bed, thereby cooling the carbon by the evaporation of the moisture. Therefore, adsorption takes place in stages 1 and 2 in series while stage 3 is being regenerated. In the second cycle, adsorption is achieved in stages 2 and 3 while stage 1 is being regenerated, and finally stages 3 and 1 are adsorbing when stage 2 is regenerating.

A three-vessel adsorption system for the removal of mixed solvents is shown in Fig. 6-32. Two of the vessels, operating in parallel, are adsorbing while the third vessel is being regenerated. The system includes distillation facilities for recovery of the adsorbed solvents.

Standardized solvent recovery systems are available for a great number of processes discharging organic pollutants into the atmosphere. Skid-mounted units, complete with accessory equipment, have rated capacities in the range of 500 to 10,000 cfm. A process flow diagram for a packaged, two-vessel adsorption system—complete with blower, condenser, water separator, dampers, and instrumentation—is shown in Fig. 6-33. The equipment arrangement for this same system is shown in Fig. 6-34.

The design of the vessels to accommodate the adsorbent is relatively uncomplicated. Sizing of the equipment is a function of the gas flow rate, adsorbent activity, desired cycle time, and pressure drop. The cost of regeneration steam and power to move the waste gas through the system represent the major operating costs. A parallel

Figure 6-32. Mixed solvents adsorption recovery system. (Courtesy of Vulcan-Cincinnati, Inc.)

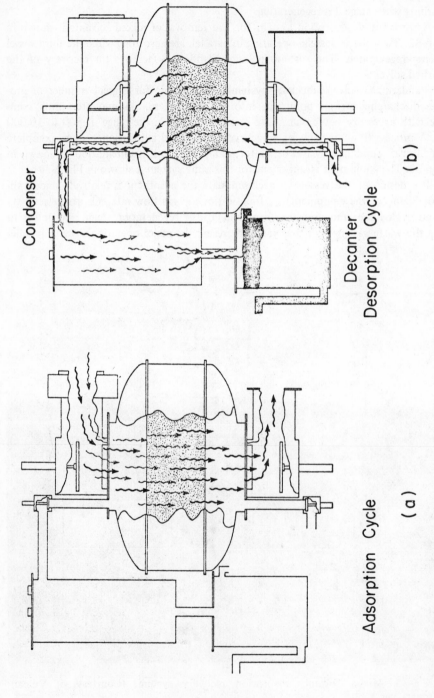

Condenser

Decanter

Desorption Cycle

(b)

Adsorption Cycle

(a)

Figure 6-33. Flow diagram for package adsorption system. (Courtesy of Vic Manufacturing Co.)

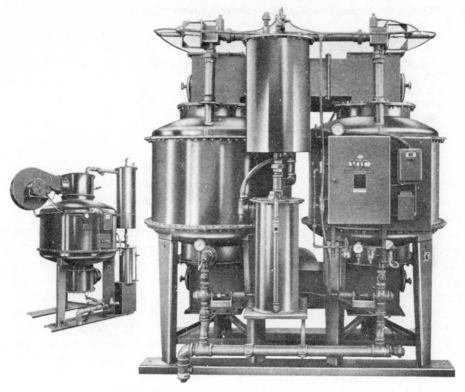

Figure 6-34. Package adsorption system arrangement. (Courtesy of Vic Manufacturing Co.)

can be drawn between the adsorption/regeneration and fabric-filtration/cleaning cycles in the determination of gas velocity, bed depth, cycle time, and pressure drop. Pressure-drop data for different activated carbon mesh sizes are shown in Fig. 6-30. The carbon-bed support grids are designed to contribute minimum resistance to the gas flow through the beds. For example, for a 4 by 10 mesh carbon adsorbent, a 10-mesh support screen is used with a wire diameter of 0.035 in. and a free opening of 0.065 in. square, yielding an estimated pressure drop of about 0.6 to 0.8 in. w.g./ft at the average gas flow rates. Physical properties of typical commercial activated carbons are shown in Table 6-9.

The key instrumentation item for a fixed-bed adsorption system is a cycle-timing mechanism. The cycle time is usually based on a specific time period that is adequate for both the adsorption and regeneration operations. Thus, the flows of waste gas and steam are automatically switched by the dampers controlled by the timing mechanism at the completion of each cycle. In some installations this switching is automatically accomplished by a vapor-sensing device located in the adsorbent bed. When that portion of the bed is fully charged, the cycle change is effected. In addition to the

Table 6-9. Typical Physical Properties of Activated Carbon

Form	Granules	Granules	Pellets	Pellets
Size, U.S. Sieve mesh	4/8,6/14,10/24	4/8,6/14,10/24	4/8	4/8
Carbon content, wt %	97	98	93	93
Ash, wt %	3	2	7	7
Bulk density, lb/ft³	32	28	34	28
Surface area, m²/g	1,100	1,400	950	1,100

Source: Nonhebel, Ref. 20.

cycle-control instrumentation, flow controls on plant air and steam are usually provided. Temperature indicators and alarms also sense conditions in the adsorbent bed to warn of potential fire risks.

In recent years, a fluidized bed approach to adsorption has been developed. In this process, the solvent-laden gas stream is continuously passed upward through the adsorber vessel, which contains a number of shallow, fluidized beds of activated carbon. The solvent is progressively adsorbed onto the carbon, the solvent-free gas being discharged from the top stage to the atmosphere through particulate collection devices for the recovery of carbon fines. The charged carbon is discharged from the bottom of the adsorber to a conveyor system, through a surge vessel, to a desorber where it is continuously and countercurrently contacted with low-pressure steam. The steam-solvent mixture is collected at the top of the vessel for recovery of the solvent while the regenerated carbon is returned to the top of the fluidized bed adsorber. The fluidized bed adsorption technique has a number of advantages over the fixed-bed method, such as: uniform adsorbent/solvent contact with very high removal efficiencies; completely automatic operation with minimum labor demands; minimum steam requirements with a more concentrated aqueous solvent product yield; and reduced operating hazards.

The continuous adsorption process is usually cost-competitive with batch operations for capacities in excess of 100,000 cfm, where the major objective is solvents recovery rather than air pollution control. Very recently a number of continuous adsorption systems, complete with recovery facilities, have been designed and installed for gas flows as low as 10,000 cfm.

Installation Features

The "package" adsorption system shown in Fig. 6-34 is shipped, as shown, on a skid mounting. It is completely prepiped and prewired so that field installation requires only the connection of ductwork between it and the process, liquor discharge piping to storage, steam and water lines for regeneration and condenser duties, and power leads to the fan and instrumentation panel. The dewatering system represented in this design is based on the collection of water-immiscible solvents so that a simple decantation method is applicable. The exhaust gas ductwork should be as direct as possible to minimize solvent condensation and attendant fire hazards. In many installations the cleaned air from the adsorption system is recycled to the process, to achieve fuel economy.

An important factor in the installation of an effective adsorption system is the design of an adequate hood and ductwork train to ensure maximum removal of the pollutants from the process. Because of the high solvents-removal efficiencies characteristic of adsorption processes, it has been found that reduced performance of the overall system is often caused by poor hood and ductwork design practices. In some applications there are unavoidable difficulties imposed by the process being served. For example, in a tank-filling operation, the collection of 100% of the offending fumes is practically impossible. Safety practices favor the use of a number of small individual hoods for the evacuation of multiple processes, rather than a single large hood. The smaller hood will maintain a higher air velocity at a particular air flow rate and prevent stratification and concentration of flammable pollutants in the ductwork. As in explosives-handling techniques, the use of multiple small hoods reduces the probability of flash fires that might occur if the solvents vapor concentration were to become critical.

Suitable atmospheric vents should be provided in any adsorption system where servicing of the equipment may involve the entry of personnel into a closed space. The atmosphere of the vessel and ductwork should be checked for the presence of any toxic or flammable pollutant, before maintenance activities are initiated.

All types of activated carbon, if raised to a certain temperature level and exposed to oxidizing conditions, will burn. The kindling point in oxygen is in excess of 350° C. However, the apparent kindling point of carbon that contains adsorbed solvents is considerably reduced because of the ignition potential of the adsorbed material. When handling "used" activated carbon, the possibility of dust explosions should not be ignored. Totally enclosed, dustproof electrical equipment should be used in the vicinity of such handling operations.

New or reactivated carbon may be stored in closed containers indefinitely without undergoing any noticeable deterioration. Precautions normally in effect for flammable materials should be taken. Because of the possibility of chemical reaction or desorption of the pollutant, used carbon should not be stored in any type of container for extended periods. Before the spent carbon is removed from the adsorbers, it must be thoroughly regenerated to remove as much of the pollutant(s) as possible, followed by cooling with air or inert gas. The carbon should then be stored outdoors in an unconfined state.

When spent carbon has been reactivated a number of times, indefinite reactivation is not practical for the following reasons. Because reactivation essentially involves some oxidation of the carbon, there is a limit beyond which most carbons can be further processed without serious impairment of their strength and retentivity. At this time in its life, the carbon adsorbent must be discarded. Disposal methods must be carefully evaluated for environmental conformity. If incineration is the chosen process, the adsorbed materials might create an air pollution problem. On the other hand, the possibility of soil or ground water contamination must be considered if land-fill disposal is proposed.

Applications

Adsorption techniques can be utilized to solve the following problems:

Solvent emission control
Odor removal
Corrosive atmospheres elimination
Process air purification

One major source of air pollution is the discharge of organic solvents from various industrial processes. A list of some of these industries and the solvent emissions that have been controlled by adsorption is shown in Table 6-10. Adsorbents used in these processes are some form of activated carbon rather than silica and alumina-base adsorbents because of carbon selectivity for organic vapors in the presence of water.

Activated carbon can adsorb vapors in concentrations from trace quantities to 100% by volume. Adsorption is most commonly applied to emissions in concentrations below the lower explosive limit, usually less than 2% by volume. In the range of 2 to 10 vol %, dilution of the waste gas stream, followed by adsorption, is favored. Above 10 vol %, other techniques discussed elsewhere, such as absorption and incineration, should be considered.

Operating data for an adsorption system to remove solvents from a printing press exhaust air stream is shown in Table 6-11. In this process a solvent mixture containing toluene, butyl acetate, and ethanol is removed from the process exhaust stream and is recovered and separated by a combination decantation/distillation train.

Table 6-10. Industrial Adsorption System Applications

Industry	*Solvents*
Acetate fibers processing	Ketones
Printing	Acetates, alcohols, hydrocarbons
Plastic coating	Ketones, acetates, alcohols hydrocarbons
Plastic production	Ketones, acetates, alcohols, hydrocarbons, ethers
Rubber production	Hydrocarbons, chlorinated solvents
Metals degreasing	Chlorinated and fluorinated solvents
Solvent extraction	Hydrocarbons, chlorinated solvents
Synthetic phenol processing	Hydrocarbons

Source: Enneking, Ref. 21.

Table 6-11. Operating Data for Printing Plant Solvents Recovery System

Exhaust volume, scfm	11,000
Exhaust temperature, °F	90
Solvents loading, lb/min	4.7
Solvents composition, wt %	
Toluene	40
Butyl acetate	14
Ethanol	46
Steam rate per adsorber, lb/hr	1,800
Steam/solvent ratio	3.5:1
Time cycle per adsorber, hr	
Adsorption	2.0
Steaming	0.5
Drying (220° F air)	0.25
Cooling (cold air)	0.25
Number of adsorbers	6
Arrangement, parallel/series	2/3
Adsorbent type	Activated carbon
Solvent adsorbed/cycle, wt %	10–14
Solvent concentration in air stream, lb/mscf	0.3–0.5
Adsorbent required, lb/ton recovered solvent	0.5–1.0
Utilities, per lb recovered solvent,	
Steam, lb	3.5–5.0
Water, gal	7–10
Power, kwhr	0.10–0.15

Source: Kohl and Riesenfeld, Ref. 14.

The application of adsorption systems for the removal of odorous or corrosive pollutants differs from that of solvent emissions control in that such air contaminants are present in lower concentrations and their recovery is not attempted. Gases and vapors such as mercaptans, acroleins, hydrogen sulfide, sulfur oxides, and similar compounds, which usually contain sulfur or nitrogen, fall into this category. Because of the low concentration of these pollutants, the load on the activated carbon is sufficiently small so that the adsorbent can be used for months before it becomes ineffective. This control system, besides achieving conformity to the "nuisance" pollution regulations, protects both plant personnel and equipment from the influence of these contaminants. For example, bread baking and photographic film processing will be seriously affected by mercaptans, and instrumentation and similar electronic equipment will malfunction in the presence of the sulfur oxides. In one installation where it was found necessary to treat the plant ventilation air, the sulfur dioxide content was reduced from 0.12 ppm to 0.002 ppm. Because large volumes of air must be handled in this type of application, very thin activated carbon beds are utilized to minimize the pressure drop across the equipment. Individual cylindrical canisters arranged in a suitable enclosure, somewhat like the multitube cyclonic separator, represent one type of design for this service. Each canister may contain from 0.5 to 1.5 lb of carbon arranged in an annular configuration, which provides a bed 3/8 in. to 3/4 in. thick. In this arrangement the carbon can be easily removed for regeneration by the suppliers of the adsorbent.

Another adsorption application involving the control of such odorous emissions as hydrogen sulfide and mixed mercaptans is represented by a pilot plant experience at the Hyperion Treatment Plant of the City of Los Angeles. In this installation a stream of ventilation air from the new sewage digestors was stripped of H_2S, skatoles, and indoles for more than 8 months in a single adsorber. The adsorbent comprised a single bed of 6 by 16 mesh activated carbon, 3 ft deep and 5 ft in diameter. The adsorber vessel was fabricated of epoxy-coated mild steel with a stainless steel grid for the adsorbent bed support. A gas flow of 1000 cfm at a humidity range of 35 to 60%, and containing sulfide compounds in a concentration of about 1 to 50 ppm, was introduced into the top of the bed at a superficial velocity of 50 fpm. During a 6000 hr test period, the hydrogen sulfide was completely removed and the remaining sulfide compounds were reduced to acceptable levels for most of this test period. During the first 4500 hr, 254 lb of H_2S were removed at an average carbon capacity of 14 wt %. According to available isotherm data for the operating conditions at this plant, the predicted adsorption potential for hydrogen sulfide should have been less than 1% of the carbon charge. As discussed under the heading "Pollutant Type," decomposition increased the adsorbate level to 14%.

The adsorption technique for the control of gaseous emissions does not seem to have kept pace with other commercially accepted methods such as thermal/catalytic incineration and absorption. The recent developments in continuous fluidized-bed adsorption systems will have very little impact on most industries where pollution emissions are usually associated with exhaust gas streams in the neighborhood of 5,000 to 50,000 cfm. As mentioned previously, economically feasible fluidized adsorption systems are in the capacity range of 100,000 cfm and upward.

The recovery of waste products from batch-operated adsorption systems is a difficult concept to promote unless the economic returns are extremely attractive. The majority of plant operators are anxious to eliminate their emissions problems without incurring the responsibility for operations in unknown technological areas. The package adsorption system illustrated in Fig. 6-34 is a step in the right direction toward the simplification of this method for air pollution control. However, this system is applicable only to a water-immiscible solvents system.

In summary, the most important operating parameter influencing the feasibility of emissions control by adsorption is that of solvents concentration. Unlike the combustion processes or wet scrubbing, adsorption performance is maintained at a high level at very low pollutant concentrations. Adsorption operating costs, comprising steam, power, and water requirements, are essentially unaffected by "low concentration" operating conditions. Combustion control costs increase considerably at low solvent concentrations because of the fuel gas necessary to heat the total process gas stream. The relative operating costs for incineration, catalytic combustion, and adsorption control processes are determined in the next section, Sample Design Calculations. It is believed that the scarcity and increasing costs of natural gas should favor the adsorption process over combustion methods, particularly at low concentration emission levels. Other techniques employing vacuum and hot air for activated carbon regeneration, now being developed, should reduce adsorption operating costs still further.

Sample Design Calculations

The design of an adsorption system will be developed for the problem previously solved by the application of thermal incineration and catalytic combustion. A restatement of the operating data follows.

Problem Statement

Process	Metal decorating
Equipment	Baking oven
Exhaust volume, scfm	6000
Exhaust temperature, °F	400
Solvents loading, lb/min	1.4
Solvents composition, wt %	
Hexane	35
Toluene	45
MEK	20
Particulate loading, grain/scf	0.08
Operating period, 24-hr days/yr	360
Steam available, psig	50
Removal efficiency (organics), %	98

Problem Solution. Data for the design of adsorption equipment must be determined experimentally. The adsorbent type, pollutant type and concentration, gas flow rate and temperature, and the desired removal efficiency are the major parameters influencing the bed dimensions and contact period.

1. *Design data:* A test vessel 2 ft 0 in. diam containing an 8 in. depth of 12 by 30 mesh activated carbon granules was operated with a sample of the exhaust gas stream defined in the Problem Statement. Values of the breakthrough value, θ_B, were determined at various values of the gas flow rate at an adsorption temperature of 150° F. The value of θ_B, which is the time required for the adsorbate to reach a given level in the adsorber effluent gas stream, was determined for a 98% removal efficiency.

Table 6-12. θ_B Values versus Gas Flow Rates

Exhaust Stream Condition Flow Rate, lb/hr-ft²	Temp, °F	Area, ft²	Depth, in.	Wt, lb.	Time θ_B, min.
		Adsorbent Mass			
100	145	3.14	8	58.6	420
150	157	3.14	8	58.6	240
200	150	3.14	8	58.6	150

Data for the three runs are given in Table 6-12.
A plot of θ_B versus W/Q is shown in Fig. 6-35, where

θ_B = breakthrough period, min
W = weight of adsorbent, 58.6 lb
Q = gas stream flow rate, lb/hr

From this curve an adsorption period of 335 min will be chosen. Design values are as follows:

Adsorbent/gas-flow ratio (W/Q) = 0.15
Gas flow rate (Q) = 391 lb/hr
Mass flow rate = 124 lb/hr-ft²
Average gas temperature = 150° F

2. Adsorbent requirements:

$$\text{Effluent gas flow rate} = 6000 \text{ scfm}$$

At 150° F,

$$\text{Flow} = 6000 \times \left(\frac{460 + 150}{460 + 60}\right) = 7040 \text{ acfm}$$

$$= \frac{6000}{379} \times 29 \times 60 = 27,600 \text{ lb/hr}$$

Adsorbent required = 27,600 × 0.15 = 4140 lb

Adsorbent capacity check. Solvents adsorbed in 335 min (5.6 hr):

$$\text{Weight} = 1.4 \text{ lb/min} \times 335 \times 0.98$$

$$= 460 \text{ lb}$$

Solvent adsorbed by carbon per cycle.

$$\text{Percent adsorbed} = \frac{460}{4140} \times 100 = 11.1$$

296

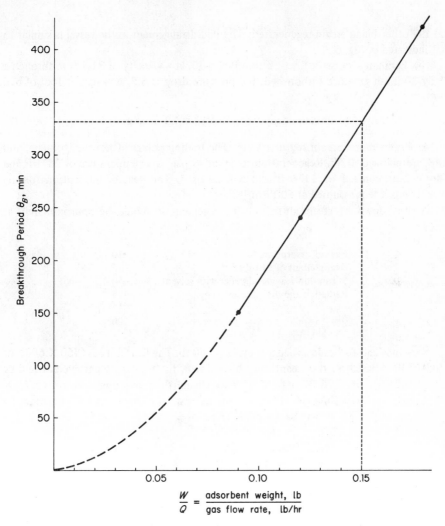

Figure 6-35. Breakthrough period at various exhaust gas flow rates.

3. *Adsorber design*

Mass flow rate (from test) = 124 lb/hr-ft^2
Superficial velocity = (124/60) × (379/29) × (610/520) = 31.7 fpm
Exhaust gas flow = 7040 acfm
Adsorbent cross section = 7040/31.7 = 222 ft^2
Total adsorbent charge = 4140 lb
Bulk density = 28 lb/ft^3
Bed depth = 4140 lb × (1 ft^3/28 lb) × (1/222 ft^2) = 0.67 ft

(a) Adsorber dimensions: 8 ft diam × 27 ft long, containing a single 8 in. bed of 12 by 30 mesh activated carbon granules. Two adsorbers are required, one adsorbing

Air Pollution

and the other being steam-regenerated. The bed arrangement in the vessel is similar to that indicated in Fig. 6-28.

(b) Estimated pressure drop: From Fig. 6-30, at a velocity of 31.7 fpm through a 12 by 30 mesh granular carbon bed, the pressure drop is 6.5 in. w.g. per foot of bed depth:

$$\text{Pressure drop} = 6.5 \times 8/12 = 4.3 \text{ in. w.g.}$$

4. *Regeneration steam requirements:* The boiling points of hexane, toluene, and MEK are below 250° F. Regeneration steam at 40 psig at a temperature of 287° F and a total heat content of 1176 Btu/lb will be used. The heat of adsorption for the solvents mixture is assumed at 600 Btu/lb.

A breakdown of steam duties during regeneration would be approximately as follows:

Heat of desorption, %	10
Heat content of adsorber, %	15
Steam flow leaving adsorber with solvents, %	70
Radiation and other losses, %	5
	100

Solvents load to be desorbed per cycle is 460 lb. The heat duty is (460 × 600), or 276,000 Btu. Based on the steam duty breakdown, the total duty per cycle would be (276,000/0.10), or 2.76×10^6 Btu. The steam flow requirements are ($2.76 \times 10^6/1176$), or 2350 lb/cycle. To maintain a reasonable load on the condenser, a steaming period of 1.5 hr will be adopted. Therefore,

$$\text{Steam flow rate} = 1,570 \text{ lb/hr}$$
$$\text{Annual steam requirements} = 3.64 \times 10^6 \text{ lb}$$

5. *Accessory equipment:* The accessory equipment for recovery of the various solvents would be a combination decantation/distillation system. The toluene-hexane mixtures, both immiscible in water, would be separated in the decanter and recovered as such. The water-soluble MEK would be separated from the decanter aqueous phase by distillation. An alternate scheme would be to re-use some or none of the immiscible solvents and sell the remainder to a solvent reclamation industry.

The exhaust gas stream leaving the oven is at a temperature of 400° F. A water-cooled heat exchanger is provided to reduce the gas temperature to 150° F.

The particulate loading value of 0.08 grain/scf represents resins and combustion by-products from the baking process. A regenerable filter must be provided ahead of the adsorbers to remove these particulates.

The design of the accessory equipment is beyond the scope of this problem. However, order-of-magnitude costs for this equipment will be developed to assess the economics of the overall system.

6. *Economic evaluation*

(a) Capital costs: The adsorber capital costs at a gas flow of 7040 acfm, complete

298

with fan, condenser, decanter, and instrumentation, is estimated at $35,000, unerected basis. This value is based on mild-steel fabrication on the assumption that the relatively low concentration of MEK will not require corrosive-resistant materials.

An order-of-magnitude cost value for the heat exchanger, filter, and distillation system would be an additional $80,000; thus, the total equipment would be $115,000 on an unerected basis. Erection costs would be estimated at 20% for the adsorption package unit and 40% for the accessory equipment, for a total installation charge of $39,000. Therefore, the total equipment cost, fully installed, would be $154,000.

(b) Operating costs: The major operating costs to be considered are steam, water, and adsorbent replacement costs.

Steam consumption was determined at 3.64×10^6 lb/yr for the adsorption process. This flow is discharged as condensate with the MEK, from the decanter to the distillation system. An additional steam load of 4.0×10^6 lb/yr is estimated as the demand for the distillation system, for a total of 7.64×10^6 lb/yr. At a steam cost of $0.80/1000 lb, the annual steam costs would be 6.2×10^3.

Cooling water is required for cooling the exhaust gases from 400 to 150° F plus that required for the adsorption and distillation condensation duties. These annual flows are:

Gas cooling	$= 6.40 \times 10^6$ Btu/cycle
Adsorption condensation	$= 4.00 \times 10^6$
Distillation condensation	$= 4.30 \times 10^6$
Total	$= 14.70 \times 10^6$ Btu/cycle

Assuming an allowable water temperature rise of 100° F, then annual water requirements would be

$$\frac{14.70 \times 10^6}{5.6} \times 24 \times 360 \times \frac{1}{(100 \times 8.33)} = 27.6 \times 10^6 \text{ gal}$$

At $0.15 per 1000 gal,

$$\text{Annual water cost} = 27.6 \times 10^6 \times \frac{0.15}{10^3} = \$4.2 \times 10^3$$

Despite the use of the regenerable filter, it would be expected that some polymerized solvents and resins from the baking operation would enter the adsorbers. Using a very conservative design basis, it will be assumed that both activated carbon beds must be replaced every 2 yr. Therefore, the total activated carbon required per year would be $(4140 \times 2)/2 = 4140$ lb.

$$\text{Activated carbon cost} = \$0.50/\text{lb}.$$
$$\text{Annual cost} = 4140 \times 0.50 = \$2.07 \times 10^3$$

(c) Amortization costs: The total installed costs for the complete system is $154,000. For an amortization period of 15 yr, the annual depreciation costs would be (154,000/15), or 10.3×10^3.

(d) Credits: The recovered solvents can be credited against the operating costs. It is assumed that 5% losses will occur in the decantation and distillation operations. Therefore, the recovered solvent value is

$$\frac{460 \text{ lb}}{5.6 \text{ hr}} \times 0.95 \times 360 \times 24 = 680 \times 10^3 \text{ lb/yr}$$

Based on a composite selling price for the particular solvents mixture of $3.00/100 lb, the annual income would be

Table 6-13. Adsorption System Design and Cost Data

Application:	*Metal decorating oven*
Exhaust volume, scfm	6000
Exhaust temperature, °F	400
Solvents loading, lb/min	1.4
Solvents composition, wt %	
Hexane	35
Toluene	45
MEK	20
Particulates loading, grain/scf	0.08
Operating period	360(24-hr day)/yr
Performance	98% organics removed.
Adsorption period	5.6 hr
Equipment type	Fixed bed adsorber
No. and dimensions	2(8 ft diam × 27 ft long)
Adsorbent bed depth, ft	0.67
Press. drop, in. w.g.	4.3
Accessory equipment	Fan, condenser, decanter, distillation system, heat exchanger and instrumentation
Equipment purchase cost, $	
Adsorber, condenser, decanter, and fan	35,000
Heat exchanger, filter, and distillation system	80,000
Total installation cost	39,000
Annual depreciation costs	10.3×10^3
Annual operating costs, $	
Steam	6.2×10^3
Water	4.2×10^3
Adsorbent	2.1×10^3
Total annual operating costs, $	22.8×10^3
Credit for recovered solvents, $/yr	20.4×10^3
Net annual operating costs, $	2.4×10^3

$$680 \times 10^3 \times \frac{3.00}{100} = \$20.4 \times 10^3$$

7. *Design Summary*. Table 6-13 summarizes the pertinent design, operating, and cost data for the adsorption recovery system.

A comparison of capital and operating costs for adsorption, as compared with thermal incineration and catalytic combustion equipment, is illustrated in Table 6-14. Power requirements for moving the process gas stream were not included in this comparison. The operating cost for this utility would favor the incineration and combustion processes, with their pressure drop of less than 1.0 in. w.g. when compared with the adsorption system pressure drop of 4.3 in. w.g. However, the overall influence of the additional operating cost would be negligible.

The total operating cost values indicate a considerable operating cost advantage in favor of adsorption. The major disadvantage is the capital cost magnitude, with the major portion of this investment value being required for accessory equipment. In actual

Table 6-14. Cost Summary for Thermal Incineration, Catalytic Combustion, and Adsorption Systems

Application: metal decorating oven
Gas flow: 6000 scfm
Gas temperature: 400°F
Duty: 98% organic removal

	Equipment Type		
Costs	*Thermal Incineration*	*Catalytic Combustion with Heat Exchanger*	*Adsorption with Recovery System*
Total capital cost, $ (erected basis)	36,000	85,500*	154,000
Depreciation, 10^3/yr	2.4	5.7	10.3
Utilities, 10^3/yr			
Gas	35.8	6.7	—
Steam	—		6.2
Water	—		4.2
Adsorbent	—		2.1
Total Operating, 10^3/yr	38.2	12.4	22.8
Credits, 10^3/yr	—	6.7†	20.4
Net operating costs, 10^3/yr	38.2	5.7	2.4

*Estimated for a total catalytic system rather than the oven-redesign unit developed at the end of Section 6.6.

†Heat recovery savings from catalytic combustion of solvents.

301

practice, adsorption would not be too seriously considered for a bake-oven emissions problem because of the high process gas temperature, particulates loading and solvents recovery requirements.

6.8. EQUIPMENT CHARACTERISTICS–WET SCRUBBING

The application of wet scrubbing to the removal of particulates was discussed in Chapter 5. The wet scrubber is unique in its ability to remove both particulate and gaseous pollutants. The principles of mass transfer in a wet scrubber, as applied to the collection of particulates, have not been theoretically established. However, gas absorption is on a more firm footing in this respect because this process has been developed as one of the basic unit operations in chemical engineering. In air pollution control, gaseous emissions abatement is usually concerned with low gas concentrations. In problems involving the simultaneous control of particulate and gaseous pollutants, the selection and design of a wet scrubber for a specific duty must be predicated on the operating principles controlling the removal of each pollutant type.

Operating Principles

Just as for particulates removal, wet scrubbers for the absorption of gases must be designed to provide maximum contact between the gas phase and the scrubbing liquor. The rate of transfer between these two phases is dependent on the gas solubility and the contact surface between both phases. For any particular gas-liquid system there are many mechanisms available to provide intimate mixing, such as atomization, impingement, spraying, and agitation. Some of the specific designs employing these mechanisms were described in Chapter 5.

One of the most simple of the impingement types of wet scrubbers for the absorption of gases is the packed tower. In this design the scrubbing liquor is introduced at the top and flows downward through the packing countercurrently to the incoming gas, which enters at the scrubber base. The packing is designed to present a maximum area of liquid film for contacting the solute gas.

The fundamental principles controlling gas absorption have been developed and known for some time [22]. In a diffusion process such as gas absorption, the rate of mass transfer is dependent upon the available contact surface of the scrubber liquor and the gas. Mass transfer from the gas to the liquid phase may be defined by

$$N = K_G aV \Delta Y \qquad (6\text{-}14)$$

where

N = solute transferred from the gas to the absorbent, lb moles/hr
$K_G a$ = gas capacity coefficient, lb moles/hr-ft^3-mole ratio
a = interfacial contact area, ft^2/ft^3 packing
V = packing volume, ft^3
Y = solute concentration ratio in gas, lb moles solute/lb mole carrier gas

The quantity a represents the effective interfacial area between the gas and the liquid. It is usually interpreted to be the surface area of the packing.

Similarly, the mass transfer from the liquid to the gas phase is represented by

$$N = K_L \, aV \, \Delta X \qquad (6\text{-}15)$$

where

$K_L \, a$ = liquid capacity coefficient, lb mole/hr-ft^3

X = solute concentration ratio in liquid, lb moles solute/lb mole solvent

Because overall coefficients vary with the gas and liquor rates, Chilton and Colburn [23] developed the concept of height of a transfer unit (HTU). This concept is useful for gaseous emissions control when the gas concentration is usually dilute. In such processes the gas film resistance most often controls when a solvent is chosen in which the solute is very soluble. Thus, the height of a gas-phase transfer unit can be expressed as

$$H_{OG} = \frac{G_M}{K_G \, aP}$$

where

H_{OG} = height of overall gas-phase transfer unit, ft

G_M = gas molar-mass velocity, lb moles/hr-ft^2

P = total pressure in system, atm

A number of empirical generalized relationships [24] have been developed for determining the height of transfer units, based on both the gas film and liquid film controlling. These formulas are based on experimentally derived operating parameters such as packing characteristics, gas and liquor flow rates, and fluid properties. Thus,

$$H_G = \frac{\alpha G'^{\beta}}{L'^{\gamma}} \left(\frac{\mu_G}{\rho_G D_G} \right)^{0.5} \qquad (6\text{-}17)$$

where

H_G = height of a gas phase transfer unit, ft

G' = gas mass flow rate, lb/hr-ft^2

L' = liquid mass flow rate, lb/hr-ft^2

$\alpha\beta\gamma$ = packing constants; see Table 6-15

μ_G = gas viscosity, lb/hr-ft

ρ_G = gas density, lb/ft^3

D_G = gas diffusivity, ft^2/hr

The expression $\mu_G/\rho_G D_G$ is known as the Schmidt number.

Similarly, the formula for calculating the transfer unit height, based on the liquid phase controlling, would be

Table 6-15. Constants for Determination of Transfer Unit Heights

Packing Type and Size	Gas Film Transfer Factors					Liquid Film Transfer Factors		
	Fluid Flow, lb/hr-ft²		Constants			Fluid Flow, lb/hr-ft²	Constants	
	G'	L'	α	β	γ	L'	φ	η
Raschig rings								
3/8 in.	200–500	500–1500	2.32	0.45	0.47	400–15,000	0.00182	0.46
1 in.	200–800	400–500	7.00	0.39	0.58	400–15,000	0.0100	0.22
	200–600	500–4500	6.41	0.32	0.51			
1 1/2 in.	200–700	500–1500	17.30	0.38	0.66	400–15,000	0.0111	0.22
	200–700	1500–4500	2.58	0.38	0.40			
2 in.	200–800	500–4500	3.82	0.41	0.45	400–15,000	0.0125	0.22
Berl saddles								
1/2 in.	200–700	500–1500	32.40	0.30	0.74	400–15,000	0.0066	0.28
	200–700	1500–4500	0.81	0.30	0.24			
1 in.	200–800	400–4500	1.97	0.36	0.40	400–15,000	0.00588	0.28
1 1/2 in.	200–1000	400–4500	5.05	0.32	0.45	400–15,000	0.00625	0.28
Partition rings								
3 in.	150–900	3,000–10,000	650	0.58	1.06	3,000–14,000	0.0625	0.09

Source: Treybal, Ref. 24.

$$H_L = \phi \left(\frac{L'}{\mu_L} \right)^{\eta} \left(\frac{\mu_L}{\rho_L D_L} \right)^{0.5} \qquad (6\text{-}18)$$

where

H_L = height of liquid transfer unit, ft
ϕ, η = packing constants (see Table 6-15)
L' = liquid mass flow rate, lb/hr-ft^2
μ_L = liquid viscosity, lb/hr-ft
ρ_L = liquid density, lb/ft^3
D_L = liquid diffusivity, ft^2/hr

The value of $\mu_L/\rho_L D_L$ is the Schmidt number, based on the solvent properties. Values of the Schmidt number for both solute and solvent phases are given for a number of gases in Table 6-16.

Both Eqs. 6-17 and 6-18 neglect each other's film resistance effect. Although in most air pollution applications the significance of the liquid film resistance is not too evident, in some systems such as ammonia-water experimental results have indicated that it does have an influence on the mass transfer. The following expression for the height of an overall gas-transfer unit, H_{OG}, is modified to allow for the influence of the liquid film resistance.

$$H_{OG} = H_G + m\,(G_M/L_M)H_L \qquad (6\text{-}19)$$

where

m = slope of equilibrium curve
G_M = gas rate, lb moles/hr-ft^2
L_M = liquid rate, lb moles/hr-ft^2

Table 6-16. Diffusion Coefficients (Schmidt Numbers) for Fluids

Gases and Vapors	Diffusion Coefficient, $\mu_G/\rho_G D_G$ in Air at 25°C	Diffusion Coefficient, $\mu_L/\rho_L D_L$ in Water at 20°C
Ammonia	0.66	—
Oxygen	0.75	558
Carbon dioxide	0.94	570
Hydrogen	0.22	196
Water	0.60	—
Chlorine	—	570
Hydrogen sulfide	—	712
Methanol	0.97	785
Ethanol	1.30	1,005
Butanol	1.72	1,310
Acetic acid	1.16	1,140
Aniline	2.14	—
Propanol	1.55	1,150
Toluene	1.84	—

Source: Perry, Ref. 25.

For most applications there are sufficient empirical data available to make it possible to utilize Eqs. 6-17 through 6-19 for design estimates.

A transfer unit is a measure of the mass transfer requirements of a specific problem. In Chapter 5 it was expressed as a function of the particulates transferred by differential contacting. In gas absorption the value of N defines the number of equilibrium stages necessary to convert the solute concentration in the gas stream entering a wet scrubber to the required exit concentration. It is based on the assumption of integral or stepwise contacting. Therefore, the height of packing required to achieve this concentration gradient can be expressed as

$$h = N_{OG} \times H_{OG} \qquad (6\text{-}20)$$

where

$\qquad h$ = height of packed section of tower, ft
$\qquad N_{OG}$ = number of overall gas-phase transfer units
$\qquad H_{OG}$ = height of an overall gas-transfer unit, ft

The number of transfer units can be expressed either as N_{OG} for the gas-film or N_{OL} for the liquid-film resistance controlling. By the selection of a high solubility solute-solvent system, the gas film resistance can be made to control. In such a case,

$$N_{OG} = \frac{\ln\left[\{1 - (mG_M/L_M)\}\{(Y_1 - mX_2)/(Y_2 - mX_2)\} + \{m(G_M/L_M)\}\right]}{1 - (mG_M/L_M)} \qquad (6\text{-}21)$$

where

$\qquad Y_1$ = mole ratio of solute in gas stream at concentrated end of
$\qquad\qquad$ countercurrent flow packed tower, lb moles solute/lb mole
$\qquad\qquad$ inert gas
$\qquad Y_2$ = mole ratio of solute in gas stream at dilute end of
$\qquad\qquad$ countercurrent flow packed tower, lb moles solute/lb mole inert
$\qquad\qquad$ gas
$\qquad X_2$ = mole ratio of solute in solvent stream at dilute end of
$\qquad\qquad$ countercurrent flow packed tower, lb moles solute/lb mole solvent

As an alternative to calculating the value of N_{OG} by Eq. 6-21, Baker [26] proposed a graphical method for determining the number of transfer units. The method is based on plotting the system equilibrium curve and an operating line on the same set of coordinates, as shown in Fig. 6-36. Equilibrium conditions for a given gas-liquid system can be determined for a fixed temperature at a number of concentration states and plotted as X-Y data. An operating line is then plotted to define the solute concentration in the gas stream and the liquid stream at the inlet and discharge of the tower. When plotted as moles of solute per mole of solvent (X) versus moles of solute per mole of gas (Y), a straight line can be drawn. The relative position of this straight operating line above the equilibrium curve is a measure of the differential between the tower-operating conditions and equilibrium. The greater the distance between the two curves, the further removed are the tower conditions from equilibrium, and the greater is the absorption driving force and the fewer transfer units

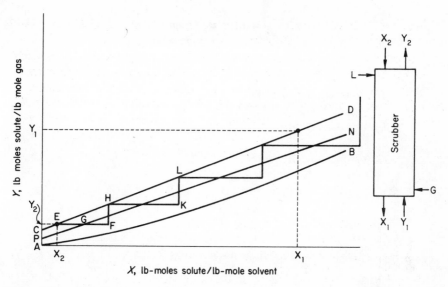

Figure 6-36. Graphical method for determination of number of transfer units. (Adapted from Baker, Ref. 26.)

required. In Fig. 6-36, the equilibrium curve is represented by line AB and the operating line by CD. Line PN is placed with ordinates midway between those of the operating line CD and the equilibrium curve AB. Assuming point E represents the operating conditions leaving the tower, line EGF is drawn so that EG is equal to GF. Then line FH is drawn, then HK, and then KL. If point L had represented the conditions at the inlet of the tower, then two transfer units (N_{OG}) would be required. To accomplish the absorption defined by Y_1, X_1 at the base of the scrubber, and Y_2, X_2 at the scrubber gas discharge, four transfer units are required.

The height of packing in a wet scrubber determines the absorption efficiency, and the cross-sectional area controls the capacity or the flow rate of gas that can be processed. In packed towers the gas and liquor flow rates are limited by a condition known as "flooding." As either the liquor or gas flow rate is increased, the liquid holdup in the packing increases and the free area for gas flow decreases. A situation known as the "flood-point" develops when most of the packing voids are filled with liquor and the gas flow falls off as the pressure drop builds up excessively. It is common practice to design packed towers for gas and liquor flow rates equivalent to pressure-drop conditions at 50 to 75% of the flood point.

Each type of packing has its own pressure-drop characteristics. This property, for any particular packing type, must be considered together with the available surface area of the packing. Thus, the absorption performance must be traded off against operating costs in terms of the power consumption required to overcome pressure drop. There are numerous packing types and shapes, varying from the fixed Raschig ring packing to the mobile hollow-sphere type. An empirical flooding correlation for various packing types can be expressed by the following formula [27]:

307

$$\frac{L'}{G'}(\rho_G/\rho_L)^{0.5} = K\,\frac{u^2\,\rho_G\,a\mu_1^{0.2}}{g_c\,\rho_L\,\epsilon^3} \tag{6-22}$$

where

L' = liquid flow rate, lb/hr-ft^2
G' = gas flow rate, lb/hr-ft^2
ρ_G = gas density, lb/ft^3
ρ_L = liquid density, lb/ft^3
K = proportionality factor
u = gas velocity, fps
a/ϵ^3 = packing factor; see Fig. 6-37

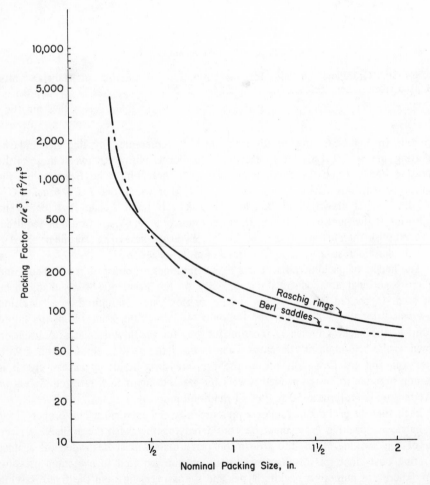

Figure 6-37. Flooding velocity factors for various packings. (Adapted from Lobo et al., Ref. 27.)

μ_1 = liquid viscosity, centipoise

g_c = gravity factor, 32.2 ft/sec^2

This relationship is plotted in Fig. 6-38 for stacked Raschig rings, dumped Raschig rings, and dumped saddles. Thus, for any L/G ratio, a knowledge of the physical properties of both fluids, and the factor a/ϵ^3 for the specific packing, the value of the gas flooding velocity u can be determined. By applying the recommended design percentage to the flooding velocity value, the operating gas velocity through the tower can be defined.

When the liquid and gas mass-flow rates have been established for a particular gas-liquid system to satisfy the desired performance, the pressure drop across the scrubber should be determined for a specific type of packing. Thus,

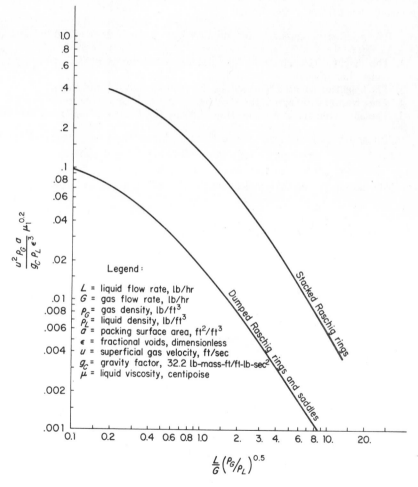

Figure 6-38. Flooding velocity correlations for various tower packings. (Adapted from Lobo et al., Ref. 27.)

$$\frac{\Delta p}{h} = m_1 \times 10^{-8} \left(10^{n_1 L'/\rho_L}\right) \frac{G'^2}{\rho_G} \qquad (6\text{-}23)$$

where

Δp = pressure drop, lb/ft^2
h = packed height, ft
n_1, m_1 = packing constants; see Table 6-17
L' = liquid mass flow rate, $lb/hr\text{-}ft^2$
ρ_L = liquid density, lb/ft^3
G' = gas mass flow rate, $lb/hr\text{-}ft^2$
ρ_G = gas density, lb/ft^3

The design sequence for a packed, countercurrent type of scrubber for a particular service would be as follows: Determine

1. The number of mass transfer units (N_{OG}) by either the method in Fig. 6-36, or Eq. 6-21.
2. The height of transfer unit H_{OG} for gas phase, liquid phase, or both phases controlling in accordance with Eqs. 6-17, 6-18, or 6-19.
3. The height of packing required, as the product of N_{OG} and H_{OG}.
4. The pressure drop across the scrubber by Eq. 6-23 and Table 6-17.
5. The allowable gas and liquid flow rates by Eq. 6-22, Fig. 6-37, and Fig. 6-38.

The foregoing relationships apply to fixed packings. More recently a mobile type of packing has been developed comprising hollow lightweight spheres, which move about in the tower under the opposing forces of the liquid and gas streams. Experimental data have indicated that considerably increased gas and liquid flow rates can be

Table 6-17. Pressure-Drop Constants for Tower Packings

Packing Type and Size	L' Value Range, lb/hr-ft²	Δp/h Range, lb/ft²-ft	m₁	n₁
Raschig Rings				
1/2 in.	300 – 8,600	0–2.6	139	0.00720
1 in.	360 – 27,000	0–2.6	32.10	0.00434
1 1/2 in.	720 – 18,000	0–2.6	12.08	0.00398
2 in.	720 – 21,000	0–2.6	11.13	0.00295
Berl saddles				
1/2 in.	300 – 14,100	0–2.6	60.40	0.00340
1 in.	720 – 78,800	0–2.6	16.01	0.00295
1 1/2 in.	720 – 21,600	0–2.6	8.01	0.00225
Intalox saddles				
1 in.	2,520 – 14,400	0–2.6	12.44	0.00277
1 1/2 in.	2,520 – 14,400	0–2.6	5.66	0.00225

Source: Treybal, Ref. 24.

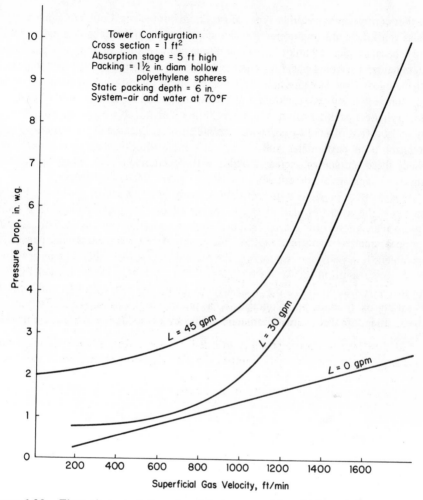

Figure 6-39. Flow characteristics of mobile tower packing. (Adapted from Douglas et al., Ref. 28.)

obtained at lower pressure drops with this packing over the fixed types. Relative fluid flow and pressure-drop values are shown in Fig. 6-39. Thus, for 1½ in. Raschig rings, an estimated pressure drop of about 3.5 in. w.g. for a 5 ft packing depth would be expected at a gas velocity and liquor flow rate of 105 fpm and 45 gpm, respectively. From Fig. 6-39 the allowable gas and liquor flows for the mobile packing would be 950 fpm and 45 gpm at this same pressure drop.

The operating principles for countercurrent-flow packed scrubbers have been discussed in some detail because semitheoretical concepts to explain absorption phenomena have been developed almost exclusively for this type of contacting device. Actually, the fixed packing type of scrubber for gas absorption, although still utilized for many industrial gas-conditioning applications, is being replaced in many air

311

pollution control areas by other types of gas-liquid contactors. Contacting mechanisms such as venturi, liquid impingement, cyclonic, spraying, and mechanical agitation are applied because such techniques can simultaneously remove gaseous and particulate contaminants. The fixed packing scrubber is somewhat limited in its ability to process particulate-laden gas streams or to handle scrubbing liquors that contain suspended solids, since it is subject to fouling and eventual plugging. On the other hand, the mobile spherical packing can handle heavy liquor slurries; this permits its application to the absorption of acidic gases with alkaline slurries. Because of the increasing need to control both particulates and gases, other interesting packing shapes have been devised; these emphasize increased voids with less tendency for plugging. Various commercial packings are described in the later section "Mechanical Design."

As discussed previously in connection with other types of pollution control equipment, the recent escalation of control regulations has made increasing demands on performance levels for gaseous removal by wet scrubbing equipment. For example, in copper smelter operations, a 75% reduction of SO_x emissions was considered acceptable a few years ago, whereas at the present time a 90% value is being enforced. The economic solution for such a typical problem can best be determined empirically through the operation of prototype equipment. However, even with this approach, the application of fundamental concepts is a prime requisite for the definition and understanding of the major parameters influencing the design of the commercial equipment.

Design Parameters

The various parameters controlling the design of a wet scrubber for gas absorption are absorption efficiency, gas flow, gas temperature and moisture content, gaseous pollutants, pollutant concentration, liquor type, liquid/gas ratio, and pressure drop.

Absorption Efficiency. In considering the various design parameters influencing absorption efficiency, some consideration must be given to the presence of particulates and the need for their removal. The absorption of a gaseous pollutant from a clean gas stream could be achieved very economically with a fixed packing scrubber, providing water or a soluble scrubbing liquor is available. However, should it be necessary to remove particulates as well, as often is the case, then a different type of scrubber must be considered with different absorption efficiency characteristics. For the absorption of chlorine, a removal efficiency up to 95% can easily be obtained by water scrubbing in a packed tower. However, if particulates are present, such as submicron aluminum chloride in an aluminum-processing effluent, then a venturi scrubber would be recommended for the relatively more difficult task of particulate removal and the absorption efficiency potential would be reduced.

Currently, the air pollution regulations for the control of gaseous emissions are not too severe. For example, for coal-fired utility boilers, the required SO_x absorption efficiencies are in the range of 75 to 90%. Even the presence of flyash does not make this control level too difficult to attain in the majority of commercial scrubber designs, providing an appropriate scrubbing liquor is employed.

The parameters defined in Eqs. 6-17 through 6-21 will permit the design of a packed tower to meet the absorption efficiency defined by the terms Y_1 and Y_2 in

Eq. 6-21. For other commercial types of scrubbers, empirical data based on test equipment or commercial installations must be relied on to establish a feasible design.

Gas Flow. Generally, gas flow determines the cross-sectional area of the wet scrubber, although both liquid and gas flows must be considered. Single scrubbers have been designed to handle flows as high as 250,000 cfm. As explained in Chapter 5, the gas flow must be corrected for temperature and moisture content before determining its saturated volumetric flow. Thus, a saturated gas flow of 250,000 cfm at 130° F through the scrubber could be equivalent to a scrubber inlet flow of 350,000 cfm at 400° F. Methods for determining this saturated gas flow are shown in the Sample Design Calculations of Chapter 5.

The gas flow through a scrubber, as related to the liquor flow, influences the operating pressure drop. In Fig. 6-39, an increase in the gas flow rate from 600 to 900 ft^3/ft^2 cross-sectional area is responsible for a gain in the pressure drop of 1.0 to 1.7 in. w.g. at a liquid flow rate of 30 gal/min-ft^2.

Gas Temperature. The temperature of the gas stream entering the scrubber influences the saturated gas flow rate and affects the sizing of the absorption equipment. An effluent gas stream from a pyrometallurgical process, at about 1600° F, undergoes a reduction in volumetric flow of about 2:1, as it is humidified and cooled to its saturation temperature in the scrubber.

The temperature also affects the absorption characteristics of a particular solute-solvent system. For example, for a hydrogen chloride-water system, the vapor pressure of HCl increases appreciably with increased temperature. Therefore, the values of Y, moles of solute per mole of gas, for the equilibrium curve in Fig. 6-36 would increase with increasing temperature. If the same performance were required as defined by the operating line, and the equilibrium curve was advanced upward in closer proximity to this operating line, then the number of transfer units would be increased. Or, as a corollary, for a specific scrubber design, an increase in the operating temperature would decrease the absorption efficiency.

As the temperature of the inlet gas is increased, at any fixed value of the gas moisture content, both saturation temperature and moisture content of the gas stream leaving the scrubber will be increased. Thus, by referring to the psychrometric chart in Fig. 5-40, it can be seen that if the inlet gas stream temperature were at 800° F rather than 400° F, at a moisture content of 0.04 lb-water/lb-dry gas, the scrubber discharge temperature would be increased from 130 to 150° F, and the moisture content would be raised from 0.11 to 0.21 lb H_2O/lb dry gas. At both temperature levels the moisture conditions cause the formation of a steam plume as the saturated gases are diffused in atmospheric air. Such plumes are formed by the moisture supersaturation conditions of the gas stream as its temperature becomes reduced by contact with the cooler atmospheric air. Loss of gas buoyancy, formation and precipitation of gaseous mist droplets in the neighborhood of the stack, and the abuse of air quality esthetics are the major charges leveled at wet scrubber steam plumes. Every scrubber operated at a saturation temperature in excess of the ambient air value will produce a plume.

There are two solutions to the steam plume problem. One involves heating the scrubber effluent gas by extraction of some of the heat content of the incoming gas through a system of heat exchangers. In the second approach, the saturation temperature can be reduced in the scrubber either by the introduction of copious quantities of cold scrubbing liquid or by the use of an external liquid-to-liquid heat exchanger, which cools the recirculated scrubbing liquor. The first method delays the formation of the plume and somewhat improves the dispersion pattern. Reheat temperature gains of 50 to 100° F are common, depending on the heat exchanger design and costs. The theoretical objective of the second method is to reduce the temperature of the discharged gas below that of the ambient air. This condition is attainable in summer climates, and stack plume appearances have been noticeably improved by this means. The degree of heat extraction is controlled by the available amount of cooling water, and by costs. In cooling the saturated gas stream, the moisture content is also reduced, with some improvement in the plume appearance even at low-temperature ambient conditions. In one paper-mill installation in northeastern Canada, the temperature of scrubber gaseous effluent from a recovery boiler was reduced from 167 to 90° F by the use of an external heat exchanger. This reduction was equivalent to a decrease in the saturation water content from 0.39 to 0.03 lb/lb dry gas.

Moisture Content. Both the inlet gas moisture content and its temperature determines the saturated gas temperature and moisture content in the scrubber. Thus, an increase in the moisture content of a process gas stream at a constant temperature will raise the gas saturation temperature and corresponding moisture content.

The gas moisture content also influences the water material balance of the scrubber, and is an important factor in the determination of the makeup scrubbing liquor flow. This effect will be demonstrated in the Sample Design Calculations for this section.

Gaseous Pollutant. The type of gaseous pollutant will decide the selection of the liquor system. Obviously, the most economical solvent is water, but despite its availability and low cost, the disposal of low pH liquors from the absorption of acidic gaseous pollutants produces a serious water quality problem.

The gaseous pollutants most commonly controlled by absorption are sulfur dioxide, chlorine, hydrogen chloride, hydrogen fluoride, ammonia, hydrogen sulfide, and water-miscible hydrocarbons. With the technology currently available, nitrogen oxides cannot be controlled at an acceptable level by absorption methods. In very few of the gaseous emissions applications is recovery of the pollutant values feasible.

One of the most active areas of technological activity is the study of SO_x absorption processes as applied to the control of coal-fired utility boiler emissions. Although the required removal efficiencies are not too stringent and commercial equipment is available to perform adequately, the most serious defect of the alkaline scrubbing system is the problem of the sulfite-sulfate liquor effluent disposal. It is this deficiency in the total system that has denied the wet scrubber complete acceptance for this application. Therefore, considerable development efforts are being expended on alternate control systems that incorporate recovery facilities. On the other hand, in the electrolytic reduction of alumina, hydrogen fluoride is evolved and adequately absorbed, together with particulates collection, in a wet scrubbing system employing a sodium carbonate scrubbing solution. The sodium fluoride liquor effluent is converted to cryolite (Na_3AlF_6), which is re-used in the process. This is an ideal pollution control-recovery situation. Thus, when a gaseous pollutant is identified, both absorption performance and recovery potential must be explored. The disposal of scrubber effluents is rapidly becoming a serious environmental problem and this trend, spurred by water quality control regulations, must be recognized in the design of gaseous emissions control equipment.

The type of gaseous pollutant and its temperature will determine the selection of materials of construction. Ammonia and hydrocarbons can be processed in mild-steel equipment, whereas the acidic gases, such as SO_2, Cl_2, etc., require the use of exotic fabrication materials, either of all-metal or polyester construction or plastic and glass-lined materials.

Pollutant Concentration. The concentration of the gaseous pollutant determines the level of absorption performance that must be achieved to comply with the applicable control regulation. In the combustion of coal having sulfur content in the range of 2 to 4%, the equivalent SO_x concentrations in the flue gas stream are 1400 and 2800 ppm. Regulatory emission requirements are commonly based on the heat-release rating of the specific boiler in question. For example, one midwestern state regulation permits a maximum SO_x emission of $0.80 \, lb/10^6$ Btu, which is equivalent to 350 ppm SO_x for one potential application. To comply with this regulation, an absorption efficiency of either 75.0 or 87.5% would be expected, corresponding to uncontrolled SO_x concentrations of 1400 and 2800 ppm.

Because, for most instances of air pollution control, the driving force for mass transfer depends on the partial pressure of the gaseous pollutant in the carrier gas stream, a reduction in concentration will cause a decrease in the absorption level. In terms of the number of transfer units (N_{OG}), lowering the pollutant concentration in the gas stream entering the scrubber is equivalent to decreasing the value of L on the operating line *CD* in Fig. 6-36. As a result of this action, the number of transfer units that must be constructed from the tower discharge conditions at E to the revised inlet conditions at L would be increased.

Liquor Types. The selection of the optimum liquor for a particular gaseous pollutant(s) should be based on a number of important properties. The first of these is that the selected liquor should have a high solubility potential for the gas to be controlled so as to improve the absorption rate and minimize the liquor makeup requirements. The liquor should not be too costly and should have sufficiently low volatility to minimize evaporation losses in the cleaned discharge-gas stream. For example, although most hydrocarbon gases are very soluble in light oils, the high volatility of these solvents preclude their use as scrubbing liquors for atmospheric pressure absorption towers.

Water is the most common solvent, and for such noncorrosive gases as ammonia and water-miscible hydrocarbons, its use is recommended. Although such gases as hydrogen chloride and hydrogen fluoride are very soluble in water and chlorine and sulfur dioxide are reasonably soluble, the use of alkaline additives in the scrubbing water is usually practiced. The reason is that, if disposal of the highly acidic effluent liquor is necessary, this stream must be neutralized before it is discharged into public waters. By introducing an alkaline liquor directly into the scrubber, a neutral effluent can be produced, the absorption performance improved, and the equipment corrosion potential reduced.

In evaluating alkaline additives, both soluble and insoluble types should be considered. For example, Na_2CO_3 is soluble and more active than $CaCO_3$. However, the price ratio in favor of the insoluble calcium base is 2:1. Because the activity difference is not significant for the more common gas absorption problems and the soluble sodium effluent liquor poses a threat to ground-water contamination, calcium base liquors are favored. This recent emphasis on the use of insoluble scrubbing liquors and/or the formation of precipitates during the scrubbing process has revised scrubber design concepts considerably. This trend has also reduced the available application areas for the fixed packing scrubber.

Special scrubbing liquors are utilized in absorption processes concerned with gas-cleaning operations where the gaseous contaminant and solvent are recovered. Thus, the ethanolamines are used for the absorption of hydrogen sulfide and carbon dioxide, which are converted to sulfur and "dry ice," respectively. Because of its cost, this solvent can be employed only in regenerable processes where the makeup scrubbing liquor requirements are minimal. Buffered aqueous solutions of potassium permanganate are being used as a scrubbing medium in many emissions control applications where oxidation of the gaseous and particulate contaminants reduce their odor level. In these operations the permanganate is chemically reduced to a manganese dioxide precipitate so that the use of a nonplugging type of scrubber becomes mandatory.

Liquid-to-Gas Ratio (L/G). For those conditions where the gaseous pollutant concentration is low and solubility is high, the gas film controls. The method for determining the number of transfer units, as shown graphically by Fig. 6-36 and algebraically by Eq. 6-21, is applicable to conditions such as those commonly representative of most emissions control problems. The slope of the operating line CD in Fig. 6-36 is $(Y_1 - Y_2)/(X_1 - X_2)$. Because the overall mass-balance equation can be written as

$$L_M(X_1 - X_2) = G_M(Y_1 - Y_2) \qquad (6\text{-}24)$$

where the definitions of terms are the same as for Eq. 6-21, it follows that L_M/G_M is the slope of the operating line. Thus, the greater the value of L/G, the steeper does the operating line become and the greater is the driving force causing absorption, as represented by the vertical distance between the operating line and the equilibrium curve. Therefore, an increase in the liquid/gas ratio will improve the absorption performance. Figure 6-40 shows the effect of L/G variation on the absorption efficiency for SO_2 as it is being removed from a power-plant boiler flue gas.

The effect of L/G, as discussed above, is based on packed tower scrubbers with countercurrent flow of fresh makeup liquid. The relationship illustrated in Fig. 6-40 was determined for a countercurrent, mobile packing type of scrubber. In this unit, typical of most commercial scrubber designs, the variation in L/G values was obtained by regulation of a recirculated liquor stream drawn from the base of the scrubber and introduced to the liquor inlet at the top. This method is not a strict interpretation of

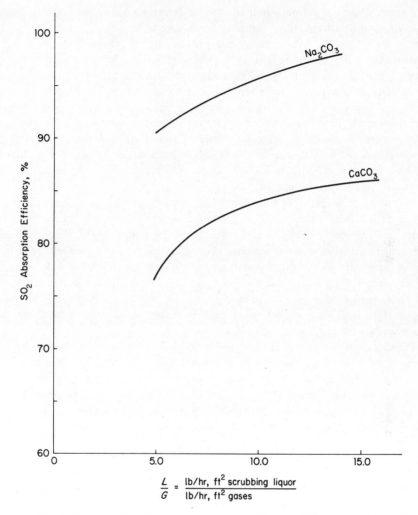

Figure 6-40. Influence of L/G on gas absorption with two different scrubbing liquors. (Adapted from Pollock et al., Ref. 29.)

the L/G concept demonstrated in Fig. 6-36 because of the solute concentration buildup in the recirculated liquor. However, because the gas film resistance is controlling, the influence of increased L/G ratios achieved by liquor recirculation is still adequately demonstrated.

In commercial scrubber operation for the control of gaseous emissions, sufficient makeup liquor must be introduced into the scrubber to satisfy the gas saturation requirements and to maintain the solute concentration in the discharged liquor (X_1) at the design value. When alkaline additives are introduced into the scrubbing medium, the makeup chemical values may be regulated by pH control of the effluent liquor.

317

Pressure Drop. A formula and data for the determination of the pressure drop in a packed countercurrent flow scrubber were shown in Eq. 6-23 and Table 6-17. Pressure drop must be overcome by the expenditure of power. This power input represents the major operating cost for the wet scrubber. An optimum scrubber design is characterized by maximum liquid/gas contact at minimum pressure drop. Presently, the use of cross-flow packed scrubber designs is being developed, with pressure drops on the gas side being about one-half those for equivalent countercurrent towers. Cross flows rather than countercurrent flows are advocated because, in the majority of pollution control situations, large quantities of liquid are used with minimum gas concentrations and because the scrubbing liquors react so readily with the gas that no equilibrium partial pressure is evident. Under these conditions, the advantages of countercurrent flow are eliminated.

The pressure drop across a scrubber is a function of the gas/liquid turbulence, necessary for effective contacting of the two phases. The power input for a specific absorption duty may be compared to the contacting power required for particulate collection. In actual practice, a scrubber is designed for a certain gas pressure drop to develop turbulence in the impingent type scrubbers or to atomize liquids, as in the case of venturi designs. Should the pressure drop fall below the design value, the absorption performance will be adversely affected. Just as for those wet scrubbers designed for particulate collection, a reliable index of absorption scrubber performance is the measured pressure drop.

Pressure-drop values for absorption service fall in the medium-range category, varying from 4 to 15 in. w.g. Where a composite service is required, involving both particulate collection and gas absorption, the pressure drop for the former duty is usually the greatest. Thus, for the removal of HCl and aluminum chloride from an aluminum processing operation, acceptable absorption performance can be achieved with an alkaline scrubbing liquor, at about 2 in. w.g. However, the particulate collection duty requires a pressure drop of 20 to 24 in. w.g.

Mechanical Design

As mentioned earlier, gaseous emissions control has only recently become a source of concern to environmental protection agencies. For a number of decades the standard item of equipment for gas absorption has been the packed tower, with the major variation in its design being in the selection of the type of packing that would provide maximum wetted surface area for minimum capital and operating costs. Usually, absorption was confined to contacting clean gases with nonviscid scrubbing solutions. Where slurry processing became necessary, low-efficiency open tower or slat packings were employed.

Air pollution control during this period was generally concerned with visible emissions involving particulates removal; therefore, practically all gas absorption operations were concerned with chemical processing. As a result of this situation, there are not too many novel wet scrubber designs specifically directed at air pollution conditions. Thus, the optimum wet scrubber design for the simultaneous removal of SO_2 and flyash, in the presence of an alkaline slurry, is still being debated. Laws are being enforced for the reduction of NO_x from power-plant effluent gas streams, but wet

scrubbers cannot presently cope with this problem because of the solubility characteristics and reaction kinetics of NO and NO_2. Such gases as HF, HCl, SiF_4, SO_2, and
NH_3 can be adequately controlled in wet scrubbers, although special designs are
required when these gaseous pollutants are accompanied by heavy particulate loadings.

Wet scrubbers for gas absorption comprise a contacting section and deentrainment
section, similar to the designs for particulate collection. A number of the more
prevalent contacting mechanisms will be discussed in connection with three common
commercial scrubbers. The deentrainment methods employed with absorption scrubbers are centrifugal, impact, and labyrinth. These devices rely on the inertia of the
solute-laden liquor droplets for their separation from the gas stream. The importance
of the removal of these droplets from the scrubber discharge gas cannot be overempha-
sized. It is not unusual for a scrubber with a malfunctioning deentrainment section to
record negative absorption efficiencies because of the loss of entrained solute in the
gaseous effluent. The centrifugal separators were described and illustrated for particulate removal wet scrubbers in Chapter 5. The impact and labyrinth types will be
discussed under the corresponding scrubber design in this section.

Fixed-Packing Scrubber. The fixed-packing wet scrubber represents the classic
design for absorption duty. It is still the most common scrubber type utilized for
processing clean gases with scrubbing solutions. By the optimum selection of a packing
having a high percentage of voids, some particulate collection can be accomplished.
This type of scrubber is almost invariably circular in cross section and is generally
designed for gas flows in the range of 1,000 to 10,000 cfm. Countercurrent flow is
achieved by introducing the gas at the bottom of the packing so that it will flow
upward, with the scrubbing liquor entering at the top and being distributed through
liquid spargers or overflow orifice trays. This type of wet scrubber was illustrated in
Fig. 5-47. The major physical design variables are diameter, height, and packing type.
They can be determined from the formulas and data given at the beginning of this
section under "Operating Principles." Average superficial gas velocities of 100 to 500
fpm and liquor flow rates of 1 to 50 gal/min-ft^2 are used as design bases. Thus, for a
gas flow of 10,000 acfm at ambient temperature, a tower 6 ft in diameter by
approximately 15 ft in height, depending on the difficulty of the duty, might be
expected. Because of the low gas and liquor flow rates, entrainment is reduced to a
minimum and a simple free-space section above the liquor distribution device may be
sufficient for its elimination.

Innumerable packing types are available for the fixed-packing scrubber. Wood
slats, size-graded aggregates, marbles, Raschig rings, saddles, Pall rings, and Tellerettes
are just a few of the forms that are used. Some of the more common of these is shown
in Fig. 6-41. The Pall ring is a recent addition to the various fixed-packing types, with
claims for increased gas and liquor flow rates at minimum pressure drop. Thus, with
1½ in. Raschig rings, a pressure drop of 0.5 in. w.g./ft of packing would allow gas and
liquor flows of about 65 fpm and 60 gal/min-ft^2, respectively. With the same size Pall
rings, the identical pressure drop can be maintained at flows of 130 fpm and 60
gal/min-ft^2. A correlation between liquid- and gas-mass flow rates and pressure drop
for an air-water system for these two packing types is shown in Fig. 6-42.

Mobile Packing Scrubber. In recognition of the increased need for a packing to

319

Raschig Ring

Sizes: ½, 1, 1½, 2 in.

Pall Ring

Sizes: ⅝, 1, 1½, 2 in.

Berl Saddle

Sizes: ½, 1, 1½, 2 in.

Intalox Saddle

Sizes: ½, 1, 1½, 2 in.

Tellerette

Size: 1 in.

Figure 6-41. Common types of fixed packings.

handle particulate-laden gas streams and viscid scrubbing liquids, a mobile packing was devised. This consists of 1¼ in. diam, lightweight plastics spheres, which are free to move randomly between a support grid and restraining grid. The turbulent motion of these spheres prevents the accumulation of solids and makes it possible to handle highly concentrated slurries for the scrubbing liquor. As illustrated by Fig. 6-39, this type of packing permits unusually high liquid and gas flows at nominal pressure drops.

This scrubber is almost solely applied to those processes where fixed packings cannot handle the solids loading. It is available in either circular or rectangular cross section, with the latter design applied to the processing of gas flows in excess of 100,000 cfm. For easily attainable absorption performance, a single contacting stage is employed, with the grids spaced from 2 to 5 ft apart, depending on the application. For more difficult contacting applications, multiple staging is used, with sometimes as many as five stages being provided in a single scrubber shell. Average gas and liquid design flow values for this type of scrubber are 750 cfm/ft^2 and 20 to 40 gal/min-ft^2 The equivalent pressure drops for various process flows are shown in Fig. 6-39.

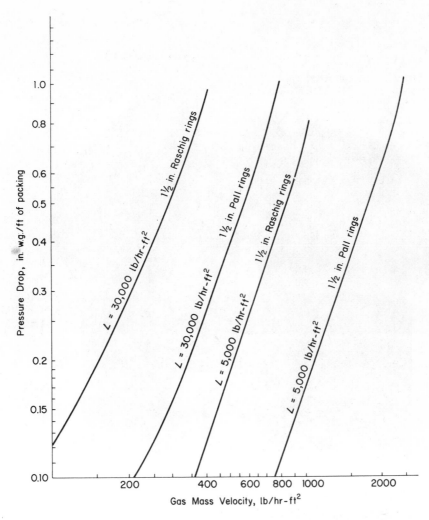

Figure 6-42. Pressure drop versus L/G for various fixed packings. (Courtesy of Norton Chemical Process Products Division.)

One recent operation area utilizing this type of scrubber is that of SO_2-flyash emissions control. This scrubber type, known as the Turbulent Contact Absorber, is shown in Fig. 6-43. This three-stage unit handles in excess of 300,000 cfm and is capable of removing 80% of the SO_2 from a power-plant flue gas stream, using a limestone slurry and 95% of this pollutant with a sodium carbonate liquor. Simultaneously, flyash can be collected at an efficiency of 99⁺%.

The flue gases at a temperature of about 350° F enter at the base of the scrubber, and are humidified and cooled to about 125° F as they flow upward through the three stages of mobile spherical packing. It is estimated that the operating pressure drop for

321

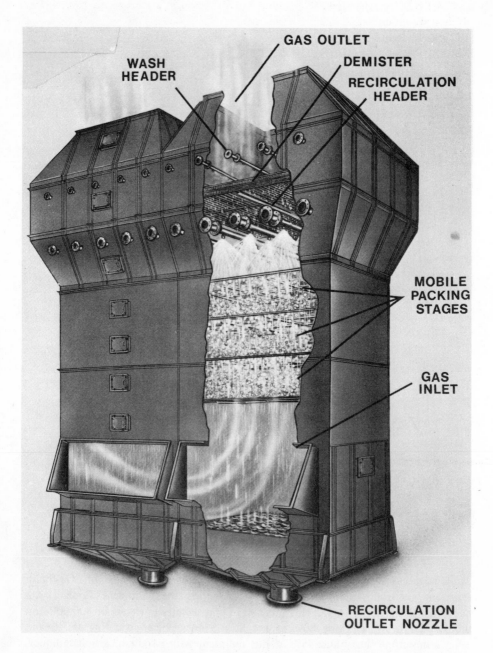

Figure 6-43.   Turbulent Contact Absorber®. (Courtesy of Air Correction Division.)

this particular service would be in the range of 6 to 8 in. w.g. The discharged liquor effluent would contain about 5% solids and be maintained at a pH in excess of 8. The deentrainment section is an impingement type of separator comprising multiple chevron blades set about 1½ in. apart. The heavily entrained liquor load strikes the blade surface area where the droplets are removed, flow along drain channels forming an integral section of the blades, and then are gravity discharged to the contacting section along the walls of the scrubber. To prevent buildup of carbonates and sulfates on the separator blades, a wash sparger located above them intermittently directs a stream of fresh wash water against the blade surfaces.

The lightweight packing requirements for this type of scrubber limits the selection of materials and wall weights for the spheres. Polyethylene and polypropylene are presently available, with some temperature and abrasion limitations.

Centrifugal Spray Scrubber. The centrifugal spray scrubber operates on the principle of finely atomized liquid spray passing through a spinning gas stream. As shown in Fig. 6-44, the gas stream carrying the gaseous pollutant enters tangentially at

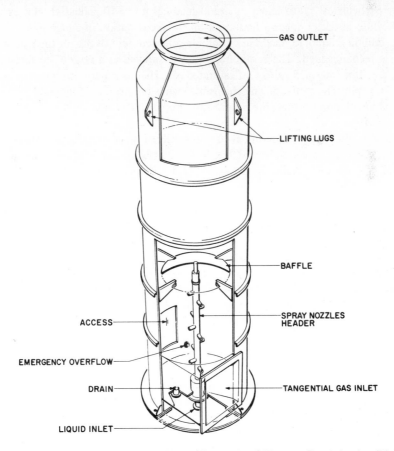

Figure 6-44. Centrifugal-spray scrubber. (Courtesy of Western Precipitation Division.)

the bottom of the scrubber and describes a spiral upward path. Liquid spray is introduced into the rotating gas stream from an axially located manifold containing several high-pressure nozzles located in the lower section of the scrubber. The atomized fine-spray droplets impinge against the rotating gas stream at the scrubber wall, thereby absorbing the gas. The scrubbing liquid and collected pollutants run down the walls and are discharged from the bottom of the unit. The cleaned gas continues to flow upward through a central core baffle, located just above the axial spray manifold, and is discharged through the top outlet. The core baffle forces the gas stream against the scrubber walls and deflects any overspray in this same direction, thus assisting in the centrifugal deentrainment action. Where very high liquid-atomization pressures are required and negligible particulate loadings are present, it is sometimes advantageous to utilize a labyrinth type of entrainment separator in the upper section of the scrubber, to achieve complete deentrainment. These separators consist of wire mesh grids or fine-filament mesh pad sections that accomplish separation of the liquor spray droplets by impaction and entrapment.

The major duty intended for this scrubber is gas absorption, although it can remove particulates 2 μ or larger in size at loadings up to 10 grains/ft^3. However, the high-pressure nozzles impose a limitation on the recirculation of a high-solids content in the scrubbing liquor; a maximum insoluble solids content of 3% for the recirculated liquor is recommended. The scrubbing liquid is sprayed at a nozzle pressure of 100 psig and a flow rate of 5 gal/ft^3 of saturated gas. The gas superficial design velocity is about 700 fpm. The contacting mechanism depends on the energy required to atomize the scrubbing liquor so that low gas-pressure drops, on the order of 1 to 2 in. w.g., are sustained.

Because this type of scrubber is essentially a single-stage contactor, the gas removal efficiency levels are rather fixed. Some suggested absorption efficiencies are 95% for SO$_2$ with an alkaline solution, 98% for HF, and 90% for HCl. The latter two values apply when water is the scrubbing liquid.

Regardless of scrubber type, the chemical characteristics and temperatures of the gaseous pollutant will determine the materials of construction. Such acidic gases as SO$_2$, H$_2$S, Cl$_2$, HCl, and HF require the use of plastic-coated, rubber, or glass-lined, fiberglass reinforced polyester, stainless steel, special high-nickel alloys, or carbon brick fabrication materials. Cost indices for some of these materials were quoted in Section 5.7 under "Mechanical Design." The various packings, both fixed and mobile, are fabricated of polyethylene, polypropylene, polyvinyl chloride, and Penton. The spherical mobile packing is limited to plastic materials, to ensure lightness and mobility, but the fixed packings are available in both metal and plastic materials.

One interesting situation that complicates the selection of construction materials is that posed by high-temperature operations combined with corrosive atmospheres. In processing chlorine or HCl in a wet scrubber at ambient temperature, the use of fiberglass reinforced polyester (FRP) is ideal. However, when the temperatures are in excess of 200° F, this material is no longer serviceable, and therefore some type of Hastelloy or other high-nickel alloy must be considered. The use of Hastelloys for scrubber fabrication imposes about a sixfold price increase over that for mild-steel construction, whereas FRP costs just twice as much. In one process where the gas enters the scrubber at 600° F, consideration was given to precooling this gas stream to its saturation temperature of about 110° F, which then allowed the use of the more moderately priced FRP. The water supply was protected by temperature controls, but unfortunately these controls failed because the technical level of the operating personnel was not considered when the decision was made to use water-protected FRP construction. As a result, the FRP scrubber had to be replaced. Regardless of such contingencies, this design decision is still a difficult one to make in very many similar situations. One alternative approach is to make use of FRP-coated mild steel so that if water failure occurs, only the lining will be destroyed, thus minimizing replacement costs.

Installation Features

The major installation considerations in the design of a wet scrubber system are the scrubbing liquor supply and disposal means for the contaminated liquor effluent. Wet scrubbers for the collection of particulates generally use water as the scrubbing medium so that its supply becomes a simple problem of determining a source of quality water consistent with the demands of the wet scrubber. The disposal of the solids-laden effluent usually involves some liquid-solids mechanical separation equipment, as shown in Fig. 5-52.

For gas absorption wet scrubbers, chemical additive equipment and instrumentation are generally required for effective performance. The disposal system must incorporate adequate equipment to deliver to public waterways a final effluent that is free from both solid material and chemical values. The disposal problem is the most difficult to overcome, and for many potential wet scrubbing systems it becomes the major stumbling block responsible for the abandonment of a proposed absorption scheme. The difficulties involved in the disposal of the $CaSO_3$-$CaSO_4$-flyash liquor resulting from the removal of SO_2 and flyash from power-plant flue gases have been cited previously. The problem is compounded by the fact that for an average size 800-mw power plant burning coal with a sulfur content of 3%, the annual slurry-disposal load amounts to almost one million tons. A typical scrubbing and liquor disposal system for power-plant gaseous effluent control is shown in Fig. 6-45. Recovery of the sulfur values would certainly make the economics of this system more acceptable. One recovery scheme now being proposed is a two-stage scrubbing system wherein the flyash is completely removed in the primary venturi type scrubber, followed by SO_2 absorption in a mobile packing scrubber employing an $Mg(OH)_2$ slurry. The concentrated $MgSO_3$-$MgSO_4$ liquor is then processed by others for sulfur recovery, and the reactivated $Mg(OH)_2$ is returned to the power plant facility as makeup scrubbing liquor.

In gas absorption installations the fans should be located upstream from the wet scrubbers to eliminate the need for costly fabrication materials. If located downstream, the combination of moving mechanical parts and a wet saturated gas, usually containing residual corrosive gases, poses a continuing maintenance burden. Fans fabricated of exotic metals are prohibitively expensive, and the use of coatings on the

325

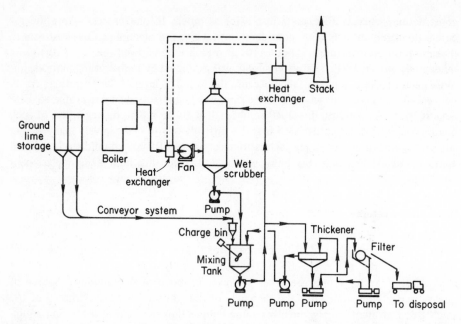

Figure 6-45. Scrubbing and liquor disposal system for power-plant gaseous effluent control.

rotating fan blades is a constant source of expensive plant upkeep and inconvenient shutdowns.

In the use of packed towers, the packing may be dumped into the column at random or may be manually stacked in some definite geometric order. Randomly dumped packings have a higher specific surface-contact area and offer greater resistance to the gas and liquor flows. Pall ring packing is normally dumped, one proposed method being to charge the packing through a manway into a flooded section of the tower. Thus, the rings are "floated" into place in a random pattern. In charging spherical mobile packing, only about a 12 in. to 18 in. static packing depth is required in each stage. A specified number of spheres is charged through a manway located between the two grids constituting a single stage, which are spaced about 4 ft apart. The spheres are thus randomly distributed under the forces exerted by the liquid and gas flows.

Effective performance in a packed tower depends on providing a maximum exposed liquid film on the packing surfaces. Maldistribution of the downflowing liquor can cause uneven coverage or nonwetting of the packing. To ensure optimum packing irrigation, the liquor should be evenly distributed through a well-designed sparger configuration. At least five distribution points are recommended for each square foot of tower cross-sectional area. The minimum liquor downflow rate should be in excess of 2 gal/min-ft^2. Low-pressure drop, spiral type nozzles can provide a variety of spray patterns for effective tower area coverage. When overflow tray distributors are employed, it is most important that they be installed level and true. For deep towers it might be necessary to prevent liquor channeling by the installation of peripheral shelf-type distributors located at definite intervals throughout the column. In randomly packed towers the liquid tends to channel toward the walls because of the lower packing density at this location. The peripheral distributors deflect the liquid flowing down the walls toward the center of the tower.

Because a packed type of wet scrubber is essentially a constant gas flow device, efforts should be made in the design of the accessory equipment to ensure as near to constant gas flow conditions as possible. In many industrial processes the gas flows are either cyclical and/or variable. For example, in an industrial steam-producing boiler, the plant process demands might impose loads as low as 25% of full rating, with a corresponding reduction in the contaminated gas flow to the wet scrubber. At this diminished gas flow, the scrubber absorption performance would be adversely affected. Although the total SO_x emission rate would also be reduced, its concentration in the effluent gas stream would be increased.

For small boilers of 100,000 lb/hr steam rating equivalent to 18,000 scfm of flue gas, or less, the use of an atmospheric damper controlled by the pressure drop across the scrubber would automatically provide a constant gas flow to the scrubber. As the boiler effluent gas flow fell off, the pressure drop across the packed scrubber would decrease and be sensed by the differential pressure controller. This instrument would then actuate the atmospheric damper, causing it to open to a position that allowed sufficient ambient air to enter the flue gas stream and satisfy the full-load pressure drop across the scrubber.

For larger boilers the power requirements necessary to process the relatively large ambient air flows needed to maintain the system in balance do not favor the use of the atmospheric damper. Instead, multiple scrubbers are used and the total effluent gas stream is split to minimize the reduction of flow to each individual scrubber. Thus, at a 50% flow reduction, a single scrubber of a dual scrubber system would accept the full flow and the second scrubber would be shut down. Above 50% rating, both scrubbers would be operated; below 50%, a single scrubber would be on stream. The major disadvantage of this system would be the requirement for some sophisticated large-scale damper systems.

The foregoing discussion pertains to both fixed and mobile types of packed scrubbers, and the solutions discussed to overcome reduced loading situations are applicable to both scrubber designs. On the other hand, the gas/liquid contacting mechanism utilized in the centrifugal spray scrubber, shown in Fig. 6-44, depends on the gas film thickness, gas retention time, and the liquor droplet size. A reduction in the gas flow rate would increase the contact time between the liquid and gas phases, and thereby improve the absorption performance. The scrubber inlet gas flow must be sufficient to maintain the centrifugal gas flow pattern up through the tower.

Wet scrubbers, whether for particulates collection or gas absorption, require minimal instrumentation. The most basic requirement is that of a simple manometer to measure the pressure drop across the scrubber. If panel-mounted instrumentation is preferred, then a draft gage is recommended. The next most important item is a multipoint temperature recorder or a controller. If gas humidification and cooling is critical to the life of the scrubber and the main water supply is unreliable, then a controller to shut down the fan and/or open a diversion damper to atmosphere and open an alternate water supply valve is recommended. Gas flows are usually determined by Pitot readings on an intermittent basis; rarely is fixed instrumentation required for this function. Makeup water requirements for the scrubber should be metered and the flow rate indicated to evaluate its influence on the quality of the liquor effluent being discharged to disposal or recovery.

For absorption service, gas monitoring should be provided to maintain a continuous record of the gaseous pollutant concentration and its total mass flow to the atmosphere. If alkaline addition is necessary, the rate of makeup chemicals can usually be controlled by a pH indicating controller. In any event the pH level of the liquor effluent being discharged into public waterways should be continuously recorded.

Applications

In the application of the wet scrubber for gas absorption, it is necessary to select an absorbent in which the gaseous pollutant is soluble. Absorbents may be classified as reactive or nonreactive; the former absorbs the gas by chemical reaction, such as in the use of NaOH for the absorption of chlorine, and the latter relies on the solubility of the gas without chemical reaction. In addition to gas absorption, wet scrubbers can be applied to the control of process effluent gas streams where the removal of both particulates and gases is required.

By-product recovery is an important consideration in the application of wet scrubbers. Thus, the use of expensive absorbents such as the ethanolamines can be justified for the control of hydrogen sulfide only because H_2S can be recovered for its sulfur values and the absorbent re-used. In a process developed by Consolidated Mining and Smelting Company [30], low-concentration SO_2 is absorbed in aqueous ammonia to form a solution of ammonia sulfite-bisulfite. This liquor effluent is steam stripped to yield a more concentrated SO_2 stream and a fertilizer grade of ammonium sulfate. The sulfur oxides are reduced from 7500 to 1000 ppm, equivalent to an absorption efficiency of 87%.

An application chart for the various types of wet scrubbing equipment is given in Table 6-18. In air pollution control applications, there has been an accelerated demand for gas/liquid contact equipment that can handle viscid fluids. This trend is prompted by the increased demand for wet scrubbers with the ability to remove both particulate and gaseous pollutants and/or accommodate slurry-type absorbents.

In the aluminum reclaiming industry [31], magnesium alloy additives are removed from junk aluminum by sparging the molten aluminum with chlorine. The gaseous effluent from the melt furnace contains aluminum chloride and aluminum oxide particulates together with free chlorine and hydrogen chloride gases. Abatement of these pollutants is achieved with a wet scrubber employing a sodium carbonate scrubbing liquor. The particulates loading is extremely heavy in this application and, on being hydrolyzed, the aluminum chloride is converted to the viscid and gelatinous hydroxide form. Because this process is classified as pyrometallurgical, the particulates are submicron; therefore, large energy inputs are required for their collection. To meet this high-energy, heavy-slurry handling problem, a six-stage mobile packing scrubber operating at a pressure drop of 28 in. w.g. is used. Particulates removal of 98⁺% has

Table 6-18. Absorption Equipment Applications

Scrubber Type	Physical Absorption		Chemical Absorption		Pressure Drop	Simultaneous Absorption and Dust Collection Capability	Typical Application
	High Solubility	Medium Solubility	Rapid	Slow			
Packed tower rings, saddles, coke	x	x	x	x	Medium	Poor	Absorb H_2S in ethanolamine
Packed tower wood grids	x	x	x		Low	Fair	Absorb HF and remove fluoride dust
Packed tower mobile packing	x	x	x	x	Medium	Good	Absorb SO_2 and remove flyash
Spray tower	x		x		Low	Good	Absorb NH_3 in water
Agitated tank	x		x		High	Fair	Reaction of H_2 with organic materials
Venturi	x		x		High or low	Good	Absorb SO_2 and remove "saltcake"
Centrifugal spray	x		x		Low	Fair	Absorb HF, and remove "some" dust
Plate impingement	x	x	x	x	Medium	Good	Absorb SO_2 and remove dust

been reported and the chlorine and hydrogen chloride have been completely absorbed. Materials of construction are rubber-lined mild steel for the scrubber shell and nickel alloy accessory equipment.

That segment of the fertilizer industry employing wet processes relies very heavily on wet scrubbers for the control of gaseous emissions. Both disposal and by-product recovery means are utilized for management of the scrubber liquor effluent. In the production of phosphoric acid, sulfuric acid is reacted with phosphate rock to yield weak product acid and gypsum. The phosphate rock contains fluorides, which are evolved as gaseous hydrogen fluoride (HF) and silicon tetrafluoride fume (SiF_4) during the reaction, filtration, and evaporation operations. The silicon tetrafluoride, on contact with water, forms hydrofluosilicic acid (H_2SiF_6) and gelatinous silica. Scrubbing water is available from ponds where the by-product gypsum is discharged for its disposal. The SiF_4 is partially submicron in size so that a medium-pressure-drop venturi scrubber is usually recommended for this service. The HF is only partially removed in this primary unit, and in many installations a secondary scrubber is employed with either fixed or mobile packing to remove residual HF. The liquor effluent is discharged to the gypsum ponds where the fluoride ion content is closely monitored to maintain its concentration at a suitable level for re-use as a scrubbing liquor. The fluorine emissions regulations are extremely severe, particularly in the Florida phosphate-processing area. Consequently, both particulate and gaseous removal efficiencies of 99[+]% are demanded. The weak phosphoric acid is separated from the gypsum on continuous vacuum filters, either the rotary or multiple-pan type. The agitation of the liquor during filtration, in addition to the reduction of pressure, causes the evolution of additional SiF_4 and HF. An installation for the collection of these pollutants is shown in Fig. 6-46. This two-stage Turbulent Contact Absorber with mobile spherical packing handles ventilation air at a flow rate of 22,000 acfm, containing low concentrations of both gaseous and particulate fluorides, from a vacuum pan filter. Removal efficiencies of 98% for both pollutants have been obtained by using gypsum pond water, with a considerable suspended solids loading, as the scrubbing liquor. The pressure drop across the scrubber is 4 in. w.g.

There are a wide variety of gaseous pollutants emitted by chemical processing, fertilizer, metal finishing, and the pigment and dye industries. A list of the gases associated with each of these operations and the estimated carrier gas flow rates is shown in Table 6-19. These effluents contain very small amounts of particulates so that fixed-packing scrubbers can reduce the noxious gas concentration to a level sufficiently low to satisfy today's gaseous emissions regulations. By-product recovery is feasible from some of these pollutants, such as the recovery of chromic acid for the metal-finishing industry and ammonia gas and phosphate dusts from fertilizer production. In some of these applications where the pollutant is highly soluble in the scrubbing liquor, a cross-flow packing configuration is used with attendant pressure-drop reductions. The Tellerette type of fixed packing, shown in Fig. 6-41, is employed in these cross-flow scrubbers. The concentration of the pollutant gases listed in Table 6-19 is less than 1% by volume. In the absence of heavy particulate loadings and with the use of water or soluble alkaline scrubbing liquors, the fixed-packing type of scrubber is the most economical control equipment available. An average pressure-drop

Figure 6-46. Turbulent Contact Absorber® in the fertilizer industry. (Courtesy of Air Correction Division.)

Table 6-19. Common Industrial Pollutants Removed by Packed Wet Scrubbers

Industry	Approx. Gas Flow, Rate, cfm	Pollutants
Chemical processing	6,000	Acid mists
	1,000	H_2SO_4 mists; NaOH and Na_2CO_3 dusts
	1,500	NH_3 gas
	250	Cl_2 and HCl gases
	3,500	HCl gas and $NiCl_2$ dust
	1,400	Cl_2 vent gases
Fertilizer		
	35,000	KCl and amine mists
	15,000	HF gas and SiF_4 fume
	14,000	NH_3 gas and ammonium phosphate dust
	12,000	Di-ammonium phosphate dust
Metal finishing	45,000	HCl, HF, HNO_3, NO_2 mists and gases
	3,500	Chrome and caustic mists
	24,000	NaOH mists and NaOH dusts
	1,000	HF and NH_3 gases HNO_3 H_2SO_4, HCl mists
	3,500	HCl gas and mist
Pigments and dyes	2,000	H_2SO_4 acid mists
	1,000	Cl_2 vent gases
	1,750	HCN gases
	10,000	Cyanuric chloride dust, NH_3 HCl, H_2SO_4 and toluene mists

Source: Ceilcote Company, Berea, Ohio.

range for the countercurrent packed scrubber for these various gas absorption applications is 2 to 4 in. w.g.

Sample Design Calculations

The problem considered in this section will be identical to that defined for the wet scrubber in Section 5-7, where the basic duty was that of particulate removal. Some discussion of the capabilities of the design for the removal of SO_2 was also presented.

In this problem, the major duty will be the removal of 90% of the SO_2 contents of the gas stream. Both particulates collection and gas absorption are to be accom-

332

plished in the same scrubber. The absorption performance requirement of 90% will be based on the average gas flow, but the influence of peak flow conditions on the SO_2 removal efficiency will also be indicated.

A restatement of the problem and its solution follow.

Problem Statement

Application	Process steam boiler
Fuel	Coal
Gas flow, acfm (avg)	168,600
Gas flow, acfm (peak)	210,000
Gas temperature, °F	400
Gas moisture, vol %	8.5
Gas analysis, vol %	
CO_2	14.2
H_2O	8.5
SO_2	0.2
Dust type	Flyash
Dust loading, grains/acf (avg)	1.1
Dust loading, grains/acf (peak)	1.8
Dust analysis, % less than 10 μ	17.0

Required Performance. SO_2 removal at 90% efficiency for average gas flow conditions is required. A dust collection efficiency of at least 98% at these same process conditions is also necessary.

The influence of the peak gas flow rate on the SO_2 performance level should be determined.

Coal at a variety of analyses and moisture contents is available at this plant. As an alternate condition, SO_2 removal should be calculated at a flue gas moisture content of 4.8 vol %.

Problem Solution. To achieve a 90% absorption efficiency for SO_2, experience dictates that an alkaline scrubbing liquor will be required. However, to illustrate the effect of such parameters as gas flow, gas temperature, moisture content, and SO_2 concentration, a preliminary solution will be undertaken, based on a countercurrent-flow packed scrubber design in which water will be used as the absorbent.

A. Scrubber Design with Water Absorbent

1. Equilibrium Data: Data for the partial pressure of SO_2 over aqueous SO_2 solutions will be determined at two different scrubber-gas temperatures corresponding to the two different moisture levels. At 400° F and a moisture content of 8.5 vol %, the saturation temperature from Fig. 5-40 is 134° F. At 400° F and a moisture content of 4.8 vol %, the saturation temperature would be 126° F.

In Fig. 6-47, vapor-pressure equilibrium curves have been determined at 134° F and 126° F. It will be assumed that sufficient makeup water will be introduced into the scrubber to maintain these temperatures throughout and to nullify the effects of heats of solution.

Because of the marked effect of temperature on the SO_2/water equilibrium data and the subsequent influence on performance, an additional curve at 115° F has also been plotted in Fig. 6-47. This temperature is based on heat exchange of the incoming gas stream at 400° F with the scrubber discharge gases so as to reduce plume opacity. It is assumed that the temperature of the gases entering the scrubber will be reduced from 400 to 250° F. At the lower moisture content of 4.8 vol % and a temperature of 250° F, the saturated temperature of the gases in the scrubber will be 115° F.

In Fig. 6-48 an X-Y plot has been drawn at 115° F from the corresponding equilibrium curve in Fig. 6-47. To illustrate the full-range effect of gas temperature variations on scrubber design, an additional X-Y plot has been developed in Fig. 6-49 for the scrubber gases at 134° F. A sample calculation for a single value of X and Y follows: For the 115° F gas scrubbing temperature, from Fig. 6-47,

$$\frac{1.5 \text{ lb } SO_2}{100 \text{ lb } H_2O} \leftrightarrow 0.294 \text{ atm } SO_2 \text{ partial pressure}$$

thus,

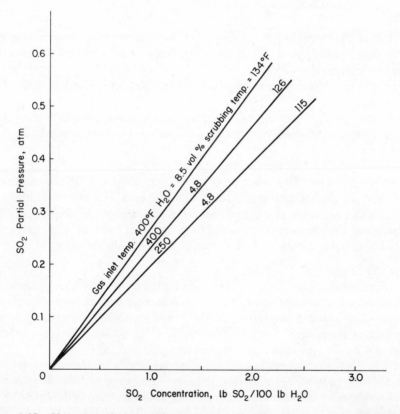

Figure 6-47. Vapor pressure equilibrium curves for SO_2 over aqueous SO_2 solutions. (Adapted from Int'l Critical Tables, Ref. 32.)

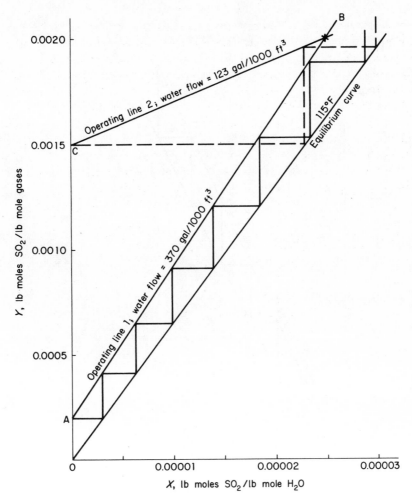

Figure 6-48. X-Y curves for SO$_2$-water scrubber design at a gas saturation temperature of 115° F.

$$X = \frac{1.5}{100} \times \frac{18}{64} = 0.00423 \text{ lb moles SO}_2/\text{lb mole H}_2\text{O}$$

$$Y = \frac{0.294}{0.706} = 0.416 \text{ lb moles SO}_2/\text{lb mole gases}$$

A number of points were plotted and the slope of the *X-Y* equilibrium curve found to be approximately 66.0. The pertinent portion of the equilibrium curve in the neighborhood of the proposed operating lines was then constructed, as shown in Fig. 6-48.

335

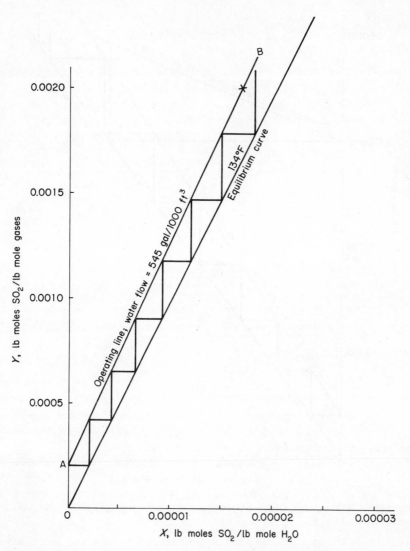

Figure 6-49. X-Y curves for SO_2-water scrubber design at a gas saturation temperature of $134°$ F.

2. Operating Lines and Determination of N_{OG}: The scrubber inlet gas contains 0.002 mole SO_2/0.998 mole flue gas. Thus,

$$Y_1 = \frac{0.002}{0.998} = 0.002004 \text{ mole } SO_2/\text{mole flue gas}$$

The scrubber discharge gas at 90% absorption contains 0.0002 mole SO_2:

$$Y_2 = \frac{0.0002}{0.9998} = 0.0002 \text{ mole } SO_2/\text{mole flue gas}$$

The value of X_1 was fixed at 0.000025 mole SO_2/mole H_2O so as to establish an agreeable slope for the operating line AB relative to the equilibrium curve:

$$X_2 = 0 \text{ mole } SO_2/\text{mole } H_2O$$

Then, for the operating line AB,

$$\text{Slope} = \frac{0.0020 - 0.0002}{0.000025 - 0} = 72 \text{ moles } H_2O/\text{mole flue gas}$$

Based on industrially acceptable values of L/G at reasonable pressure drops, as shown in Fig. 6-42,

$$\begin{aligned} \text{Water flow} &= 1010 \text{ moles/hr-ft}^2 \\ &= 18{,}180 \text{ lb/hr-ft}^2 \\ \text{Gas flow} &= 14 \text{ moles/hr-ft}^2 \\ &= 406 \text{ lb/hr-ft}^2 \end{aligned}$$

where

$$L/G = 1010/14 = 72.0$$

The liquor rundown rate is rather high for a packed tower design, being equivalent to a ratio of 370 gal/1000 ft^3 of flue gas. A more reasonable value of 123 gal/1000 ft^3 will be chosen, with the corresponding operating line CB shown in Fig. 6-48. This reduction in the water downflow results in an SO_2 content of the discharged gases of 0.0015 mole/mole flue gas, equivalent to an absorption efficiency of 25%.

Reading directly from Fig. 6-48, at a scrubber temperature of 115° F and a 90% absorption efficiency, there would be required seven transfer units (N_{OG}) and an irrigation rate of 370 gal/1000 ft^3 of flue gas. Reducing the liquor rundown to 123 gal/1000 ft^3 reduces the number of stages to two at a corresponding absorption efficiency of 25%.

For the 134° F equilibrium curve, the operating line slope was determined and plotted in Fig. 6-49. The values of 0.002 for Y and 0.000017 for X were determined for the upper end of the operating line AB. The line was then drawn in Fig. 6-49 to obtain 90% removal of SO_2. The operating line slope is 106, equivalent to a liquor flow rate of 545 gal/1000 ft^3 of flue gas. Seven transfer units would be required.

A summary of absorption efficiencies, liquor flow rates, and transfer unit requirement at both temperature levels are summarized in Table 6-20.

3. Height of Transfer Unit (H_{OG}) and Packing Height: Conditions are 115° F scrubbing temperature and 90% absorption efficiency. From Eq. 6-17,

$$H_G = \frac{\alpha G'^\beta}{L'^\gamma} \left(\frac{\mu_G}{\rho_G D_G} \right)^{0.5}$$

337

Table 6-20. Summary of Transfer Unit Calculations

Gas Inlet Conditions		Scrubbing Temp., °F	Operating Line Slope	Absorption Eff., %	Liquor Rate, gal/1000 ft³	N_{OG}
Temp., °F	Moisture, vol. %					
250	4.8	115	72	90	370	7
250	4.8	115	24	25	123	2
400	8.5	134	106	90	545	7
400	4.8	126	—	—	—	—

The packing will be 2 in. Raschig rings:

$$L' = 18,180 \text{ lb/hr-ft}^2$$
$$G' = 406 \text{ lb/hr-ft}^2$$

From Table 6-15,

$$\alpha = 3.82, \qquad \beta = 0.41, \qquad \gamma = 0.45$$

From Table 6-16,

$$(\mu_G/\rho_G D_G) = 0.90 \text{ (estimated)}$$

$$H_G = \frac{3.82 \times 406^{0.41}}{18,180^{0.45}} \times 0.90^{0.5} = 0.47 \text{ ft}$$

From Eq. 6-18,

$$H_L = \phi\left(\frac{L'}{\mu_L}\right)^{\eta} \left(\frac{\mu_L}{\rho_L D_L}\right)^{0.5}$$

From Table 6-15,

$$\phi = 0.0125, \eta = 0.22, \mu_L = 2.42$$

From Table 6-16,

$$(\mu_L/\rho_L D_L) = 550 \text{ (estimated)}$$

$$H_L = 0.0125\left(\frac{18,180}{2.42}\right)^{0.22} \times 550^{0.5} = 2.1 \text{ ft}$$

From Eq. 6-19,

$$H_{OG} = H_G + m\left(\frac{G_M}{L_M}\right) \times H_L$$

where

m = equilibrium line slope = 66
G_M = 14 moles/hr-ft^2
L_M = 1010 moles/hr-ft^2

H_{OG} = 0.47 + [(66 × 14 × 2.1)/1010] = 2.40 ft

From Eq. 6-20,

$h = N_{OG} \times H_{OG}$
= 7 × 2.40 = 16.8 ft of packing

4. Pressure Drop: From Eq. 6-23,

$$\frac{\Delta p}{h} = m_1 \times 10^{-8} \ (10^{n_1 L'/\rho_L}) \left(\frac{G'^2}{\rho_G}\right)$$

where

h = 16.8 ft packing height
m_1 = 11.13 and n_1 = 0.00295 for 2 in. Raschig rings (from Table 6-17)
ρ_L = density of water, 62.4 lb/ft^3
ρ_G = gas density at 115° F sat conditions, 0.066 lb/ft^3

$$\frac{\Delta p}{h} = 11.13 \times 10^{-8} \times 10 \left[\exp\left(0.00295 \times \frac{18,180}{62.4}\right)\right] \times \frac{(406)^2}{0.066}$$

Δp = 33.8 lb/ft^2 = 6.5 in. w.g.

5. Flooding/Operating Gas Velocities: From Eq. 6-22,

$$\frac{L'}{G'}\left(\frac{\rho_G}{\rho_L}\right)^{0.5} = K \frac{u^2 \rho_G a}{g_c \rho_L \epsilon^3} \mu_1^{0.2}$$

where

L' = 18,180 lb/hr-ft^2
G' = 406 lb/hr-ft^2
ρ_G = 0.066 lb/ft^3
ρ_L = 62.4 lb/ft^3

$$\frac{L'}{G'}\left(\frac{\rho_G}{\rho_L}\right)^{0.5} = \frac{18.180}{406} \times \left(\frac{0.066}{62.4}\right)^{0.5} = 1.46$$

From Fig. 6-38 for dumped Raschig rings,

$$\mu_1^{0.2}\left(\frac{u^2}{g_c} \times \frac{\rho_G}{\rho_L} \times \frac{a}{\epsilon^3}\right) = 0.012$$

339

From Fig. 6-37 for dumped 2 in. Raschig rings,

$$a/e^3 = 75$$

$$u^2 = \frac{0.012 \times 32.2 \times 62.4}{1 \times 0.066 \times 75} = 4.86$$

$$u = 2.22 \text{ ft/sec}$$

Operating gas velocity taken at 75% of flooding value:

$$\text{Velocity} = 2.22 \times 0.75 = 1.63 \text{ ft/sec}$$

6. Scrubber Sizing

Gas flow = 168,600 acfm at 250° F, 4.8 vol % H_2O
Spec. volume at 250° F, 4.8 vol % H_2O = 18.7 ft^3/lb dry gas
Spec. volume at 115° F, saturated = 16.2 ft^3/lb dry gas
Corrected gas flow = 168,600 × 16.2/18.7 = 146,000 acfm
Cross section area = 146,000/(1.63 × 60) = 1490 ft^2

Choose twelve 12 ft, 6 in. diam × 36 ft high scrubbers.

$$\text{Gas velocity} = \frac{146,000}{12 \times (12.5)^2 \times \pi/4 \times 60} = 1.63 \text{ fps}$$

$$\text{Gas mass flow} = \frac{146,000}{1490} \times \frac{1}{379} \times \frac{520}{575} \times 60 = 14 \text{ moles/hr-ft}^2$$

Liquor flowdown rate = 1010 moles/hr-ft^2

$$\text{Liquor flow/scrubber} = \frac{1010 \times 18}{8.33 \times 60} \times \frac{1490}{12} = 4500 \text{ gpm}$$

7. Effects of Peak Gas Flow on Scrubber Operation:

Peak flow = 210,000 acfm at 250° F
= 182,000 acfm sat at 115° F

Performance
L/G would decrease to 57 so that the revised operating line in Fig. 6-48 would indicate a drop in performance to 70% and N_{OG} would be 4. The value of H_{OG} would be reduced from 2.40 to 1.85 so that the height of packing required would be reduced to (1.85 × 4), or 7.4 ft. The performance could be restored to the 90% value by increasing the liquor rundown rate so as to maintain L/G at a value of 72. However, this scrubbing liquor increase would probably increase the pressure drop to a prohibitive value.

Pressure Drop

Increasing the gas flow to 210,000 acfm at 250° F would cause the superficial velocity through the scrubber to be increased to 2.12 ft/sec or within 5% of the flooding velocity at the original liquor flow rate of 370 gal/1000 ft³. This is too close. Assuming flooding does not take place, then the pressure drop would be increased from 6.5 to 10.2 in. w.g., in accordance with Eq. 6-23.

8. Particulates Removal: With this type of fixed-packing scrubber, using counter-current flow of flue gas and water absorbent, it would be difficult to meet the required particulates collection duty. Actually, the packing would plug after a short operating period. Therefore, it would be recommended that the packed scrubber be operated in series with and downstream from the impingement type scrubber designed for partic-ulates removal; see Fig. 5-45.

9. Scrubber Specifications and Costs

Gas flow, acfm	168,600
Gas temperature, °F	250
Gas flow (sat), acfm	146,000
Gas discharge temperature, °F	115
Scrubber type	Fixed packing
No. of scrubbers	12
Packing, size and type	2 in. Raschig rings
Packing height, ft	17
Liquor downflow/scrubber, gpm	4,500
Pressure drop, in. w.g.	6.5
Performance, % SO_2 removal	90
Scrubber dimensions, diam × hght	12′6″ × 36′0″
Fabrication material	Type 316 stainless
Cost, unerected bases (each), $	120,000
Total cost, unerected basis, $	1,440,000

B. *Scrubber Design with Alkaline Liquor*

The system recommended for this application is a countercurrent flow type of scrubber with mobile packing. To obtain a 90% SO_2 absorption efficiency, a sodium carbonate scrubbing liquor would be required. The major advantages of this alkaline system are the reduced scrubbing liquor flows and minimal contacting equipment requirements.

In absorption, for mass transfer to occur, the partial pressure of SO_2 in the carrier gas must be greater than the partial pressure of the SO_2 above the saturated solution. The latter, representing a counterdriving force, is considerable for aqueous SO_2 solutions, as illustrated in Fig. 6-47. However, in an alkaline solution the vapor pressure of SO_2 above the scrubbing liquor is essentially zero, being a function of scrubbing solution temperature and pH. In such solute/absorbent systems involving gas film, liquid film and chemical reaction resistances, reliable masstransfer values are not too readily available. Instead, empirical data are employed for commercial scrubber designs. There follows the basic design information obtained from the operation of a 1000 cfm pilot, mobile packing scrubber [29].

1. Design Data

Gas velocity, ft/sec	12.5
Liquor rate, gal/1000 acf	40
Pressure drop, in. w.g./stage	2.0
Stage definition	12 in. packing; 4 ft grid spacing
Packing	1¼ in. diam polypropylene spheres; wt. = 4.5 g
Number of stages	3
Liquor type	5% Na_2CO_3 solution
Performance, % SO_2 removal	90

2. Scrubber Sizing

146,000 acfm saturated at 115° F
Area = 146,000/(12.5 × 60) = 195 ft²
Tower dimensions: 12.5 ft × 16.0 ft × 54.0 ft high
Gas inlet: 5 ft × 12 ft

3. Liquor Recirculation

40 gal/1000 ft³ gas flow
146,000/1000 × 40 = 5840 gpm

This flow will be recirculated from the base of the scrubber to the liquor inlet at the top.

4. Pressure Drop

$$\text{Gas velocity} = \frac{146,000}{12.5 \times 16} = 730 \text{ fpm}$$

$$\text{Liquor flow} = \frac{5840}{12.5 \times 16} = 30 \text{ gal/min-ft}^2$$

From Fig. 6-39, prorated pressure drop for a 4 ft high absorption stage is 2.0 in. w.g., and the total pressure drop is 2.0 × 3, or 6.0 in. w.g.

5. Performance

$$G' = (146,000 \times 60)(29/379)(520/575)(1/200) = 3020 \text{ lb/hr-ft}^2$$

$$L' = 30 \times 8.33 \times 60 = 15,000 \text{ lb/hr-ft}^2$$

$$L'/G' = 15,000/3020 = 5.0$$

From Fig. 6-40, SO_2 absorption with Na_2CO_3 = 91%
Particulate collection [29] = 99%

6. Peak Flow Conditions
For 210,000 acfm at 250° F ⇆ 183,000 cfm sat at 115° F
Press. drop per stage at 910 ft/min and 30 gal/min-ft² = 2.8 in. w.g.

Total press. drop = 2.8 × 3 = 8.4 in. w.g.

7. Chemical Makeup and Water Balance

Chemical balance:

SO_2 absorbed = $168,600 \times 0.002 \times 0.90 \times \dfrac{64}{379} \times \dfrac{520}{710} \times 60 = 2250$ lb/hr

The amount of Na_2CO_3 required, assuming 95% purity and 10% excess used, is

$$2250 \times \frac{106}{64} \times \frac{1}{0.95} \times 1.10 = 4300 \text{ lb/hr}$$

Assuming scrubber pH control maintains effluent liquor at a value of 8 to 9, so that discharged liquor will contain 100% Na_2SO_3. Thus,

Na_2SO_3 discharged = $3720 \times \dfrac{126}{106} = 4420$ lb/hr

Excess Na_2CO_3 discharged = 4300 − 3720 = 580 lb/hr
Total chemicals discharged = 5000 lb/hr

Flyash loading at peak conditions:

Flyash collected = $210,000 \times \dfrac{1.8}{7000} \times 60 \times 0.99 = 3220$ lb/hr

Evaporation Water

Inlet gas water content = 4.8 vol %

Water entering = $168,600 \times 0.048 \times \dfrac{18}{379} \times \dfrac{520}{710} \times 60 = 16,900$ lb/hr

Gases leave saturated at 115° F; water content = 0.07 lb/lb DG
Water leaving = 168,600 cfm × (1 lb DG/18.7 ft³) × (0.07 lb H_2O/lb DG) × 60
= 37,800 lb/hr
Evaporation water required = 37,800 − 16,900 = 20,900 lb/hr

Total Water Requirements

Total solids discharged = 5000 + 3220 = 8220 lb/hr

Assuming 10% solids in liquor effluent stream, then

Water required = 8220 × (0.90/0.10) = 74,000 lb/hr
Evaporation water required = 20,900 lb/hr
Total makeup water required = 94,900 lb/hr

Thus, with Na_2CO_3 added as 5% solution, the water required is

4300 × (0.95/0.05) = 81,700 lb/hr

Wash water for entrainment separator, washdown, etc., is then

94,900 − 81,700 = 13,200 lb/hr

The overall water balance is summarized as

In, lb/hr		Out, lb/hr	
Flue gas,	16,900	Evaporation,	37,800
Na₂CO₃ sol,	81,700	Slurry,	74,000
Wash service,	13,200		
Total,	111,800	Total,	111,800

Therefore,

Recirculation liquor rate = 5840 gpm
Makeup water rate = 94,900/(8.33 × 60) = 190 gpm

8. Scrubber Specifications and Costs

Gas flow, acfm	168,600
Gas temperature, °F	250
Gas flow (sat), acfm	146,000
Gas discharge temperature, °F	115
Scrubber type	Mobile packing
Number of scrubbers	1
Number of stages	3
Packing depth and size	12 in.–1¼ in. diam
Liquor downflow, gpm	5,840
Water makeup, gpm	190
Pressure drop, in. w.g.	6.0
Performance;	
% SO₂ removal	90.
% flyash removal	99.
Scrubber dimensions (see Fig. 6-43), ft	12.5 × 16 × 54
Fabrication material	Type 316 stainless
Total cost (unerected basis), $	210,000

REFERENCES

1. "Profile of Air Pollution Control in Los Angeles County Air Pollution Control District," Los Angeles County APC, January 1969.
2. "Rules and Regulations," Los Angeles County Air Pollution Control District, December 1969.
3. G. L. Brewer "Odor Control for Kettle Cooking," J-APCA, Vol. 13, No. 4 (1963), p. 167.
4. *North American Combustion Handbook,* 1st ed. Supplement 78, North American Manufacturing Co., Cleveland, Ohio.
5. "Control Techniques for Particulate Air Pollutants," U.S. Dept. of Health, Education, and Welfare, Public Health Service, National Air Pollution Control Administration, Washington, D.C., January 1969.
6. J. P. Tomany (private communications), Air Correction Division, Darien, Conn., 1967.
7. *Handbook of Industrial Loss Prevention.* Factory Mutual Engineering Division, McGraw-Hill Book Co., Inc., New York, 1959.
8. J. A. Danielson, ed., *Air Pollution Engineering Manual,* 2d Ed., Air Pollution

Control District, County of Los Angeles. Environmental Protection Agency, Research Triangle Park, N.C. May 1973.

9. W. C. Verner and M. S. Edsall, "Analysis of Economic Factors in Catalytic Air Treatment," *Air Engineering* (October 1965).

10. W. B. Krenz, R. C. Adrian, and R. M. Ingels. "Control of Solvent Losses in L. A. County," Air Pollution Control Association Meeting, St. Louis, Mo., June 1957.

11. H. R. Suter, "Range of Applicability of Catalytic Fume Burners," *J-APCA*, Vol. 5, No. 3 (November 1955), pp. 173–175, 184.

12. J. Leonard, R. MacPhee, and R. J. Byran, "Analysis and Testing of Organic Solvents," Technical Services Division of APCD, County of Los Angeles, November 1966.

13. L. D. Decker, "Incineration Techniques for Controlling Emissions of Nitrogen Oxides," Air Correction Division, 1968.

14. A. L. Kohl and F. C. Riesenfeld, *Gas Purification.* McGraw-Hill Book Co. Inc., New York, 1960.

15. M. Polanyi, *Z. Elektrochem.* Vol. 26 (1920), p. 370.

16. M. M. Dubinin and L. V. Raduschkevich, *Compt. Rend. Acad. Sci. U.S.S.R.,* Vol. 55, No. 4 (1967), p. 327.

17. W. K. Lewis, E. R. Gilliland, B. Chertow, and W. P. Cadogan, "Pure Gas Isotherms," *Ind. Eng. Chem.,* Vol. 42 (1950), p. 1326.

18. I. Edeleanu, "The Recovery of Gasoline from Field and Refinery Gases with Special Reference to the Bayer Charcoal Process," *J. Inst. Petrol. Tech.,* Vol. 14 (1928), pp. 286–317.

19. O. A. Hougen and W. R. Marshall, Jr., "Adsorption from a Fluid Stream Flowing through a Stationary Granular Bed," *Chem. Engrg. Prog.,* Vol. 43 (1947), pp. 197–208.

20. G. Nonhebel, ed., *Gas Purification Processes for Air Pollution Control,* 2d ed. George Butterworth and Co., Ltd., London, 1972.

21. J. C. Enneking, "Adsorption Control of Air Pollution," *Plant Engineering* (December 25, 1969).

22. M. Leva, *Tower Packings and Packed Tower Design,* 2nd ed., The United States Stoneware Co., December 1953.

23. T. H. Chilton and A. P. Colburn, *Ind. Eng. Chem,.* Vol. 27 (1935), p. 255.

24. R. E. Treybal, *Mass Transfer Operations.* McGraw-Hill Book Co., Inc., New York, 1955.

25. J. H. Perry, ed., *Chemical Engineers Handbook,* 3d ed. McGraw-Hill Book Co., Inc., New York, 1950.

26. T. C. Baker, "Distillation and Absorption in Packed Columns," *Ind. Eng. Chem.,* Vol. 27 (1935), p. 977.

27. W. E. Lobo, L. Friend, F. Hashmall, and F. Zenz, "The Limiting Capacity of Dumped Tower Packings," *Trans. A.I.Ch.E.,* Vol. 41 (1945), p. 693.

28. H.R. Douglas, I.W. Snyder, and G.H. Tomlinson, II, "An Evaluation of Mass and Heat Transfer in the Turbulent Contact Absorber," Chem. Engrg. Prog., Vol. 59 (1963), pp. 85–89.

29. W. A. Pollock, J. P. Tomany, and G. Frieling, "Removal of Sulfur Dioxide and Fly Ash from Coal Burning Power Plant Flue Gases," A.S.M.E. Meeting, November 1966.

30. R. A. King, *Ind. Eng. Chem.,* Vol. 42 (1950), p. 2241.

31. J. P. Tomany, "A System for Control of Aluminum Chloride Fumes," *J-APCA,* Vol. 19 (1969), p. 420.

32. National Research Council, *International Critical Tables of Numerical Data,* Vol. III. McGraw-Hill Book Co., Inc., 1926, p. 302.

Chapter 7

AIR POLLUTION CONTROL
FOR SPECIFIC INDUSTRIES

7.1. INTRODUCTION

Having developed some background knowledge of the various pollutants and their emission rates, the control regulations limiting these emissions into the atmosphere and the type of control equipment available for their abatement, consideration will now be given to some actual industrial air pollution situations. In this chapter the case histories of specific existing plant operations will be discussed, with regard to their pollutant emissions, the pertinent regulations, and the actual or proposed controls to achieve compliance with the standards.

The solutions of air pollution problems associated with existing industrial facilities fall under three categories; discontinuation of the operation, process modifications, and/or installation of control equipment. The decision as to the optimum approach is largely economic. Every air pollution situation can be relieved or eliminated, but both capital and operating costs are required. The economic impact of air pollution controls on the foundry industry [1], for example, is sufficiently severe to cause a number of the smaller plants to discontinue operations. Control equipment annual costs for gray iron foundries amount to almost 60% of profits before taxes for these small firms.

Such process modifications as fuel substitution or process equipment redesign must be economically justified. Thus, the substitution of electric furnaces for gas-fired models by the glass industries to reduce NO_x and particulate emissions is not an inexpensive solution. The application of control equipment usually represents a nonreturnable investment except in some applications such as solvents emissions control, where by-product recovery is feasible.

The problems encountered by existing manufacturing facilities have been chosen for treatment in this chapter because of the practical value inherent in the consideration of real situations and solutions. However, the plight of the "new plant" should not be minimized. As evident from the regulations of the Los Angeles Air Pollution District (Appendix I), the more stringent section of each rule is directed at new facilities. Every reasonably sized community in the United States requires the review and approval of proposed industrial installations by its APCD group, to assess the potential emissions impact on air quality standards. Process equipment costs for new plant construction include sizable expenditures for environmental control systems to comply with air, water, noise, and waste product disposal standards. For their own

346

protection, many industrial firms have found it advisable to conduct an ecological survey at the plant site before construction of their own production facilities, so as to establish an environmental reference point. Their own estimated emissions are then evaluated in the light of existing air quality conditions and the current regulation requirements to determine the possibilities and costs of compliance.

The economic burden imposed by both the installation and operation of air pollution control systems is a heavy and continuing one. A controlled and acceptable emissions level today may be under surveillance and contested next month or next year because of further revisions of the code. The costs of air pollution and air pollution control will be discussed further in Chapter 9.

7.2. EMISSIONS ABATEMENT REQUIREMENTS

The degree of abatement required for a specific pollutant emission is a function of both the current discharge level and the maximum, allowable emission permitted by the regulations. Regulations do change, however, continuously and in the direction of increased severity. Regardless of the existing emissions requirements in various parts of the country, it is believed that eventually, probably within a year or two, all particulates discharged into the atmosphere will be evaluated by visual standards. A nonplume condition for both dark and light particles should be met by an emissions concentration of 0.02 grain/acf. This value can be considered as a realistic criterion for assessing plant emission control problems. For some industries, such as aggregates-processing plants where the discharged particulate concentration may be as great as 40 grains/acf, this emission standard would require an abatement efficiency of [1 − (0.02/40)]100%, or 99.95%. This is a very difficult collection-efficiency level to sustain with currently available control equipment. In the Waste Wood Combustion Process subsequently discussed in this chapter, the maximum particulates emission concentrations is about 1.5 grains/acf, so that if compliance with the "0.02" standard becomes necessary, a control equipment performance of 98.7% would be required. Control equipment manufacturers would underwrite this latter performance rating without too much hesitation, but might be somewhat reluctant to even propose equipment for the 99.95% duty. There would be real reason to consider various process modifications to reduce the relatively heavy particulate emissions in the aggregates-plant example.

The most common gaseous pollutants are sulfur oxides. Again, as in the case of particulates, the regulations differ considerably across the nation because of the variations in local industrial concentrations and meteorological conditions. Because regulations are never downgraded, it is estimated that the acceptable SO_x emission level will gradually be reduced in the very near future to a value of 200 ppm. This is not too severe a standard for power plant operators relying on coal- and oil-fired combustion. A coal containing 3% sulfur would emit SO_x concentrations of approximately 2100 ppm; thus, an emissions abatement of about 91% would be required for compliance. In today's control equipment market this would require a wet scrubbing system. An alternate approach would be fuel substitution. In this case, a source of coal would have to be found with a maximum sulfur content of about 0.3 wt %. The current regulations in Los Angeles County under Rule 62 (see Appendix I) limits the sulfur content of any liquid or solid fuel to 0.5 wt %.

7.3. PROCESS MODIFICATIONS

Although gas cleaning by the application of control equipment is the most common method of reducing particulate and gaseous emissions, the possibility of process modifications should be considered. Almost invariably, any major process alteration undertaken to achieve compliance with air quality standards can be assumed to add to the plant-operating costs. Rarely do such actions result in financial benefits. Generally, process modification is considered as a substitute for unavailable or economically prohibitive emissions control equipment. In such cases the total costs of varying process equipment, raw material, fuels, or procedures must be balanced against the capital and operating cost requirements for control equipment.

One of the most common examples of process modification is in the use of low-sulfur oil and coal fuels as a substitute for flue gas cleaning to effect SO_x emissions control. The major disadvantage in the use of such fuels, aside from their continuously increasing costs and uncertain supply, is that they may just barely meet the current regulatory demands. Any further tightening of the SO_x standards would exclude this method of control, whereas wet scrubbing control techniques have a wider range of effectiveness.

Many air pollution problems require a combination of process modification and control equipment. The major source of gaseous pollutants such as carbon monoxide, hydrocarbons, and nitrogen oxides in the principal cities of the world is the automobile. The currently favored control technique is catalytic conversion, oxidation of the carbon monoxide and hydrocarbons, and reduction of the nitrogen oxides. However, the tetraethyl lead additive present in most gasolines, to improve their performance characteristics in the automobile engine, also acts as a catalyst poison. It has therefore been concluded that any effective system for the control of automobile engine-exhaust pollutants requires the elimination of the gasoline lead contents. This process modification would impose a prodigious economic strain on gasoline-refining operations, according to automotive authorities. Federal emission standards for the automobile demand compliance by 1975–1976. Automotive manufacturers are attempting to delay this schedule in the hope that a noncatalytic solution to the problem can be devised which would allow the continued use of leaded fuels.

The required investment levels and operating costs for some categories of control equipment make it mandatory to evaluate the complete process or, as mentioned earlier, to adopt a total systems attitude. This approach is illustrated by a study undertaken by the San Francisco Bay Area Air Pollution Control District [2], which assessed the various control methods available to the foundry industries. In the control of cupola emissions, the quantitative capture of the total dust can be achieved by a high-energy scrubber or a fabric filter in addition to thermal incineration for destruction of the gaseous pollutants. As an alternate, a medium-energy scrubber might remove about 90% of the particulates. By processing only clean scrap, the need for incineration equipment could be eliminated; by such alternate melting methods as electric heating, the emission of gaseous pollutants and submicron particulate would be avoided. Once again, classified scrap represents an operating cost extra, and the "alternate" melt method requires a considerable investment and annual charges. Both incremental operating costs must be evaluated against the combined particulate collection and gaseous pollutant control system.

One simple process variation that has been utilized to improve emissions abatement in many batch operations is that of processing rate. Prior to the current decade, marked by an increasing awareness of the pollution characteristics of various processes, a batch period was determined by the desired production rate, labor costs, and optimum equipment utility. Short periods and intense processing rates were the rule. However, the emission rates for many batch processes operated under these conditions follow a typical probability-curve shape, with the peak rate often in excess of economically reasonable control equipment capacities. A typical emissions curve for

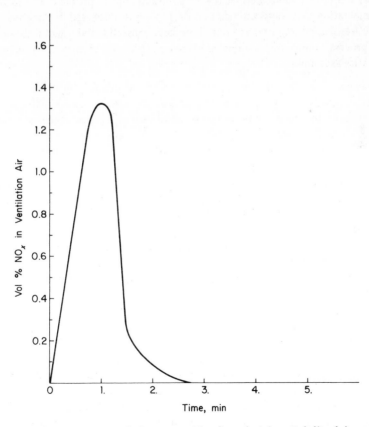

Figure 7-1. Typical release of nitrogen oxides from batch metal-dissolving process.

the evaluation of nitrogen oxide fumes from a metals-dissolving process is shown in Fig. 7-1. By extending the batch period, the curve shape can be flattened, approaching a straight line as the process nears a continuously operated condition. Thus, the sharp and uncontrollable maximum pollutant concentrations can be brought within the performance range of the emissions control system. Labor costs and reduced equipment capacities are the economic factors that must be traded off against the increased probability of complying with air quality requirements.

7.4. SYSTEMS APPROACH TO AIR POLLUTION PROBLEMS

When a limit is imposed on materials being discharged into the atmosphere by a specific industry, a qualitative and quantitative definition of the problem must be established. Control regulations, similar to those defined in Chapter 3, will describe the particular pollutants and specify their acceptable emissions level. Source tests, such as those described in Chapter 4, must then be performed to identify the various pollutants and quantify their emission rates. The difference between the regulation limits and source test values defines the air pollution control problem. It is at this time that consideration of various emission control systems, discussed in Chapters 5 and 6, and/or a fresh appraisal of the overall process must be undertaken. Some of the specific process factors that should be examined from the viewpoint of air pollution emissions are as follows:

Replacement or Modification of Equipment. Motivated by the need to meet emissions standards, many of the process industries have evaluated their process and have determined the need for improved or additional equipment. Such process alterations, representing improved production practices in some cases, have partially offset the control equipment costs. Thus, the addition or replacement of an induced draft fan in an aggregates plant operation has made it possible to increase the overall production rate by 20% while increasing the negative pressure on the process train sufficiently to prevent dust leaks that contributed to the depreciation of the air quality in the neighborhood of the plant.

Substitution of Fuels or Raw Materials. The substitution of electric power for gas fuel to achieve reduction of particulate and gaseous emissions from high-temperature melting operations has been previously discussed. Actually, this action represents a shift in air pollution potential and the need for controls from the processing plant to the power plant. Ultimately, nuclear power plants will represent the source of power most free of air pollution—but not just yet. A comparison of the various fuel types for their ability to provide pollutant-free energy is shown in Table 7-1.

Raw material substitution efforts are slightly more complex. Keeping in mind that most processes have evolved to their present status in an economically motivated climate, serendipitous raw material substitutions are not too frequent. The use of unleaded gasoline for automobiles is still being warmly contested because of the increased costs. However, some segments of the paint industry have eliminated phosphorus in their formulations, to permit the use of catalytic control devices;

350

Table 7-1. Pollutant Emissions Levels for Various Energy Sources

Energy Source	Particulate Emissions. lb/10⁶ Btu Input	Gaseous Emissions	
		SO_x ppm/% S	NO_x lb/10⁹ Btu Input
Nuclear	0*	0	0
Hydroelectric	0	0	0
Geothermal	0	0	0
Gas (no controls)	0.02	0	370
Oil (no controls)	0.07	0	680
Coal, 90% flyash removal	0.67	700	840
Coal, 99.5% flyash removal	0.03	700	840

Source: Adapted from Refs. 3 and 4.
*Disposal of radioactive wastes presently poses a strongly debated problem.

aluminum processors are considering the use of nitrogen rather than chlorine for their molten-metal fluxing operations, to eliminate the evolution of such noxious pollutants as $AlCl_3$, HCl, and Cl_2; the glass industry is attempting to find a substitute for salt cake in its batching operations because of its reduction in the process to submicron particulate emissions; and the paper industry has expended untold wealth in its efforts to develop economically a nonsulfate pulping process.

Product or By-Product Recovery. In dry processing, there are usually many emission sources that represent product losses. In feed and grain processing, the danger of spontaneous ignition and health hazards has resulted in the early application of such dry collection equipment as cyclonic separators and fabric filters, with the return of the collected material to the process. Recently, in the evaluation of a grass-seed manufacturing operation, the emissions from a drying operation were calculated as a loss equivalent to a 4 yr write-off period for the necessary control equipment.

By-product recovery seems often to be cast into the role of a fond dream. For years, power plants have been striving to develop a use for the prodigious quantities of flyash they collected. Although some of this material has value as aggregates filler, the major portion is still used as ground fill. In the combustion of vanadium-bearing fuel oils, the collected flyash does contain some amounts of the metal oxide. Some firms are purchasing this material for vanadium recovery. In both the aluminum and fertilizer industries, fluoride emissions collected in wet scrubbers do have by-product value; in the former the fluorides are converted to cryolite ($Na_3 AlF_6$) and recycled to the electrolytic reduction cells, while the hydrogen fluoride and silicon tetrafluoride compounds emitted by fertilizer operations are sometimes utilized as hydrofluosilicic acid by the glass industry. However, such by-product recoveries are exceptions. In the application of particulate collection equipment, added expenses are usually incurred for disposal of the collected material.

This "having of cake and eating it as well" aspect of air pollution control is most apparent in the various processes presently being developed for the control of sulfur oxides from the combustion of fossil fuels. The control system closest to commercial realization is based on wet scrubbing with calcium-based alkaline slurries, which convert the sulfur to a mixture of insoluble sulfites and sulfates in vast quantities that has very little commercial value. This has been one of the greatest limitations of the wet scrubbing process. A recent innovation involves the use of magnesium-based scrubbing liquors, which form an easily decomposable sulfite compound. Thus, the collected by-product can be reprocessed for the recovery of sulfur values, thereby releasing the magnesium compound for re-use as an alkaline scrubbing agent.

One by-product that should be carefully considered, using the term broadly, is heat recovery. In the application of thermal or catalytic combustion systems for the control of gaseous pollutants, a plant's overall heat balance generally discloses some process heat savings; see the Sample Design Calculations for catalytic combustion in Section 6-6. A very novel and economically prudent heat-recovery system was installed on a wet scrubber for the control of particulate emissions from a sulfate pulp recovery boiler. To reduce the scrubber plume and take advantage of heat savings, a heat exchanger was installed in the recirculating scrubbing liquor line, thereby reducing the temperature of this stream by about 75° F. The process water used as the coolant in the exchanger was heated sufficiently to be utilized in the log de-barking operation. Based on its equivalent steam-producing capacity, the scrubbing system amortization period was of short duration.

Good Practice and Proper Operation. These terms cover many plant-operating functions such as reasonable application, a good installation, and effective maintenance. In a recent survey of an aggregates plant, many of the emissions problems could have been effectively solved by reasonable attention to these three categories. Open trucks were gravity-charged with extremely fine and dry particulates by payloader operation, and open conveyors were employed for the transport of this same material—clearly a problem of misapplication of equipment. The installation was strung out so that those items of equipment making the greatest contribution to the air pollution situation were located at a considerable distance from the engineering and service centers. Instrumentation was misapplied and not maintained, particularly draft gages that could have monitored the existing control equipment performance. The cyclonic separators were estimated to be performing at about 40% of their designed level. The downstream wet scrubber was consequently overloaded and improperly applied for the rather stringent regulations being enforced at the time. What obviously was required at this particular plant was a survey of the existing operations, some modifications of operating procedures, an overall plant material balance to determine where and why the major emission losses occurred, and sustained efforts to upgrade the existing control equipment. Only then should consideration of additional control equipment be undertaken.

Thus, the approach to an air pollution problem should be a broad one and not limited to the design and installation of control equipment. In the illustrated problems and solutions that have been developed for a number of specific industries in the following sections, such a systems attitude has been maintained.

7.5. AIR POLLUTION PROBLEMS AND THEIR SOLUTIONS FOR SPECIFIC INDUSTRIES

First Industry—Waste-Wood Combustion

Introduction. The wood products industry has always been confronted with the problem of wood-wastes disposal. Sawdust, shavings, cuttings, and bark that constitute hogged fuel were incinerated in the tepee burner prior to 1960. The low cost and simplicity of operation inherent in this type of incinerator made it an ideal solution for the combustion-disposal of wood wastes. With the advent of air quality regulations, it was soon realized that the variability of the flue gas flows and the inaccuracy of combustion controls made it impractical to consider the addition of an emissions control system.

The phasing out of the tepee burner due to environmental considerations was responsible for the rapid acceptance of the hogged fuel boiler, which essentially adapts fossil-fuel boiler technology to the combustion of waste-wood fuel. As in modern boiler design, the production of process steam and controlled flue gas rates and composition made it possible to utilize the same general type of emissions control equipment. Thus, this major process modification created a more reasonable environment-economic situation, in which the disposal of a waste product, production of process steam, and air standards compliance balanced favorably with a more sophisticated and expensive type of incinerator and the need for emissions control equipment.

The following discussion concerns such a waste-wood combustion process operated in the state of Montana [5].

Process Description. The hogged-fuel boiler in this study can burn either natural gas for standby conditions, or hogged fuel. This unit has the capability of producing process and generative steam at the maximum rate of 150,000 lb/hr, at 600 psig and 750° F. The primary fuel charged to this boiler is hogged fuel, available at nearby lumber camps and sawmills. The fuel has an average moisture content of 40 wt % and a heating value of about 5300 Btu/lb, wet basis. The hogged fuel is trucked into the mill and dumped into a pit where it is transferred to a three-compartment, live-bottom, surge bin at the rate of 65,000 lb/hr by a 30 in. wide conveyor belt traveling at about 400 fpm. Each of the three compartments is provided with five variable-speed screw conveyors that transfer the fuel into the furnace injection chute. The fuel is injected onto the furnace hearth by an air blower rated at 2600 cfm. The furnace has an under-over fire combustion chamber with a porous bed for underfire fluidization of the hogged material. The forced-draft fan supplies a maximum of 59,000 cfm of air at 105° F to the underfire and overfire air inlets. Combustion gases leaving the economizer section were at one time discharged through cyclonic separators and then vented through the stack to the atmosphere.

Revision of the Montana air pollution control regulations made it necessary to install a secondary wet scrubber type of collector. Accessory equipment for disposal of the discharged scrubber liquor, comprising slurry-pumping stations and clarifier, were included with the scrubber installation. The concentrated slurry was transferred by sludge pumps to land-fill disposal sites. The general arrangement of the boiler, pollution control, and accessory equipments is shown in Fig. 7-2.

Emissions and Regulations. The emissions from hogged-fuel boilers are flyash particulates only. There is no sulfur in the fuel and therefore sulfur oxide gaseous emissions are nonexistent. Because of the relatively low combustion temperatures, the formation of nitrogen oxides and their emissions are negligible. The flyash from this type of boiler is considerably different from that produced by coal firing in that it contains a relatively high proportion of carboniferous combustible material or char, as compared to the noncombustible coal flyash. The material agglomerates readily and is usually light and fluffy in texture. Approximately 30 to 80% of the particles are less than 10 μ in size, but because of its lightness in the agglomerated state, a lower collection efficiency is obtained than would be expected for this size distribution. Entrained with this flyash is a varying amount of sand, the quantity depending on the methods by which the original wood was logged and delivered to the mill. A series of particulate source tests indicated that average emissions to and from the cyclonic separators were 2.46 and 0.84 grains/scf at 12% CO_2, respectively, at the rated boiler capacity. Thus, this collector was operating at an efficiency of 66%.

In 1963, when the boiler and cyclonic separator were initially installed, rigorous regulations were not in effect for the control of particulate emissions. In March 1969,

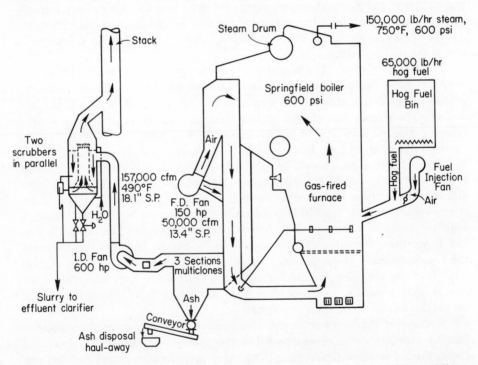

Figure 7-2. Hogged-fuel boiler control equipment arrangement. (Adapted from Effenberger et al., Ref. 5.)

the Montana State Board of Health adopted a series of regulations that practically outlawed the tepee burner and at the same time established particulate emission standards for all fuel-burning equipment, which included hogged-fuel boiler installations. Based on the rated heat input of 188×10^6 Btu/hr for the existing boiler, and categorizing the subject boiler as a "new fuel burning installation" because of some process alterations introduced after enactment of the ruling, the allowable particulate emissions were defined as 0.27 lb/10^6 Btu total heat input (see Fig. 7-3). A Ringelmann 2 reading was also held to be in effect.

Emission Control Methods. The proposed emission control methods were to be predicated on the required degree of abatement. The source tests performed at the rated boiler capacity of 150,000 lb/hr of steam, equivalent to an input duty of 188×10^6 Btu/hr, indicated the cyclone discharge concentration to be 0.84 grain/scf at a CO_2 content of 12% in the flue gas. The total flue gas flow was 155,100 acfm at 490° F, having an average CO_2 content of 8.5%. Converting this gas flow rate to the source test conditions at 70° F, 760 mm Hg, and 12% CO_2, the corrected volumetric flow becomes $155,100 \times (530/950) \times (690/760) \times (8.5/12.0)$, or 55,500 scfm at 12%

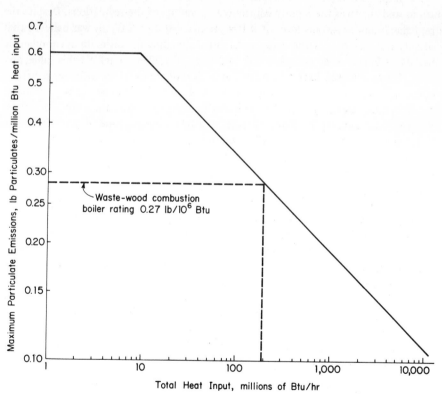

Figure 7-3. Montana State Board of Health maximum particulate emissions rate for new fuel-burning installations (Regulation No. 90-001; revised 7/10/70).

CO_2. The particulates mass emission rate is calculated as $0.84 \times 55,500 \times (60/7000)$, or 400 lb/hr. The equivalent boiler heat-input emission value would be (400/188), or 2.13 $lb/10^6$ Btu heat input. The regulation demands a maximum emission value of 0.27 $lb/10^6$ Btu, which meant that a control device would be required in series with the cyclonic separator to provide an additional removal efficiency of $[1 - 0.27/2.13] 100$, or 87.3%. The overall collection efficiency for the system from the existing collector inlet to the atmosphere discharge point would be 95.7%.

These collection efficiencies are not too difficult to attain. The required atmospheric discharge concentration would be 0.84 (1 - 0.873), or 0.104 grain/scf at 12% CO_2. Some consideration was given to various in-series cyclone collectors, but, at best, the estimated overall collection efficiency would barely meet the code. An anticipated plant expansion might require an additional boiler. This possibility would allow the existing boiler to handle infrequent peak loads so that an in-series cyclonic collector could probably perform sufficiently well at the reduced boiler capacity to meet the regulations. The new boiler would be provided with an efficient afterburner section, which would reduce the char emissions rate sufficiently to allow the use of a single cyclonic collector.

One important factor discouraged the adoption of any of these various marginal solutions and that was the rapidly continuing upgrading of the regulations. Despite the current maximum emissions level of 0.104 grain/scf at 12% CO_2, it was believed that eventually the majority of the codes would be demanding the ultimate in compliance, arbitrarily estimated at an emissions concentration of 0.02 grain/acf. At this value, any in-series collector would have to perform at a level of about 97.5% efficiency. After a careful evaluation of these various alternates, it was decided to install a wet scrubber. Mechanical and performance specifications for the two collectors are as follows:

1. Cyclonic Separator: Three parallel, multiple, cyclone type of centrifugal collectors were installed, similar to that shown in Fig. 5-9. Each section contains eighty 9 in. diam cast-iron collector tubes bolted to and supported by a steel tube sheet contained in a mild steel shell. Each section is provided with an ash hopper from which the collected flyash is discharged through a rotary air lock into a screw conveyor. The three sections have a combined capacity of approximately 200,000 acfm at 500° F at a maximum pressure drop of 3.0 in. w.g. The estimated collection efficiency for flyash measuring 30% $< 10 \mu$ should be in the range of 70 to 75%.

2. Wet Scrubber: Two impingement type medium-energy wet scrubbers were installed in parallel; see Figs. 5-45 and 5-46. These scrubbers were fabricated of Type 316 stainless steel, although subsequent assays of the discharged slurry indicated that less expensive Corten or mild-steel materials would have been adequate. As shown in Fig. 7-2, the gases from the cyclonic separator were drawn into an induced-draft fan and delivered through a vertical duct to the two scrubbers, arranged in parallel. The furnace draft was adjusted by dampers on the fan inlets.

Performance specifications for *each* scrubber are as follows:

Number of scrubbers	2
Dimensions, ft diam × ft hgt	12.25 × 23.25
Gas flow, acfm	80,000
Gas temperature, °F	490

Inlet dust loading, grain/dscf at 12% CO_2	1.0
Makeup water, gpm	60
Slurry-discharge concentration, wt %	1.0
Collection efficiency, %	93
Pressure-drop range, in. w.g.	5.0–7.0

3. Accessory Equipment: A 200 ft diam clarifier with a 10 hp drive, a slurry sump and centrifugal pump, and a reciprocating type of sludge pump were required for disposal of the scrubber liquor effluent. The discharged slurry was slightly alkaline, with the pH varying from 6.8 to 7.4. The rather large clarifier was installed to accommodate the total plant waste products. Similar equipment to handle the effluent from both scrubbers at a flow rate of about 50 gpm would be approximately 20 ft in diameter.

Performance and Economics. The total emissions control system exceeded all expectations in its performance. The final emissions concentrations for both collectors in series were determined for a total of three runs. For reference, the performance of the cyclonic separator alone was also tested during three additional run periods. The results, summarized in Table 7-2, show that in runs 1, 2, and 3 the emissions from the cyclonic separator of 1.30, 1.09, and 3.48 lb/10^6 Btu heat input were indeed in violation of the Montana State Board of Health regulation allowing a maximum atmospheric discharge value of 0.27 lb/10^6 Btu. However, in runs 4, 5, and 6, with both collectors in service, the emissions were reduced to values of 0.081, 0.060, and 0.097 lb/10^6 Btu, well below the allowable level. In fact, in all three of these runs, where total control was achieved, the emission concentrations were less than the arbitrarily set, ultimate regulated value of 0.02 grain/acf. The test results also indicate the attainment of an average collection efficiency by the wet scrubber alone of 96%.

To confirm the theory of wet scrubber contacting power, developed in Section 5.7, a quantitative assessment of the influence of the pressure drop across the scrubber on performance was explored. The gas flow resistance in the scrubber was varied by adjusting the level of the liquor reservoir relative to the point of gas impingement on its surface. Pressure drops of 7.5, 8.0, and 8.5 in. w.g. were thus obtained at varying boiler capacities, and their effect on the flyash emissions values are plotted in Fig. 7-4. As predicted by the theory, at any single boiler-loading value, increasing pressure drops yielded improved collection efficiencies, as reflected by decreasing scrubber emission concentrations.

As mentioned in Chapter 5, one mounting criticism of wet scrubbers is the discharge of a saturated steam plume, often an esthetic objection. One solution to this problem is to reheat the scrubber effluent gas. In an examination of the data in Table 7-2, it is apparent that the wet scrubber is performing far in excess of that level required to satisfy the regulations. Although operating at an average collection efficiency of 96%, the allowable emissions value of 0.27 lb/10^6 Btu could be attained at a scrubbing efficiency of 86%. Therefore, a portion of the hot flue gases discharged from the cyclonic collector could be made to bypass the scrubber and be introduced into the saturated plume stream as a source of heat. For this particular situation it is estimated that about 10 vol % of the gas flow, or about 16,000 acfm at an average

Table 7-2. Summary of Emissions Control System Data for Hogged-Fuel Boiler

Run No.	Date	Emissions Source	Boiler Load, lb/hr steam	Flue Gas Flow		Flue Gas Temp, °F	Discharge Loadings			Emissions, lb/10⁶ Btu	
				acfm	scfm		Grain/acf	Grain/scf	lb/hr	Actual	Allowed
1	3-23-71	Cyclonic separator	140,000	158,000	95,000	425	0.18	0.30	245	1.30	0.27
2	3-23-71	Cyclonic separator	140,000	158,000	95,000	425	0.15	0.25	204	1.09	0.27
3	3-25-71	Cyclonic separator	108,000	167,000	98,000	442	0.46	0.78	655	3.48	0.27
4	3-24-71	Wet scrubber	140,000	110,000	93,500	163	0.016	0.019	15.2	0.081	0.27
5	3-24-71	Wet scrubber	140,000	110,000	93,500	163	0.012	0.014	11.2	0.060	0.27
6	3-25-71	Wet scrubber	105,000	113,000	96,500	162	0.019	0.022	18.2	0.097	0.27

Note. Opacity check: Ringlemann reading was estimated by a smoke inspector to have a value of 1.5 or less.

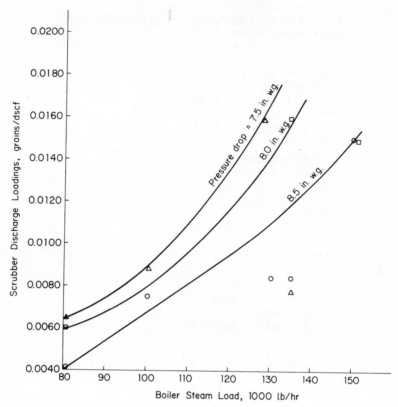

Figure 7-4. Hogged-fuel flyash emission loadings at various scrubber pressure drops. (Adapted from Effenberger et al., Ref. 5.)

temperature of 430° F, could be made available for reheat duty. The combined atmospheric emissions comprising 90% of the gas stream at an average flyash concentration of 0.26 grain/scf, which is treated in the scrubber at a 96% collection efficiency; and the untreated residual 10%, at this same concentration, would be at a final concentration of $[(0.90 \times 0.26 \times 0.04) + (0.10 \times 0.26)]$, or 0.035 grain/scf, which would easily meet the code.

The economic evaluation of the project shown in Table 7-3 is strictly an "after the fact" effort and is made to indicate the pertinent and major factors to be considered in the overall cost balance. Thus, the annual depreciation charges for the installed equipment were estimated from approximate equipment costs, while labor and utilities were presented on an order-of-magnitude basis.

Obviously, the process is economically attractive so that the condemnation of tepee incinerators, which did not provide for heat recovery, did not represent a serious hardship for the wood products industry. The only other possible alternative solution to the problem would be land-fill disposal such as is practiced for elimination of the furnace ash. The estimated annual cost for such a disposal means would be approximately $350,000.

Table 7-3. Economic Evaluation of Hogged-Fuel Boiler Emissions Control System

1. Estimated Equipment Costs: Installed Basis
   Hogged fuel boiler and accessories
     at $6 lb/hr steam $900,000
   Cyclone separator and wet scrubbers
     at $1.10/acfm of flue gas   180,000
   Accessory equipment   50,000
       Total $1,130,000

2. Annual Operating Costs
   Equipment depreciation based on a
     15 yr amortization period $ 75,500
   Labor costs for 4 men at $15,000/yr   60,000
   Utilities   12,000
   Ash disposal costs, 20,000 ton/yr at
     $1.50/ton   30,000
       Total $177,500

3. Credits
   Credit for process steam estimated at
   $0.50/1000 lb for 8000 operating hours
   per year. Annual credit value,
   150,000 × 8,000 × (0.50/1000) $600,000

4. Net Annual Income Gain,
   600,000–177,500 $422,500

Second Industry—Aggregates Processing

Introduction. Aggregates processing involves mining, crushing, screening, classification, drying, and transferring stone and sand products. Before the serious enforcement of emission standards, the use of simple, large cyclones for the control of particulates from the drying operation was considered adequate. More recently, dust collection from screening, crushing, classification, and, in some instances, at conveyor transfer points must be effected to meet recently enacted APCD regulations. The use of single, large cyclones or multiple-tube collectors for the control of particulates emitted by the dryer are not sufficiently effective to meet the more stringent regulations. Practically all aggregates plants operating in the United States today have either installed or are planning to install a secondary collector for dryer emissions control. This collector is either a fabric filter or wet scrubber type. Because of the advantages of collecting and recycling the aggregate fines in the dry state, and because of the recent advances in its technological development, the selection of the fabric filter for this service is favored.

The major source of air pollution in an aggregates plant is the dryer, which is either oil- or gas-fired. In addition to the problem of dust discharge, gaseous emissions are now being recognized. Standards for both NO_x and CO emissions are not being too seriously enforced at the present time, and burner design modifications and fuel/air ratio regulation are the current means used to achieve limited control of these gaseous

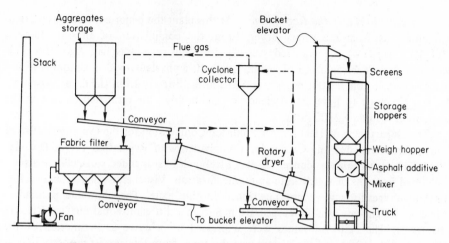

Figure 7-5. Typical aggregates plant equipment flowsheet.

contaminants. In some plants, using a high sulfur-content fuel oil, SO_x emissions must be recognized and possibly controlled.

The following actual plant experience is concerned with the installation and performance testing of a fabric filter collector in an aggregates plant located in Pennsylvania.

Process Description. In this particular plant, aggregates of all sizes up to 2½ in. are processed to be used in hot-mix asphalt paving. The coarse aggregates are crushed stone, crushed gravel, or naturally occurring limestone-based aggregates. The fine aggregates usually consist of natural sand with additive materials such as crushed stone, slag, or gravel. In this operation the aggregates are fed to the dryer batchwise, predicated on the size of the asphalt charge being produced. The batch size is approximately 6000 lb, equivalent to a production rate of asphalt-paving mix in the range of 250 to 300 tons/hr. The batch period during the source testing program rarely exceeded 1 hr. The hot, dried, classified aggregates are withdrawn from the dryer and transferred by an elevator conveyor to a weight hopper and then charged into the asphalt-mix tank. Petroleum-based asphalt cement, comprising about 5% of the product composition, is introduced into the mix tank with the aggregates, the batch is agitated, and then the charge is dumped into waiting trucks for transport to the paving site. Actually, the plant is operated on a demand basis, each complete operating cycle being timed to coincide almost exactly with the arrival of the asphalt transfer trucks.

The flue gases leaving the rotary dryer pass through a cyclone, which removes the larger dust particles, and then continue to the fabric filter for final cleaning. An induced draft fan draws the gases through both collectors and discharges them to the atmosphere through a 3 ft diam, 20 ft high stack. The collected material from both cyclone and fabric filter is conveyed by way of screw conveyors and a bucket elevator to the asphalt-mix tank. The equipment arrangement for such a typical plant is shown in Fig. 7-5.

361

The Emissions and the Regulations. At this plant the major concern was particulate emissions from the rotary dryer. From the compilation of emission factors, (Ref. 10, Chapter 4), the expected emissions from this type of asphalt aggregates plant was estimated at 17.5 lb/ton of hot-mix asphalt, equivalent to a maximum emission rate of (17.5 lb/ton × 300 ton/hr), or 5250 lb/hr, directly from the dryer. Assuming the large cyclone to have an efficiency of about 60%, the estimated loading to the fabric filter would be 2100 lb/hr.

The Department of Health's Air Pollution Commission for the Commonwealth of Pennsylvania amended the Guides for Compliance with Regulation IV on Dec. 19, 1969. This regulation was concerned exclusively with "suspended particulate matter," as defined in terms of weight-rate and concentration. Visual emissions criteria such as Ringelmann readings were not mentioned in this issue. The allowable particulate emissions weight-rate was developed from Sutton's diffusion equation (see Ref. 10, Chapter 2). The emission values were expressed by a curve as a function of the effective stack height and the distance of the stack from the nearest property line, as

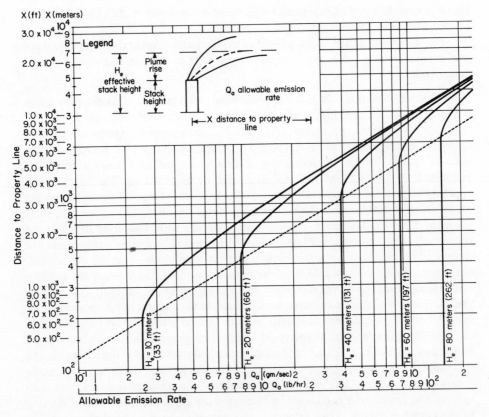

Figure 7-6. State of Pennsylvania allowable particulate emissions (Regulation IV; amended Dec. 19, 1969).

shown in Fig. 7-6. The amended regulation also requested the determination of the concentration values of particulate emissions, both at the source and in ambient air. However, limitations were not specified for either value.

To determine the allowable emissions level from Fig. 7-6, it was necessary to calculate the effective stack height. The following formulas for this determination were cited in the regulations:

$$H_e = H + \frac{(1.5 V_s d + 4.09 \times 10^{-5} Q_h)}{u} \qquad (7\text{-}1)$$

where

H_e = effective stack height, m
H = actual stack height, m
V_s = stack gas ejection velocity, m/sec
d = internal diameter of stack at the top, m
u = wind speed, assumed at 3.8 m/sec
Q_h = heat emission rate of stack gas relative to ambient atmosphere, cal/sec

The heat emission rate Q_h is computed from the expression

$$Q_h = Q_m \times C_{ps} \times \Delta T \qquad (7\text{-}2)$$

where

Q_m = mass emission rate of stack gas, g/sec
C_{ps} = specific heat of stack gas, cal/g-°K
ΔT = stack gas temperature minus ambient air temperature, °K; assume the latter value at 238° K

The value of H_e was calculated for each of the four source test runs summarized in Table 7-4. A sample calculation for run No. 1 follows:

$$H_e = 6.05 + \frac{(1.5 \times 26.8 \times 0.92) + (4.09 \times 10^{-5} \times 453,000)}{3.8}$$

$$= 20.8 \text{ m}$$

With an effective stack-height value of approximately 21 m and the stack located within 1000 ft of the property line, the allowable emission rate from Fig. 7-6 is 8.0 lb/hr.

If we consider the estimated dryer emissions value of 5250 lb/hr obtained from the Emissions Factor Compilation to be valid, then the required overall collection efficiency to meet the Pennsylvania code would be [1 − (8/5250)] 100, or 99.85%. Actually, the process emission rates will be verified during the source tests performed to rate the efficiency of the installed equipment.

Emissions Control Methods. Because particulates only were the major concern in meeting the regulations, either a wet scrubber or fabric filter could be selected for the

Table 7-4. Fabric Filter Performance Test Results for Aggregates Dryer Effluent

Run No.	Date	Gas Flow, acfm	Gas Temp, °F	Press. Drop, in. w.g.	Dust Load, grain/acf Inlet	Dust Load, grain/acf Outlet	Collect'n Eff, %	Atmosphere Emissions Grains/scf	Atmosphere Emissions lb/hr	Effective Stack Height, meters	Allowable Emissions, lb/hr
1	8-15-72	36,000	260	2.0–2.4	15.7	0.022	99.86	0.029	6.7	20.8	8.0
2	8-15-72	37,000	280	2.2–2.5	10.9	0.024	99.78	0.033	7.5	20.1	7.7
3	8-16-72	39,000	330	2.4–2.6	18.2	0.020	99.89	0.030	6.8	22.6	9.4
Average		37,300	290	2.2–2.5	14.9	0.022	99.85	0.031	7.0	21.6	8.4

Source: J. P. Tomany, Private Communications, Ref. 6.

duty. If the presence and control of sulfur oxides were a consideration, as found necessary in processing many sulfur-bearing aggregates, then the application of a wet scrubber would be favored. However, because of the following advantages, the fabric filter was selected.

1. Dry collected dust could be re-used without additional processing.
2. Wet processing facilities for handling aggregate slurries were unavailable at this plant.
3. The effluent gas temperature was in a suitable range for fabric filtration.
4. Dewpoint and condensation considerations were not a problem.
5. The required filtration pressure drop was very low, thereby minimizing operating costs.
6. The physical properties of the aggregates dust were conducive to the use of high filtration ratios, with attending low capital costs.
7. The equipment arrangement, available space, and elimination of the need for wet processing accessory equipment contributed to the reduction of fixed-cost requirements.

Two additional factors are often involved in the decision to go "wet" or "dry" in this type of application. Fabric filters are notorious for their high maintenance requirements; this limitation was discussed in some detail in Chapter 5. Much of the difficulty in this area was caused by the use of mechanical shaking devices, which imposed severe wear on the fabric. Today, however, practically every manufacturer of fabric filtration equipment can offer air backflow systems for removing the collected dust from the individual bags. In some designs this cleaning technique approaches a condition of continous operation so that the stop-and-start effect and the strain on heavily loaded bags have been markedly reduced.

There has been an increased tendency to look with disfavor on the wet scrubber when fabric filtration might be equally effective. This attitude has been encouraged by many APCD agencies in an effort to produce a clear invisible plume, but it is no more than a naive but comfortable approach to the solution of emission problems. In fact, in some areas the control authorities have discouraged the potential "emitter" from considering the wet scrubber approach. However, for this particular application, the fabric filter is ideally suited. Process specifications for this type of dry collector are as follows:

Collector type	Fabric filter
Number required	1
Dimensions, lgth × width × hgt, ft.	36 × 10 × 18
Gas flow, acfm	42,000
Gas temperature, °F	350
Dust type	Aggregates
Estimated inlet loading, lb/hr	2100
Filter ratio, cfm/ft^2	6.0
Promised collection efficiency, %	99.7
Pressure drop, in. w.g.	2.5

The fabric collector is a multiple bag, reverse air-cleaned type containing 600 felted wool bags of 6 in. diam by 8 ft length. The bag assembly is supported in a mild-steel enclosure suitably designed for negative pressures up to 100 in. w.g. and furnished with 60 deg collection hoppers. A reverse action, pulsed air-flow cleaning device is incorporated into the design, being furnished complete with timers, relays, and control cabinet. The pulse duration can be set in the range of 0.2 to 0.8 sec, with a recommended interval for this specific service of 15 sec. Operation is completely automatic. The unit is shipped preassembled except for installation of filter bags and connection of utilities. The compressed-air source for bag cleaning must be furnished by other suppliers; the requirement is 40 cfm at 50 psig.

Performance and Economics. In the process specifications a promise of a 99.7% collection efficiency was made at an assumed inlet loading of 2100 lb/hr of aggregates dust.* Thus, the final atmospheric emissions weight rate would satisfy the Pennsylvania code requirement of approximately 8 lb/hr; i.e., $2100 \times (1 - 0.997)$, or 6.3 lb/hr. Interestingly, for this particular installation, a source test was not performed prior to the control equipment design. Rather, nameplate data were relied on for the prediction of the specified gas flow rate of 42,000 acfm at 350° F. Published source emissions data plus very approximate material balances and prior experience were used to predict the inlet loading value of 2100 lb/hr.

A series of three equipment performance test runs were made; the results are summarized in Table 7-4. For the plant-operating conditions, the APCD regulations allow a maximum average atmospheric emissions rate of 8.4 lb/hr. At the conditions of gas flow and temperature, the equivalent loadings would be $(8.4 \times 7000/60) \times (1/37,300 \times 750/530)$, or 0.037 grain/scf. This value is very close to the ultimate 0.02 grain/acf which may soon become the universal value, for a time, in the United States. The fabric filter performed very well, achieving an average collection efficiency of 99.85%, an increase over the promised value of 99.7. The design bases for the filter were not too reliable. The average gas flow rate was lower than expected by about 13%, which is not too serious a discrepancy. However, the inlet loading to the filter was predicted at 2100 lb/hr, whereas the average value determined in the source test was about $(14.9/7000 \times 37,300 \times 60)$, or 4750 lb/hr. Actually, the estimated loading was based on a very conservative assumption of 60% cyclone collection efficiency, whereas 65 to 70% can be expected usually. The discrepancy between 2100 lb/hr and 4750 lb/hr might be explained either by the possibility of a malfunctioning cyclone or the plant's being operated in excess of its capacity. For example, if the plant were operated at 30% excess capacity and the cyclone efficiency was reduced to 50%, then the source test value of 4750 lb/hr and the suggested source emission value of 2100 lb/hr could be reconciled.

As discussed earlier, a wet scrubber was considered for the solution of this problem. However, the required collection efficiency of 99.8+% is a little difficult for the wet scrubber to attain. The contacting power required for this level of performance and in this type of service is considerable. Thus, the aggregates dust would probably analyze at about 60% $< 10 \mu$ so that the pressure-drop requirements for the scrubber might be as high as 30 in. w.g. In some plants the equivalent power requirement to attain this pressure drop often exceeds the transmission line and

*Based on the emissions factor value of 5250 lb/hr from the dryer and a 60% cyclone collection efficiency.

switch-gear capacities; therefore, costs for expansion of the power facilities must be represented as wet scrubber installation charges. In many aggregates plants, the required pumps, settling ponds, adequate water supplies, and experience in wet processing are available, but such was not the case for this particular application and so it was necessary to include these additional costs in the overall economic evaluation.

Table 7-5 shows an economic evaluation of the fabric filter and wet scrubber systems. This comparison demonstrates the lower capital cost of the wet scrubber and its associated equipment—$65,500 as compared with $102,500 for the fabric filter system on an installed basis. The overall annual operating costs of $42,850 for the fabric filter is less than the annual $50,480 estimated for the wet scrubber, mainly because of the power requirements necessary to develop the high pressure-drop requirements of the scrubber. An adjustment was made in labor costs for the filter, to indicate the extra maintenance charges associated with this collector.

There is an additional operating cost that has not been included in this evaluation. The fabric filter can claim total credit for the collected fines because they are directly returned to the batch weighing station and asphalt-mix tank. However, in the wet scrubber system, the settling pond effluent must be concentrated before it can be reclaimed and sold as filler material to the concrete and building material industries.

Table 7.5.   Economic Evaluation of Aggregates Plant Emissions Control Systems

Function	*Fabric Filter*	*Wet Scrubber*
Gas flow, acfm	42,000	42,000
Gas temperature, °F	350	350
Dust loading,		
Grains/acf	14.9	14.9
Lb/hr	4,750	4,750
APCD permissible emission, lb/hr	8.0	8.0
Required collection efficiency, %	99.84	99.84
Pressure drop, in. w.g.	2.5	30.0
Fan drive, hp	100	500
Collector dimensions, ft	40 long × 10 wide × 18 high	9 diam × 20 high
Collector shipping weight, lb	28,000	11,000
Collector capital cost, $	48,000	12,000
Fan and motor capital cost, $	7,500	13,500
Accessory equipment capital cost, $	3,000	8,200
Expanded power facilities, $	0	6,800
Installation costs, $	44,000	25,000
Total installed cost, $	102,500	65,500
Operating costs, $/yr		
Power	6,000	21,000
Water	0	1,100
Labor	30,000	24,000
Depreciation	6,850	4,380
Total	42,850	50,480

Thus, the ability of the fabric filter to recover dry aggregate fines for blending into the final product is another advantage favoring the use of this collector by the asphalt aggregates industry.

REFERENCES

1. "Economic Impact of Air Pollution Controls on Gray Iron Foundry Industry," U.S. Dept. of Health, Education, and Welfare, Public Health Service, Environmental Health Service, National Air Pollution Control Administration, Raleigh, N.C., Nov. 5, 1970.
2. "Air Pollution in the Bay Area, Technical Report and Appraisals," 2d ed., Bay Area Air Pollution Control District, San Francisco, 1962.
3. "Control Techniques for Particulate Air Pollutants," AP-51, U.S. Dept. of Health, Education, and Welfare, Public Health Service, Consumer Protection and Environmental Health Service, National Air Pollution Control Administration; Washington, D.C., January 1969.
4. "Control Techniques for Nitrogen Oxide Emissions from Stationary Sources," AP-67, U.S. Dept. of Health, Education, and Welfare, Public Health Service, Environmental Health Service, National Air Pollution Control Administration, Washington, D.C., March 1970.
5. H. K. Effenberger, D. D. Gradle, and J. P. Tomany, "Hogged Fuel Boiler Emissions Control—A Case History," paper presented at the Environmental Division Conference of TAPPI at Houston, Texas, May 1972.
6. J. P. Tomany (private communications), Western Precipitation Division, Los Angeles, Calif., 1971.

Chapter 8

AUTOMOTIVE POLLUTANT EMISSIONS

8.1. INTRODUCTION

The greatest single source of pollutant emissions in the United States is the automobile. In Table 1-1, the total annual particulate and gaseous emissions in this country from both stationary and mobile sources for the year 1969 were estimated to be 281.2 million tons. Of this amount, transportation sources—essentially automobiles, trucks, and buses—contribute 144.4 million tons. The degree of contribution to the air pollution situation made by the automobile in any single community is a function of the relative automobile to industry population. Thus, in Los Angeles County, with its 4½ million automobiles, this source is responsible for discharging into the atmosphere about 42% of the total particulates and an average of 80% of the total gaseous pollutants comprising hydrocarbons, carbon monoxide, and nitrogen oxides. By contrast, in St. Louis, Missouri, with an automobile population of about one million cars, its relatively large industrial concentration reduces the particulate emissions contribution of the automobile to about 3%.

The two automotive vehicle types responsible for pollutant emissions are the gasoline-powered internal combustion engine and the diesel engine. Practically all automotive pollutants are emitted by gasoline-powered vehicles because of their numbers. The overall concentrations of gaseous emissions from the diesel are lower than the gasoline engine, although its exhaust is responsible for higher particulate emissions and has an offensive odor level.

Because of their marked influence on air quality, automotive emissions have been the subject of more legislation for their control than those for any single stationary source. There is considerable confusion in the various regulations being enacted by the federal and state agencies in view of the unusual environmental economic impact of any actions directed at the automotive industries. There were approximately 90 million registered passenger cars in the United States in 1972 and obviously even minor proposed modifications can result in prodigious national expenditures. It has been estimated [1] that a charge of $10.00 per new car for an emissions control system amounts to an expenditure of $100 million per year for the country's car buyers. Such figures as these have caused both technical and legislative authorities to pause and reflect before initiating any extreme control measures. The Environmental Protection Agency's (EPA) first concern is with the hydrocarbon and carbon monox-

369

ide emissions, calling for a 90% reduction of these two pollutants from 1970 standards by 1975. (These standards have since been postponed and are being re-evaluated.) EPA is insisting that by 1976 the emissions of nitrogen oxides be reduced by at least 90% of the average levels from uncontrolled 1971 vehicles. These are rigorous requirements and are being actively opposed by the automotive industries. Under pressures by this industry, the EPA has recently reduced some of its demands for hydrocarbons, CO, and NO_x emissions standards for heavy-duty diesel equipment by 1973.

The present technological status of automobile pollution control devices is not too healthy. The generally accepted control approach involves the application of catalytic elements, which are considered particularly unreliable when entrusted to the care and supervision of the average automobile owner. Unconventional automotive power-plant development activities are being supported by the federal government; electric power, liquefied natural gas, steam, and hybrid engines are some of the power-plant concepts being investigated by private industry, with support from EPA. Mass transportation, automobile utility taxation, and urban travel restrictions are some of the nontechnical options presently being evaluated by private and governmental development groups to reduce automobile-generated air pollution.

8.2. EMISSIONS PROBLEM OVERVIEW

The ever-increasing proliferation of the automobile would indicate that, if uncontrolled, gaseous exhaust products could increase without limit. It is estimated that each private passenger vehicle annually emits on an uncontrolled basis 0.15 tons of hydrocarbons (HC), 1.06 tons of carbon monoxide (CO), 0.053 tons of nitrogen oxides (NO_x), 0.005 tons of sulfur oxides (SO_x), and 0.004 tons of particulates, the latter containing some lead compounds. The projected registration of passenger cars in the United States up through 1985 is shown in Fig. 8-1. Based on the operation of 90 million cars in the United States in 1972, the total uncontrolled pollutant emissions for that year would be 13.5×10^6 tons of HC, 95.4×10^6 tons of CO, 4.8×10^6 tons of NO_x, 0.5×10^6 tons of SO_x, and 0.4×10^6 tons of particulates. If viable control devices are not available by 1985 and the current registration trend continues, there could be expected a twofold increase in the discharge of automotive air pollutants into the atmosphere. The contribution that each of these pollutants makes to the overall deterioration of the nation's air quality is indicated by Table 1-1. Transportation sources are responsible for the emission of 53% of the total HC, 74% of the CO, 50% of the NO_x, 3.3% of the SO_x, and 2.3% of the particulates.

The vast automobile population in the United States is concentrated in urban centers. Unlike emissions from stationary plants, those from the automobile represent a variable source of air pollution, which is strongly dependent on traffic flow patterns and more directly on automobile commuting activities. Thus, weekday pollutant concentration peaks disappear almost entirely on Sundays and assume an intermediate value on Saturdays. A plot of the daily traffic density in Washington, D.C., on a single day in September 1964 is shown in Fig. 8-2 (a). Superimposed on this same curve is a plot (Fig. 8-2b) of measured CO concentrations. The curves are in agreement as

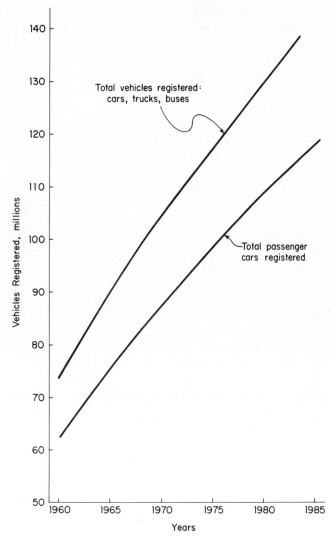

Figure 8-1. Projected registration of motor vehicles in the United States. (U.S. Department of Commerce, Ref. 2.)

expected for the early morning period, but the pollutant intensities seem to fall off during the evening traffic surge. Varying meteorological conditions are believed responsible for this dispersion effect. The exposure of the automobile-confined commuter is usually more severe than indicated by stationary monitoring stations. Table 8-1 lists exposure of commuters to CO in five major urban centers during the approximate half-hour required to travel to work. Commuter proximity to the source is reflected by the fact that these levels are from two to seven times the station-

371

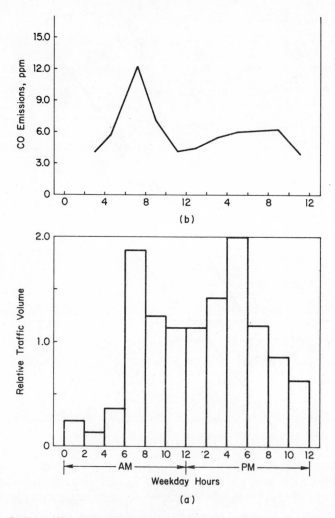

Figure 8-2. Daily traffic pattern and carbon monoxide emissions. (Adapted from Ott et al., Ref. 3.)

monitored values. As a point of reference for reviewing these values, it is interesting to note that the California Department of Public Health has established maximum CO concentrations at 30 ppm for an 8 hr exposure and 120 ppm for 1 hr exposure.

The effects of hydrocarbons in the atmosphere are not too well known except for their photochemical reaction with the nitrogen oxides to produce ozone and PAN, as discussed earlier. Most of the particulates from automotive exhausts are submicron. Particles less than 0.1 μ serve as condensation nuclei that may adsorb gaseous pollutants. Lead is a known industrial poison. The tetraethyl lead compound used as a

**Table 8-1.   Automobile Commuter Exposure to Carbon Monoxide in Various Major Cities
(Based on 20–30 minute runs in traffic)**

City	Year	Route Type	No. of Runs	Exposure during Runs			Brief Peak Exposure		
				Min	Avg	Max	Min	Avg	Max
Los Angeles	1966	Expressway	87	9	29	62	23	52	102
		Arterial	15	24	38	60	45	74	147
		Center city	17	27	40	60	44	68	111
New York	1967	Expressway	17	5	21	39	16	45	141
		Center city	30	9	27	42	16	50	119
Washington, D.C.	1967	Expressway	35	2	12	33	5	27	76
		Arterial	17	10	20	30	22	40	86
		Center city	34	8	26	63	13	45	116
Chicago	1967	Expressway	34	3	20	35	7	43	86
		Arterial	29	1	16	41	3	32	68
		Center city	10	19	34	53	45	70	109
Atlanta	1966	Expressway	10	12	23	35	50	73	92
		Arterial	9	21	30	40	37	62	91
		Center city	7	13	21	27	33	49	75

Source: U.S. Dept. of Commerce, Ref. 2.

Notes: (1) Concentrations expressed in parts per million (ppm); (2) generally accepted maximum carbon dioxide "healthful" levels are 120 ppm for 1 hr and 30 ppm for 8 hr.

gasoline additive has become a major contaminant in the atmosphere. Each gallon of gasoline contains about 2.5 cu cm of tetraethyl lead, and about 70 to 80% of this amount is discharged from the automobile exhaust. Thus, of the total estimated annual uncontrolled emissions of 0.4×10^6 tons of particulates from the nation's automobiles, lead compounds contribute about 20% as submicron particles, which remain suspended in the atmosphere for some time. There is a lack of toxicological data on the body's lead levels of the general population or the contribution to this level made by automobile emissions. These unknown areas make it difficult to establish air quality criteria for lead and its compounds.

Because automobile emissions have been the major cause of the smog problem in Los Angeles, the California Air Resources Board has taken some bold steps to restrict emissions from new vehicles (defined as 1966 and later models) [4]. In every pollutant category the emission standard was exceeded according to surveillance test data developed by the Board. Thus the HC limit for 1966 to 1969 models was exceeded by 2.5%, and the CO standard for 1966 to 1967 models was exceeded by 15% but was met by the 1968 and 1969 models. In their attempts to reduce HC and CO emissions in 1968–1970 the automobile manufacturers prescribed an approach that caused an increase in NO_x of about 50% over the pre-1966 vehicles. This marked increase was believed responsible for the nine ozone concentration alerts that occurred in Los

373

Angeles County during 1970, the highest number in any year since 1956. These problems attest to the difficulties involved in establishing and maintaining control of automotive emissions on the more recent model cars.

Those cars without emission control, 1966 and older, are estimated to number about 55 million in the United States at the present time. Considering that they emit as much as five times more hydrocarbons and three times more carbon monoxide than the 1971 models, they represent the largest single source of air pollution in the country.

8.3. INTERNAL COMBUSTION ENGINE EMISSIONS

Emission Sources and Conditions

The automobile emissions sources and the estimated discharge of the various pollutants from each source are as follows:

Crankcase; 25% HC, 25% particulates
Fuel tank and carburetor; 20% HC
Exhaust
    55% HC
    99% CO
    99% NO_x
    99% SO_x
    75% particulates

Leakage of gases past the pistons and piston rings into the crankcase is termed "blowby." This loss occurs during both the compression and power strokes, and is a function of air-fuel mixture flow to the engine. Thus, emissions are at a maximum at heavy load conditions. About 85% of this loss is the air-fuel charge and the remainder comprises combustion products. The hydrocarbon losses result from the quenching action of the cylinder walls on that portion of the combustible mixture farthest from the flame source. The hydrocarbon concentration in blowby gases is usually quite high, about 10,000 ppm, measured as hexane. Before controls were imposed, the standard method for discharging crankcase fumes was by the use of a draft tube that vented them to the atmosphere.

Fuel evaporation losses occur in decreasing order of their magnitude at the fuel tank and the carburetor. The loss of hydrocarbons from the fuel tank comprises over half the total evaporation losses. The tank is vented to the atmosphere so that such losses are a function of gasoline volatility, ambient temperature, and the volume of fuel in the tank. During car operation the losses from the carburetor are at a minimum. This is because the carburetor bowl, which serves as a fuel reservoir, is usually under a slight vacuum when the engine is running. The major carburetor evaporation losses occur when the engine is shut down and the ambient temperature in proximity to the motor might be increased by as much as 70° F above normal by release of the stored engine heat.

The composition and quantities of the exhaust products from an internal combustion engine is a function of the mode of driving. Thus, cruising, acceleration or maximum load and deceleration, or idling cause several changes in the fuel combustion variables. Cruising produces high air/fuel ratios, moderate-to-low manifold vacuum, and moderate exhaust gas flows. Because the formation of nitrogen oxides is favored by the presence of sufficient oxygen for reaction with nitrogen and high peak pressures and temperatures, the concentration of this pollutant is found to increase at cruising conditions. The air/fuel ratios at which maximum nitrogen oxides are formed are in the range of 15.5:1 to 17:1. Further fuel dilution causes decreasing peak cycle temperatures and is reflected in lower emitted NO_x concentrations. Acceleration demands moderate air/fuel ratios and is marked by low manifold vacuum and high exhaust gas flows. A low air/fuel ratio in the combustion zone causes an increased formation of carbon monoxide so that the emission of this contaminant is lowest at cruising and medium during acceleration. Idling is accompanied by low air/fuel ratios, a high manifold vacuum, and low exhaust flows. The carbon monoxide levels are greatest at these conditions because of the very lean gas mixture.

The hydrocarbons emitted in the exhaust differ from those resulting from evaporation losses because they include various oxygenated and partially oxidized compounds. These emissions vary with air/fuel ratio and engine load conditions. Their concentrations are highest when the engine is running rich in fuel during warmup and deceleration. Such factors as the amount of residual fuel dilution at the start of the compression cycle and wall-quenching effects also influence hydrocarbon exhaust emissions. The discharge of hydrocarbons is strongly dependent on the manifold vacuum.

At low values in the range of 0 to 7 in. Hg with maximum engine load, up to intermediate values of 18 to 20 in. Hg at cruising conditions, hydrocarbon concentrations are low and fairly constant. However, for deceleration, when the vacuum values exceed 21 in. Hg, relatively high hydrocarbon losses are experienced, which may amount to as much as 20% of the fuel furnished to the engine. This relationship is not unexpected, since engine operating conditions such as temperature, pressure, and air/fuel ratio are optimized for complete combustion at the low and medium manifold vacuum values. In general, increased hydrocarbon and nitrogen oxide concentrations can be expected at the same engine operating conditions.

The automobile is not considered a serious source of sulfur oxides pollutant emissions. As mentioned previously, the transportation industries are responsible for only 3.3% of the SO_x atmospheric pollution level. Therefore, the control of SO_x is of little concern at the present time. More realistically, it is believed that if one more pollutant is added to the present burdensome package of unsolved emission problems faced by the automobile industry, they might quietly give up or completely rebel.

Particulates discharged from automotive exhausts consist of lead compounds, carbon particles, motor oil, and nonvolatile reaction products. The latter are probably high molecular-weight olefins and carbonyl compounds formed by the degradation of motor oil in the combustion zone. Particle sizes are estimated to be about 70% by count in the range of 0.02 to 0.06 μ. Particles in blowby gases consist almost entirely of lubricating oil. The amount of particulates emitted from the crankcase is approximately one-third to one-half the amount discharged in the exhaust. Although lead compounds represent only an estimated 20% of the total particulate emissions, they

375

create the most concern, not only because of their toxicological impact but also because of their interference with the application of effective catalytic control devices. The amount of lead discharged in the exhaust of an automobile engine seems to be a function of the present and past driving modes to which the engine has been subjected. Thus, for normal in-city cruising conditions, the quantities of lead emitted vary from a low value of 20 to 25% of the total input for prior extended severe engine service to a range of 30 to 60% for similar but moderate engine activity [5]. It is believed that the severity of demand on the engine determines the degree of lead deposition in the exhaust system, and the position of the equilibrium loading determines the relative retention/discharge ratio.

The diesel engine utilizes a compression ignition system and represents the major source of power for heavy trucks and buses. The diesel operates with a considerable excess of combustion air so that carbon monoxide emissions are essentially eliminated. The excess oxygen promotes a more complete oxidation of the hydrocarbons, while the resultant lower peak-temperature limits formation of the nitrogen oxides. At all engine-operating conditions except deceleration, the gaseous emission concentrations are less for the diesel than for the spark-ignition engine. At idling and deceleration the nitrogen oxides emissions are almost the same for both engine types. Particulate loadings from diesel engines run higher than for the gasoline engine.

Diesel smoke discharges, accompanied by noxious odors, vary in shade from white to black, depending on combustion conditions. White smoke consists largely of unburned hydrocarbons, and black smoke comprises carbon particles discharged under incomplete combustion conditions, usually under heavy load demands. When operated under optimum air/fuel ratios and temperature levels, the diesel engine emits small amounts of smoke. The reason for diesel odor is not completely understood at the present time, and research efforts are being expended to determine its cause and develop methods for its abatement. Because diesel-powered vehicles represent less than 1% of the automobile population and their gaseous emissions are considerably less than for the automobile, their contribution to the overall air pollution problem is not receiving as much attention as that given to the automobile. Standards have been established by a number of states for diesel exhaust odor and smoke opacity.

Pollutant Concentrations and Discharge Rates

The pollutant concentrations and total emission rates are a function of the various driving modes discussed earlier. The influence of cruising, acceleration, and deceleration driving conditions on the formation and exhaust discharge of hydrocarbons, carbon monoxide, and nitrogen oxides can be seen in Table 8-2. The values in this tabulation have been determined from dynamometer test runs based on a cyclical, standard mode, test procedure. This test system was derived from average traffic pattern surveys and is specified for various engine load conditions, with the engine dynamometer directly connected to the flywheel. Vehicle accelerations are simulated using dynamometer flywheels for inertia loadings, and cruise loads are achieved with constant energy-absorbing mechanical systems. During the entire test period, continuous recordings are made of HC, CO, and NO_x concentrations. These values are

Table 8-2. Automobile Exhaust Emissions Test Results

Mode	Engine Air/Fuel Ratio	Exhaust Rate, acfm	Hydrocarbons Test A mol %	lb/hr	Test C mol %	lb/hr	Test D mol %	lb/hr	Carbon Monoxide Test A mol %	lb/hr	Test C mol %	lb/hr	Test D mol %	lb/hr	Nitrogen Oxides Test A ppm	lb/hr	Test B ppm	lb/hr	Test C ppm	lb/hr
Idle	11.9	6.8	1.11	0.17	0.54	0.08	0.28	0.04	6.69	1.97	6.56	1.94	6.59	1.95	30	0.001	–	–	–	–
Cruise, mph																				
30	13.3	24.6	0.75	0.42	0.30	0.17	–	–	3.35	3.57	1.90	2.03	–	–	905	0.16	1465	0.37	1208	0.21
40	13.6	35.6	0.56	0.45	–	–	0.13	0.11	2.51	3.88	–	–	1.43	2.22	–	–	–	–	–	–
50	13.9	48.5	0.78	0.86	0.26	0.29	–	–	1.90	4.00	1.42	3.00	–	–	1625	0.56	–	–	1273	0.44
Accel. open throttle, mph																				
0–60	12.7	125	0.70	1.98	–	–	–	–	4.83	26.2	–	–	–	–	–	–	–	–	–	–
0–50	–	105	–	–	–	–	0.14	0.33	–	–	–	–	5.76	26.3	–	–	–	–	–	–
20–45	12.7	90.9	–	–	–	–	–	–	–	–	–	–	–	–	1185	0.77	506	0.38	700	0.45
Accel. part throttle, mph																				
15–30	13.3	42.4	0.73	0.70	–	–	–	–	3.09	5.66	–	–	–	–	1700	0.51	–	–	–	–
Decel. free at start of 50 mph	11.9	6.8	3.60	0.55	–	–	1.47	0.23	6.69	1.97	–	–	–	–	60	0.003	–	–	–	–
Decel. mph, brake at start of																				
40	11.8	6.8	2.19	0.34	–	–	–	–	6.70	1.98	–	–	–	–	–	–	–	–	–	–
30	11.9	6.8	1.99	0.31	–	–	–	–	6.68	1.97	–	–	–	–	–	–	–	–	–	–

Source: Stern, Ref. 5.

averaged and weighted across repetitive cycles, and the emissions for these three pollutants are expressed as illustrated in Table 8-2.

These average values confirm the theoretical relationship between the driving mode, combustion conditions, and pollutant concentration. Thus, the carbon monoxide and hydrocarbon concentrations are at a maximum at idling and deceleration, corresponding to minimum air/fuel ratios and poor combustion quality, whereas the nitrogen oxide concentrations are greatest for cruise and acceleration when the maximum oxygen is available for their formation. The total emission values are a product of the concentration and exhaust flow rate and thus, even for medium concentrations during acceleration, the total carbon monoxide mass discharge rate is at a maximum.

Although lifetime emissions for an automobile are usually based on the life of the vehicle, it has been found that engine deterioration rates are nearly linear after the first few thousand miles. Engine deterioration and abuse, and unique driving practices, are heavily weighted factors in the prediction of total lifetime exhaust-emission values.

The average uncontrolled emission concentrations for automobiles have been estimated to be 900 ppm for hydrocarbons, 3.5 mol% for carbon monoxide, and 1500 ppm for nitrogen oxides. These are weighted averages and are based on the various driving modes and emissions associated with them. These average values fall in the same range as the emissions data compiled in Table 8-2. For example, the 1500 ppm for the nitrogen oxides corresponds to values of 30, 905, 1625, 1185, 1700, and 60 ppm for the various driving modes.

These average values are compared with those for the diesel engine in Table 8-3. Because diesel heavy-haul vehicles are almost invariably driven under heavy-load conditions, the nitrogen oxide concentrations have been adjusted upward accordingly and are in excess of the automobile source levels. The extremely low hydrocarbon and carbon monoxide diesel emissions as related to the automobile values are the major reason that federal regulations for controlling diesel emissions are lagging behind automobile source controls. Additional accurate exhaust measurement methods are being developed for diesel exhaust emissions.

The determination of automobile emission discharge rates are a function of concentrations and exhaust volume, the latter being proportional to fuel consumption. Pollutant outputs for the "average" car have been estimated by the Current Automotive Systems Subpanel of the Automobile and Air Pollution study [2]. The basic values for the "average" car were established at an annual mileage of 10,000 miles and

Table 8-3. Comparison of Gaseous Emission Levels for Automobile and Diesel Engines

Pollutant	Automobile	Diesel
Hydrocarbons, ppm	900	150–500
Carbon monoxide, vol %	3.5	0.2
Nitrogen oxides, ppm	1500	2000–3000

Source: U.S. Dept. of Commerce, Ref. 2.

Table 8-4. Estimated Annual Pollutant Emissions for "Average" Automobile

Pollutant	Crankcase			Evaporation			Exhaust			Total, Lb/yr
	Source, %	Lb/yr	G/mile	Source, %	Lb/yr	G/mile	Source, %	Lb/yr	G/mile	
Hydrocarbons	25	130	—	20	90	—	55	240	11.0	460
Carbon monoxide	0	–	–	0	–	–	100	760	80.0	1760
Nitrogen oxides	0	–	–	0	–	–	100	90	4.0	90
Total		130			90			2090		2310
Particulates										
Total									0.1–0.4	

Source: U.S. Dept. of Commerce, Ref. 2.

379

a gasoline consumption rate of 14.27 mi/gal. These values are equivalent to the combustion of 700 gal or 4900 lb of gasoline per year. Based on this fuel consumption rate and average emission values for the various driving modes, a group of automotive design authorities have estimated the pollutant discharge rates from the crankcase, fuel tank plus carburetor, and exhaust of the vehicles. An overall summary of the estimated total pollutant annual emission rates for the "average" automobile is shown in Table 8-4. Emissions expressed in grams per mile are the equivalent federal values, established by the Environmental Protection Agency as "precontrol" norms upon which to base emissions control goals.

8.4. AUTOMOBILE EMISSIONS REGULATIONS

State Regulations

The State of California was the first to realize the role of the automobile as the major source of gaseous pollutants. Enactment of the California Pure Air Act of 1968 established emission requirements for new vehicles through the 1974 year model. The California State Air Resources Board followed with the definition of standards for 1975 and later model vehicles. In addition, the most recent air quality legislation demands the assembly-line testing of all gasoline-powered vehicles intended for sale in California, starting with the 1973 models. California continues to be the most actively concerned state involved with the development of a motor vehicle emission-control program.

The terms of the Federal Clean Air Act of 1967 reserved the right to enact automobile emissions control legislation for federal agencies. Because of its pioneering activities in the control field, California was permitted to continue its active development of emission standards, but only with the approval, at each standard enactment level, of the Environmental Protection Agency. The State has appealed to the federal government a number of times to be permitted to apply automotive emissions standards at variance with those proposed by the government [6]. These appeals have been based on the recognition of the automobile-generated photochemical smog problem existing in many areas of California.

Emission standards have been defined by the State of California for gasoline-powered vehicles for crankcase ventilation losses, tailpipe emissions of hydrocarbons, carbon monoxide and nitrogen oxides, and fuel tank and carburetor evaporation losses. For diesels, standards have been adopted for smoke emissions, enforceable by roadside inspections. The early California standards for new vehicles, starting with the 1966 model, defined reduced exhaust-emission values for hydrocarbons, carbon monoxide, and nitrogen oxide. More stringent standards were proposed for 1970 cars at this time. Both sets of emission standards, compared with the average typical uncontrolled values, are shown in Table 8-5.

In August 1968, California passed a revised emissions code for cars and trucks under the Pure Air Act of 1968. These values applied to the 1970 through 1974 and later model year cars and are specified in Table 8-6. These regulations were at variance with, and more severe than, the federal HEW-EPA agency values and therefore it was

Table 8-5. California Automobile Exhaust Emission Standards, 1966

Pollutant	Uncorrected Emissions	Emission Standards for Car Model Years	
		1966	1970
Hydrocarbons	900 ppm	275 ppm	180 ppm
Carbon monoxide	3.5 vol %	1.5 vol %	1.0 vol %
Nitrogen oxides	1500 ppm	350 ppm*	350 ppm*

Source: U.S. Dept. of Commerce, Ref. 2.
*These standards to become effective when at least two suitable control devices are made available.

necessary to obtain a waiver from the federal government exempting them from provisions of the U.S. Air Quality Act of 1967. The federal Act provides that a state may be granted such a waiver if it can demonstrate that a compelling and extraordinary need exists for the enactment of stricter regulations and that the standards proposed by the state are both technologically and economically feasible. The waiver was granted.

In 1963, New York adopted legislation requiring an approved crankcase ventilation system for all cars manufactured after July 1963. The standard demanded that the control system must reduce atmospheric emissions from this source by 80 wt %, based on measurements of a representative number of automobiles in each specified vehicle class. In 1966, additional legislation was enacted to define emission standards that were to include exhaust and evaporation losses for every motor vehicle, starting with 1968 models. In establishing such standards, two factors were considered:

1. Exemptions were permitted where a practical control device was unavailable.
2. Due consideration was to be given to the federal regulations and standards.

The State of New Jersey enacted a law directing the Air Pollution Control Commission to establish test procedures and develop inspection standards for the determination of motor vehicle emissions in compliance with federal regulations.

The majority of other states are now establishing standards based on emission limitations developed by their own air pollution control agencies or health boards. However, in all these efforts, the state defers to the federal standards proposed by EPA. Because of the unique air pollution problem in many of its cities, and in view of its knowledgeable background in emissions evaluation and control, California is the only state to actively develop its own vehicle exhaust standards while continuously negotiating with the Department of HEW for their acceptance.

In another type of action, a total of 18 states are presently bringing suit against the four largest automobile manufacturers to force them to install pollution control devices on some 85 million old cars [8], model years 1953 to 1967. Realizing the pollution potential of the early model cars, the California Air Resources board has

381

Table 8-6. 1968 California Pure Air Act Allowable Vehicle Emissions for Various Model Years

I. Gasoline-Powered Motor Vehicles under 6001 Pounds Manufacturer's Maximum Gross Vehicle Weight Having an Engine Displacement of 50 Cubic Inches or Greater.

| | *A. Exhaust Emissions, Grams per mile* | | | | | *B. Evaporative Loss* |
	Uncontrolled	*1970*	*1971*	*1972, 73*	*1974, Later*	*1970 and Later Model Year*
Hydrocarbons	11	2.2	2.2	1.5	1.5	6 Grams of Hydrocarbons per Test, where a test is defined as average 20 minute trip
Carbon Monoxide	80	23.	23.	23.	23.	
Nitrogen Oxides	4.0	–	4.0	3.0	1.3	

II. Gasoline-Powered Truck, Truck Tractor, or Bus over 6001 Pounds Manufacturer's Gross Vehicle Weight.

| | *Exhaust Emissions* | |
	1970, 1971	*1972, Later*
Hydrocarbons	275 ppm	180 ppm
Carbon Monoxide	1.5 vol %	1.0 vol %

Source: California Air Resources Board, Ref. 7.

382

required the installation of control devices on all 1955 to 1967 cars that are resold by September 1972. These controls must achieve the reduction of HC emissions by 40%, CO by 20%, and NO_x by 34%.

Federal Controls

The need to control motor vehicle pollution was recognized by Congress when it adopted the 1965 amendments to the Clean Air Act. Under this legislation, HEW was given the authority to establish and enforce standards to control air pollution from new motor vehicles. The California standards of 1966 were adopted for the nation by the federal government for new gasoline-powered passenger cars and light trucks, both American and imported types, beginning with the 1968 model years. Standards for hydrocarbons and evaporation losses from fuel tanks and carburetors were also proposed by HEW. Evaporation losses were tentatively set for the 1970 model year at 2 grams per test, wherein a test was defined as being equivalent to an average 20 min trip.

In the Clean Air Act Amendment of 1970, a 90% control of hydrocarbon and carbon monoxide emissions, based on 1970 levels, was required by Jan. 1, 1975. Similarly, a 90% control of nitrogen oxides was the goal set for Jan. 1, 1976, while the allowable evaporation losses were increased to 6.0 g per test. These standards are shown in Table 8-7. The amended Act also allowed for the possibility of the EPA director's granting a 1 yr extension to the automobile manufacturer for meeting these requirements. This option has since been exercised.

There have been some violent objections to these standards expressed by the auto industries, usually based on the fact that the control technology to accomplish these emission reductions is not currently available. The majority of automobile manufacturers take the position that even with a one-year delay there is "no technical basis whatsoever" for believing that these standards can be met.

Table 8-7. Federal Automobile Emission Standards, 1970 Clean Air Act

Pollutant	Pre-control Value grams/mile	Permissible Emissions, grams/mile			
		1968	1974	1975	1976
Hydrocarbons (exhaust)	11.0	3.4	3.4	0.41	0.41
Carbon monoxide	80.0	34.0	39.0*	3.4	3.4
Nitrogen oxides	4.0	–	3.0	3.1	0.4
HC Evaporation loss†	–	–	2.0	2.0	2.0
Particulates	0.1–0.4	–	–	–	0.03

Source: Environmental Protection Agnecy, Ref. 9 and HEW, Ref. 10.

*Revised stringent testing procedures announced Nov. 10, 1970.[11] Modification of proposed test procedure announced July 15, 1970, *Federal Register,* Vol. 35, No. 136–II.

†Units are grams per test.

The existing federal legislation also requires that automobile manufacturers test representative models and submit these test results for government review and approval. A typical emissions control performance statement, included by an automobile manufacturer as a section of the vehicle specifications for a 1973 passenger car, is shown in Table 8-8. This statement defines emissions control compliance as demanded by the State of California.

Some of the difficulties encountered in meeting the federal emission standards are acknowledged by EPA authorities. In mid-January of 1973 the federal government maintained that to comply with the automobile emission standards defined by The Clean Air Act of 1970, implementation plans would require drastic gasoline rationing. This announcement was particularly aimed at the Los Angeles area to combat the smog situation during the months of May through October. Based on the Los Angeles conditions, it was estimated that an 82% reduction in gasoline consumption would be necessary. This ruling was to become effective in 1975, but has since been revoked.

Testing Procedures

Compliance to any air quality standard is measured by a specified testing procedure. The automobile emissions data presented in Table 8-2 were based on dynamometer tests for a defined system of driving modes, and this test method has been adopted in principle as the basis for evaluating automobile exhaust emissions.

To comply with California's allowable vehicle emissions, a test procedure referred to as the California Cycle was established. This procedure, as defined in Table 8-9, presumes to simulate an average trip of 20 min in a metropolitan area. It comprises four tests based on a seven-mode warmup cycle, test number 5, which is not recorded, and three tests during a seven-mode hot cycle. The average concentrations of the warmup and hot cycles are properly weighted to yield a reported overall average value. The hydrocarbons are usually measured by a hexane-sensitive, nondispersive, infrared analyzer. The basis for this cycle was a driving-pattern survey made in Los Angeles in 1965. Although it has been contested that this cycle is not representative of a realistic, universally applicable driving mode, it has provided a reasonable procedure for comparing the performance of various exhaust emission control systems.

In November 1970, HEW issued amendments [10] to its earlier July 15 rules [9]

Table 8-8. Typical Manufacturer Emissions Control Performance Statement for a 1973 Chevrolet Camaro Model

Conditions: This vehicle has been tested under and to California assembly line test requirements.

	Hydrocarbons	Carbon Monoxide	Nitrogen Dioxide
California standard,* gram per mile	3.2	39.0	3.0
Certified values, gram per mile	2.7	30.0	2.5

*These values represent the 1968 standards listed in Table 8-6 as amended in 1971 to reflect a revised testing procedure.

Table 8-9. California Test Procedure for Motor Vehicles Exhaust Emissions Control Evaluation

Mode No.	Mode, (speed), mph	Rate of Acceleration, mph/sec	Time in Mode, sec	Cumulative Time, sec	Weighting Factor, $\Sigma=1.000$
1	Idle	–	20.0	20.0	0.042
	Acceleration				
2	0–25	2.2	11.5	31.5	0.244
	25–30	2.2	2.5	34.0	Omitted
	Cruise				
3	30	–	15.0	49.0	0.118
	Deceleration				
4	30–15	–1.4	11.0	60.0	0.062
	Cruise				
5	15	–	15.0	75.0	0.050
	Acceleration				
6	15–30	1.2	12.5	87.5	0.455
	30–50	1.2	16.5	104.0	Omitted
	Deceleration				
7	50–20	–1.2	25.0	129.0	0.029
	20–0	–2.5	8.0	137.0	Omitted

Source: U.S. Dept. of Commerce, Ref. 2.

Notes: (1) This seven-mode sequence is to be repeated for a total of seven cycles: four warmup cycles and three hot cycles. (2) The fifth cycle of the seven-cycle series is not to be read.

regarding automobile emissions testing procedures. Emission standards and dynamometer test methods were specified for application to 1972, 1973, and 1974 model year light-duty vehicles. The revised standards were intended to ensure reductions in emissions of 80% for exhaust hydrocarbons and 69% for carbon monoxide. The test procedure was based on a simulated average 7.5 mile trip in a urban area from an initial vehicle temperature-conditioned state. A nonrepetitive driving cycle of about 23 min was specified with a proportional fraction of the exhaust flow being drawn into a sample bag. The bag contents are analyzed after completion of the test cycle, with total hydrocarbons being measured with a flame ionization detector and carbon monoxide with a nondispersive infrared analyzer.

8.5. EMISSIONS CONTROL SYSTEMS

Following the total systems concept for the reduction of automobile-generated air pollution, the solution to the problem can be considered under three broad action categories.

1. Application of control devices
2. Alternative engine design and fuel substitutions
3. Curtailment of automobile travel by the development of mass transportation systems, traffic pattern control, and other alternatives

Application of Control Devices

Evaporation Losses. Evaporation loss control can be achieved by a number of systems that have been proposed. The most common is to vent the fuel tank to the carburetor, where such losses are at a minimum because of the slight negative pressure maintained on the carburetor. Returning the gasoline vapors to the engine during vehicle operation involves the risk of upsetting the carburetor adjustment on which optimum combustion conditions and exhaust emissions control depend. In many cases this method for evaporation control has caused an increase in the exhaust pollutant concentrations. Esso Research has devised an adsorption-desorption system using an activated carbon canister and a pressure balance valve. Both carburetor and fuel tank are vented through the adsorbent canister so that when the engine is off and heat soaks into the fuel tank and carburetor, all gasoline vapor is adsorbed by the charcoal. When the engine is operating, the pressure-balance valve allows the return of the desorbed vapors to the carburetor only during those modes of engine operation, usually acceleration and high-speed cruise, when a slightly enriched fuel mixture is acceptable. Other evaporation control systems include pressurized fuel tank, vapor condenser, tank insulation, and reduced fuel volatility. Evaporation emissions are considered so easy to control that the standards enacted for this air pollution source are readily attainable.

Crankcase Emissions. The main objective of crankcase control is to reduce hydrocarbon emissions. All automotive industries use the same approach for their control, which involves recycling the gases from the engine oil sump to the combustion chamber. This system, known as positive crankcase ventilation, is the basis for over 60 devices that have been approved for new vehicles in California. During the early development stages, some difficulties developed because of the deposits of sludge at the valves controlling the flow of blowby gases into the engine intake system. Through valve redesign and the use of improved lubricating oils with additives, this problem has been largely solved.

Exhaust Emissions. Exhaust emissions control is the most difficult to accomplish. Control devices for hydrocarbons and carbon monoxide exhaust abatement are based on total oxidation of these compounds as they enter the exhaust manifold, by excess air injection, direct flame afterburners, or catalytic converters. In the more advanced air injection systems, exhaust manifolds have been sufficiently enlarged to convert them into reactors, which are insulated so as to provide the necessary retention time and temperature to achieve complete combustion. One system of this type had yielded hydrocarbon emissions of 50 ppm and carbon monoxide values of 0.5% during a continuous 50,000 mile test. A similar system has been tested in which sufficiently rich air-fuel mixtures are supplied to the engine so as to reduce the nitrogen oxide emissions to a 500 ppm level. However, at these conditions, high hydrocarbon and carbon monoxide concentrations are emitted. The latter two pollu-

386

tants are then oxidized in the manifold-reactor. This system is capable of reducing all three pollutant emissions, but does result in poor fuel economy.

The direct-flame afterburner is located in the exhaust system and provides for the oxidation of unburned fuel and carbon monoxide. The afterburner must be furnished with its own air and fuel supply and an ignition means to initiate the flame. Generally, the combustible mixtures discharged from the engine are too lean to maintain the oxidation reaction and therefore carburetor enrichment must be used or additional fuel added directly to the afterburner for that purpose.

Air injection and flame combustion methods require high operating temperatures. In an attempt to attenuate the problems of poor fuel economy and exotic material requirements attending high-temperature combustion, considerable efforts were undertaken to develop catalytic combustion methods for oxidation of the hydrocarbon and carbon monoxide exhaust emissions. However, catalytic devices are not compatible with exhaust gases from the combustion of gasolines containing lead-based additives. The lead poisons the catalyst materials and limits their life to less than a 15,000 mile service span. To realize the benefits of catalytic conversion of the combustible components of the exhaust stream, the lead contamination must be eliminated. The most direct expedient for this accomplishment is the production of lead-free gasolines.

One industry pressing for the acceptance of unleaded gasolines is the Universal Oil Products Company, which develops and licenses petroleum processes and manufactures catalysts. In tests conducted in early 1970 with a single catalyst converter for the control of unleaded gasoline exhaust products, this company claims to have reduced hydrocarbon emissions by 73% and carbon monoxide by 94% while cutting back the nitrogen oxides emissions from 746 ppm to 72 ppm [11]. The NO_x reduction was made possible by an oxygen feed-back system for the carburetor which maintained near-stoichometric air-fuel conditions.

In connection with these tests, it has been estimated that the additional per-gallon cost for the production of unleaded gasolines at the refinery would be on the order of $0.01 or less. The use of a single catalyst for the reduction of the three major automobile exhaust pollutants would seem to have some advantages over the combination of control systems requiring extra fuel inputs. However, even with unleaded gasoline, there are still some reservations in many circles with regard to catalyst life. Various reports and data have indicated catalyst life periods varying from under 25,000 miles up to 50,000 miles. Federal emission-control device testing procedures include a 50,000 mile durability test, with intermediate and final exhaust emissions to be determined as a rating index of performance.

Nitrogen oxide emissions can also be controlled by recycling exhaust gases, withdrawing them from the exhaust manifold, and introducing them into the engine. By recycling about 10 to 20% of the exhaust gas stream, the peak-cycle combustion temperatures can be reduced to a level sufficient to curtail the formation of NO_x considerably. Emission levels below 350 ppm have been attained by this method, but at the expense of vehicle performance. The recirculation rate is very critical to hydrocarbon and carbon monoxide emissions. The lean mixture conditions necessary for the reduction of these combustible pollutants are not compatible with the exhaust recirculation technique.

Except for the advantages that currently seem to be associated with the use of unleaded gasolines, there has been very little effort directed at fuel modifications. Because it has been proved that automobile emissions are directly responsible for smog formation, it is reasoned that a gasoline mixture containing a reduced level of reactive and/or volatile hydrocarbons might have a beneficial effect on the overall air quality. However, the simplicity and efficacy of blowby elimination and evaporation control techniques, and the complexities of petroelum refining modifications, seem to have diminished development efforts in the latter direction.

Engine Design Changes and Fuel Substitutions

Engine modifications and alternative designs with and without fuel substitution are presently being explored to reduce pollutant exhaust emissions. The majority of automobile manufacturers have adopted various engine modifications, while alternative engine designs and fuel substitutions are still considered to be in the development stage. A discussion of the more important technical advances under each category follows.

Engine Modifications. The stratified charge engine concept has been under investigation for some time. The Japanese-produced Honda CVCC (Compound Vortex Controlled Combustion) engine utilizes this operating principle and its manufacturers claim to be able to meet the 1975 EPA emission standards.

The CVCC engine utilizes a precombustion chamber, connected to the main cylinder by a small throat. A system of valves charges the chamber with a rich fuel/air mix while a very lean mixture is simultaneously introduced into the cylinder. The rich mixture is fired by a spark in the prechamber, and the hot combustion gases, entering the cylinder in a swirling motion, cause the combustion of the lean mixture. In a Honda engine tested at the EPA Ann Arbor Laboratory, HC, CO, and NO_x emissions were about 0.20, 1.7, and 0.9 gram/mile, respectively. The 1975 federal standards (see Table 8-7) demand corresponding emissions for these three pollutants of 0.41, 3.4, and 3.1 gram/mile.

Engine design improvements utilizing various combinations of more precise carburetor mixtures, spark-timing adjustments, and air injection into the exhaust manifold for incineration of the combustibles are currently incorporated in most gasoline-powered vehicles, starting with the 1968 model year.

Alternative Engine Designs. The rotary engine, gas turbine, Rankine cycle engine, and a hybrid drive train with an electric-motor wheel drive are the most promising engine designs being investigated under the impetus of emissions control requirements.

Although the rotary engine was actually developed and demonstrated, and is currently being marketed, its customer appeal is based on vehicle performance rather than pollutant abatement. Exhaust emission from the present version of the rotary spark-ignition engine does not compare favorably with the reciprocating engine. Oil consumption is high and blowby losses are considerable, particularly at lower engine speeds. However, considerable emissions reductions are promised for future models.

The regenerative gas turbine was primarily considered as a solution to the utilization of a variety of fossil fuels. Engine load and speed have far less influence on the uncontrolled emissions from the turbine engine than that exerted by these same variables on the exhaust levels of the internal combustion engine. Its exhaust emission characteristics are greatly superior to the spark-ignition reciprocating engine at a given energy output. A number of vehicle manufacturers have concluded that turbine pilot production for automobiles is about 5 yr away, with possible mass production capabilities scheduled for early 1980.

The Rankine cycle steam engine is dependent on a continuous combustion device to convert the chemical energy of the fuel into the thermal energy content of the steam. In this approach, it is assumed that continuous combustion burners can be designed with a cleaner exhaust than the present typical Otto and diesel cycle engines [12]. The major elements in the design of a suitable continuous combustor for automotive application are precise air/fuel controls and optimum fuel atomization so as to maintain maximum flame performance at all power-demand levels. Proper interfacing of the combustor and steam engine, rapid startup to full power for the system, overall compactness, and reasonable costs are some of the criteria to be considered for the successful adaptation of this combustor-engine system to the automobile. A reciprocating steam engine rated at 100 hp output would represent a practically sized power plant for automobile utility.

Hybrid drive trains—consisting of a prime mover, a parallel energy-storage device, and wheel drive—are being investigated. Even the standard combustion engine, which would be operated under constant load in the hybrid system, is capable of 50 to 70% reduced emissions under these conditions. A combination presently receiving some attention is a Stirling engine connected to a generator which acts as the prime mover, with a battery energy-storage system and electric motor drive. One present limiting factor in this type of hybrid system is the battery power-delivery characteristics. All batteries deliver less energy at increased power demands. In the Stirling hybrid system, power requirements for acceleration and grade climbing must be provided by the engine while the battery-electric motor combination takes over under light-load conditions. There is also a need for batteries capable of delivering higher energy densities, and considerable development efforts are being expended in the investigation of new battery types to fit the special demands of electrically driven cars.

Besides the hybrid arrangement, various versions of an all-electric drive system with zero emissions are being evaluated. These rely on either fuel cells or external "charging" facilities for their energy source. Both the hybrid system with electric motor drive and the all-electric automobile would seem to have ideal emissions characteristics, but both must await the development of suitable storage batteries. The present commercially available rechargeable battery cannot provide the range and power required by a vehicle that must match the performance characteristics of internal combustion engines. One other serious limitation of the electric engine is the prodigious demand that it would place on central power-generating facilities should it be necessary to resort to external battery-charging methods. In this case the problem of automobile pollutant emissions would be transferred to the power plant, which is presently under siege for the control of its own emissions.

Fuel Substitution. The idea of substituting alternative fuels for gasoline and diesel oil has been considered for some time. It is a known fact that the photochemical reactivity of the hydrocarbon exhaust products from the combustion of propane is about 10 to 20% of that from gasoline-fueled engines. When operated with the same air/fuel ratios as for the automobile engine, there is very little improvement in the carbon monoxide and nitrogen oxide emission concentrations from propane-fired engines.

Recent laboratory and automobile engine tests were performed at the Bartlesville

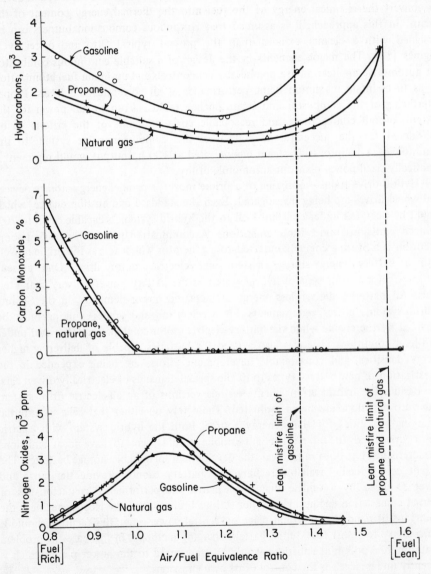

Figure 8-3. Comparison of automobile exhaust emissions for propane, natural gas, and gasoline fuels. (Adapted from Fleming et al., Ref. 13.)

Energy Research Center of the Bureau of Mines, to evaluate propane as a low-pollution fuel and to provide information regarding the adjustment of engine parameters for the advantageous use of propane as an automotive fuel [13]. The results of this study indicated that engines using propane as compared with gasoline can operate over wider air/fuel ratio ranges with minimum hydrocarbon and carbon monoxide emissions. Propane and natural gas emission levels were similar, and no significant difference was found in the nitrogen oxide emissions for the three fuels. Under simulated city-driving conditions with propane, emissions were markedly reduced by retarding ignition timing and increasing the air/fuel ratio to approach the lean misfire limit. However, serious power loss accompanied those engine adjustments that were most compatible with minimum emissions. A comparison of the three pollutant emission levels for propane, natural gas, and gasoline at various air/fuel ratios is shown in Fig. 8-3. These results were obtained with a single-cylinder engine operated at a speed of 1000 rpm, with the ignition timing set at 30° BTC (before top center). The air/fuel equivalence ratio value of 1.0 corresponded to stoichiometric ratios of 15.7 for propane, 16.9 for natural gas, and 14.6 for gasoline. Since propane and natural gas do not contain lead additives, catalytic treatment of the exhaust gases becomes feasible, if further reduction of the three pollutant emissions becomes necessary.

Should the use of propane or liquified petroleum gas (LPG) prove feasible, there would be required major shifts in petroleum refining and fuel distribution practices. Natural gas and/or LPG has become a much valued commodity for industrial combustion processes because of its minimal emission characteristics, and is currently in very short supply. The technology to manufacture LPG from petroleum stock is available, but the lead time required for a processing shift of this magnitude might be on the order of 8 to 10 yr.

Curtailment of Automobile Travel

Air pollution in general and the contributions made to it by the automobile are dependent on population distribution. All urban areas in the United States with populations in excess of one million are beset by air pollution problems. This relationship between population and air quality is illustrated in Table 8-10 [14].

As our knowledge of air pollution technology has expanded, the role of the automobile as the major cause for the deterioration of our atmosphere is becoming more evident. Various technological approaches for the abatement of automobile pollutant emissions have been defined and explained in this chapter. However, a prognosis of the time-benefit relationship for many of these palliatives was not too favorable, so that in a moment of despair or lucidity it could be concluded that the automobile "must go"—just as the demise of a number of marginally operated industrial operations has been arranged because of their inability to comply with emission standards.

Unfortunately, the automobile's position in the United States economy and its development as a way of life would make its elimination not only inconvenient but also catastrophic. Still, with the realization of its air pollution potential, some consideration must be given to the curtailment of its use. It seems so obvious that some 2 million-plus cars, each with its single occupant-driver, cannot continue to

Table 8-10. United States Urban Population and Air Quality

Particulate Pollutants

Population, Millions	Suspended Particles, * $\mu g/m^3$
0.01 – 0.10	83
0.10 – 0.40	101
0.40 – 0.70	113
0.70 – 1.00	119
1.00 – 3.00	154
> 3.00	176

Gaseous Pollutants

Population, Millions	Concentration,** ppm	
	SO_2	NO_2
< 1	0.02	0.045
> 1	0.08	0.09

Source: Herber, Ref. 14.
*1957–1961 geometric mean values.
**1961 averages.

choke the freeways of Los Angeles every weekday morning at average speeds of 15 to 25 mph without someone devising a better way.

Admittedly, any constraints imposed on a private United States citizen in the use of his automobile is anathema. To counter this effect, economically motivated restrictions of automobile traffic must be sought, and a number of possibilities are suggested below.

Mass Transportation Systems. The need for a revival of mass transit facilities is apparent. It is the only potential competitor of the automobile, but it must be tailored to certain conditions where it can truly compete. Patronage of transit systems has been steadily declining. In the face of a remarkable growth in suburban population and the attendant need for the daily transportation of people to work locations in the city, mass transit patronage declined from 12.9 billion passengers in 1951 to 7.2 billion ten years later. During this same period, passenger car registrations climbed from approximately 35 to 65 million.

A consideration of mass transportation for the abatement of air pollution becomes a basic exercise in determining its influence on emissions reduction. For example, the San Francisco Bay Area Rapid Transit System has the potential of eliminating 100,000 automobile trips per day. Assuming a daily round trip from suburbia to the center of the city to average 40 miles, and estimating the total exhaust emissions characteristic of the inefficient driving modes during rush-hour traffic to be

200 g/mile- car, then the total emissions eliminated by the use of rapid transit in this instance would be 100,000 × 40 × (200/454), or 1,760,000 lb/day. A mass-transit system would also be instrumental in eliminating traffic congestion during the most troublesome periods of the day, 7 to 9 A.M. and 4 to 6 P.M.

There is a need for enlightened attitudes and innovative ideas among public officials to make public transportation attractive to the American automobile driver. Some interesting policies to be considered are:

1. Creation of express bus lanes or commuter railroad highway median strips. The rail service from downtown Cleveland, Ohio, to the airport requiring only 15 min travel time is an excellent example of effective mass transportation.

2. Park-and-ride rail facilities of the type that enjoy popular patronage levels in Boston, Massachusetts, and other large eastern cities.

3. Adoption of a nationwide four-day work week with staggered days and hours spread over the full week span. This action would immediately reduce the weekly commuting load by 20% and remove the twice-daily peak emission loads that accompany morning and evening mass automobile commuting.

4. Establishment and enforcement of schedules for the admission of automobiles to metropolitan city centers.

5. Gradual reduction of automobile parking accommodations at public facilities, to encourage mass transportation. This trend seems to be evolving naturally at New York City's J. F. Kennedy Airport.

6. Money management education to convince the public of the true costs in health, time, and monies involved in their use of the automobile for commuter service. The airlines advertised a very effective cost analysis of air versus auto travel, which convinced many people to fly rather than drive from New York to Miami Beach.

7. Solicitation of the support and active involvement of the automotive and petroleum industries in alternate travel modes by economic trade-off studies. A parallel might be drawn between the use of tobacco and the use of the automobile, the latter under certain conditions. The problems involved in the use of tobacco were responsible for the vast diversification program undertaken by the tobacco interests.

8. Public subsidization of mass transportation, under certain regulated conditions. Police and fire protection, even waste-solid removal services are supported by the community. A very popular case could be made for subsidized rail transportation to an early morning commuter sitting in his car on a congested highway in any major city of the world.

9. Improvement of the existing modes of public transportation by an infusion of federal financial aid. In those areas where new equipment, improved stations, and adequate parking facilities have been offered, the public has responded favorably by increased patronage.

10. Expanded practice of toll refunds to those motorists accompanied by two or more fellow commuters. This scheme recently adopted in the San Francisco Bay area has encouraged more effective transportation by private means.

In recognition of the enormity and needs of the mass transportation problem, the federal government formed an Urban Mass Transportation group in 1961, assigned to the Department of Housing and Urban Development. A program has been developed by this agency to investigate, develop, and demonstrate novel and effective transportation media. It is also empowered to extend capital assistance to any deserving urban area troubled by transit problems.

More recently, Congress, in its continuing recognition of the pollution potential of the automobile, has found it expedient to specify the use of a portion of released highway funds for mass transportation improvements and implementation.

Traffic Pattern Control. In the design of highways, no special recognition has been given to differences in vehicle size or speed range. Thus, the bus, the foreign compact car, and the standard six-passenger automobile—each with its different speed, driving modes, and pollutant emission characteristics—travel along the same highway during the same time intervals. One approach that has been proposed would aim to change this traffic mix, particularly during peak commuting hours. The proportion of both buses and compact cars would be increased, which would effect a considerable reduction in the emissions per passenger and thus reduce the overall pollutant mass discharged into the atmosphere during these critical periods of the day.

The restriction of standard automobile entry to the major traffic arteries might be accomplished by computer control, which would supervise the optimum proportion for the various vehicle classes. This system would undoubtedly lead to some delays for automobiles awaiting their turn to enter the traffic stream. This inconvenience, coupled with the improved bus transport, might entice some automobile users to switch to bus transportation.

Taxing Policies. One method to regulate the wanton use of the automobile is by levying taxes on its purchase and/or operation. Some of these suggested tax policies are:

1. An increased federal excise tax on new automobiles or extension of the present tax.
2. A tax structure to encourage the purchase and operation of compact cars with reduced gasoline demands.
3. A graduated tax or insurance premium on multivehicle ownership.
4. Graduated federal and state taxes to reflect total pollutant emission potential of the vehicle as specified in the control performance statement presently issued with new cars.
5. Encouragement of innovative design and improved engine performance by granting special federal tax concessions to manufacturers, as a development/production incentive.
6. An air pollution tax to be based on miles driven or on the fuel consumption rate, with the tax collectible at the gasoline pump.
7. A tax on gasoline lead additives, to be paid by the refiners.

Planning Design and Urban Development. Aside from the generally growing conviction that there are too many environmentally unplanned superhighways in the United States, which encourage the proliferation of automobiles and travel mileage, highway engineers are evaluating their own contributions to the air pollution problem. In an interesting approach to the planning, design, and operation of transportation systems and urban development, the Highway Research Board of the National Research Council has compiled a number of case examples to illustrate the relationship of

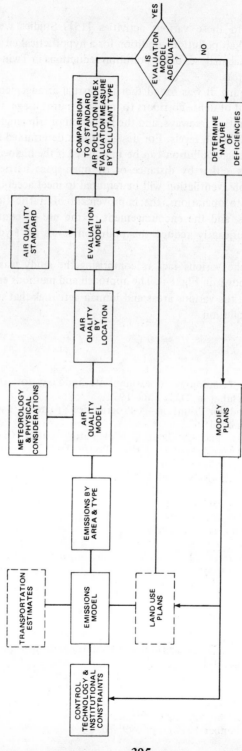

Figure 8-4. Process for incorporating air pollution considerations in planning transportation and urban development systems. (Adapted from Bellomo et al., Ref. 15.)

air pollution emissions to these various activities [15]. Studies were made of a simulated work-travel and air pollution condition for a hypothetical city of 2½ million people, based on individual analyses of air pollution reductions in Twin Cities, Seattle, Montgomery, and other areas.

With respect to planning, it was found that the spatial arrangement of economic activities, the development of dense corridors to increase transit usage, the diversion of more travel to freely operating freeways, and the reduction of trip length have positive effects on reducing air pollution levels. For design, it was determined that concentrations resulting from vehicular pollution can be minimized if the highway is segregated from adjoining structures, either by distance or by open-space barriers. When these separations are not feasible, ventilation will be required to meet acceptable air quality standards. With respect to operations, the improvement of traffic flow, the use of transport pricing policies, and the encouragement of the use of compact cars with smaller engines can significantly reduce the pollution levels for hydrocarbons and carbon monoxide.

The evaluation of the various factors comprising the study was effected by a process similar to that shown in Fig. 8-4. The approach and methods employed in this study focus attention on the various areas and human activities that must be further evaluated to reduce air pollution.

REFERENCES

1. W. E. Jacobsen, "A Technology Assessment Methodology for Auto Emissions," The Mitre Corp., Washington, D.C., June 1971.
2. "The Automobile and Air Pollution—A Program for Progress," Part II, U.S. Dept. of Commerce, December 1967.
3. W. Ott, J. F. Clarke, and G. Ozalins, "Calculating Future Carbon Monoxide Emissions and Concentrations from Urban Traffic Data," U.S. Dept. of Health, Education, and Welfare, Public Health Service, National Air Pollution Control Administration, Durham, N.C., June 1967.
4. "Profile of Air Pollution Control—1971," County of Los Angeles Air Pollution Control District, Los Angeles, Calif.
5. A. C. Stern, ed., *Air Pollution*, vol. II. Academic Press, New York, 1962, chap. 20.
6. *California Air Resources Board Bulletin*, Los Angeles, Calif., Vol. 1, No. 3 (January 1968).
7. *California Air Resources Board Bulletin*, Sacramento, Calif., Vol. 1, No. 10 (September-October 1968).
8. Los Angeles Times, Part I, p. 5, April 25, 1972.
9. "Control of Air Pollution from New Motor Vehicles and New Motor Vehicle Engines," Dept. of Health, Education, and Welfare, *Federal Register*, Vol. 35, No. 136, Part II (July 15, 1970).
10. "Control of Air Pollution from New Motor Vehicles and New Motor Vehicle Engines," Dept. of Health, Education, and Welfare, *Federal Register*, Vol. 35, No. 219, Part II (Nov. 10, 1970).
11. U.O.P. "Announced Breakthrough in Auto Emissions Control," Universal Oil Products Co., News Release, Des Plaines, Ill., March 1970.
12. W. A. Compton, J. R. Shekleton, T. E. Duffy, and R. T. LeCren, "Low Emissions from Controlled Combustion from Automotive Rankine Cycle Engines," *J-APCA*, Vol. 22, No. 9 (September 1972), pp. 699–705.

13. R. D. Fleming, J. L. Allsup, T. R. French, and D. Eccleston, "Propane as an Engine Fuel for Clean Air Requirements," *J-APCA*, Vol. 22, No. 6 (June 1972), pp. 451–458.
14. L. Herber, *Crisis in Our Cities*. Prentice-Hall, Inc., Englewood Cliffs, N.J., 1969, pp. 44–45.
15. S. J. Bellomo, A. M. Voorhees and Associates, Inc., and E. Edgerley, Jr., Ryckman Edgerley Tomlinson Associates, Inc., "Ways to Reduce Air Pollution through Planning Design and Operations," Highway Research Record No. 356, Highway Research Board, National Research Council, Division of Engineering, Washington, D.C., 1971.

Chapter 9

SOCIAL AND ECONOMIC
ASPECTS OF AIR POLLUTION

9.1. SOCIAL AWARENESS

There has been increasing recognition of and concern for the problem of air pollution. Its pervasive nature causes it to have impact on every individual, from the housewife frustrated with the demands of frequent house-cleaning chores, to the automobile-driving commuter who, if he resides in a heavily populated urban center, must arrive at work every day slightly asphyxiated by carbon monoxide and not quite totally relaxed after his driving ordeal.

Although the majority of the population does not fully understand the causes or some of the far-reaching consequences of air pollution, they recognize its immediate effect when they can smell it, feel it, and see it. The public's reaction to the abuse of its air supply is usually strong—and it should be—but some understanding of the reasons for the deterioration of our air quality might incite some constructive action as an expression of that anger and frustration. One inevitable consequence of air pollution is that each individual must ultimately pay in some manner for his parcel of clean air. That price is presently being determined.

The feeling of being the aggrieved victim of this modern plague is mildly compensatory, and one can attenuate his inner tensions by reproaching the various industries for their disregard of health and property values. However, as one air pollution control officer expressed it, if the average city dweller were made fully aware of the contribution to the air deterioration being made by his automobile, would he be willing to embrace the most effective solution and retire his automobile? There is some reason to doubt.

With the public's increased awareness of the problem, there is more need than ever for improved community relations and understanding between the public and industrial segments of our society. The newspapers and various magazines do not always take the most enlightened viewpoint of the air pollution situation. Reports of the incidents at Donora, Pennsylvania, and the Meuse Valley in Belgium, and London's various smog/fog episodes created exciting visions of total societies being decimated by the insidious specter of air pollution. These were isolated experiences created by uncontrollable meteorological and topographical conditions and are far from being representative of the current condition. However, when sensibly appraised, such events do point out the need for constructive action to arrest a worsening situation. Various concerned public groups have recently learned the meaning, if not the spirit, of the word "ecology" and are defending the "status quo" with a vengeance. Unfortunately, they often find themselves in the untenable position of denying our citizen masses the right to the better life by taking to the wilderness "in their own way," in an effort to maintain the forests for endangered species and posterity.

With regard to those problems strictly concerned with air pollution, it is strongly urged that the public's social awareness be broadened to encompass the total picture. The following statements of air pollution facts represent a summary of current conditions and are meant as a guide for an honest assessment of the situations.

1. The uncontrolled emission of pollutants into our limited air reservoir is increasing at such a rate that, unless means for their control are achieved in the near future, the quality of our air can deteriorate to an "unacceptable level."

2. The definition of an "unacceptable level" for the various pollutants, based on firm toxicological background data, is far from being established. This need to determine healthful ambient air quality is one of the major obstacles in the establishment of valid emission standards.

3. The automobile is the major contributor to air pollution, but the development of emissions control devices will be difficult and slow. The most immediately effective control method is the curtailment of gasoline-powered vehicle travel.

4. The enactment and enforcement of pollutant emission regulations have generally proceeded most efficaciously. In some politically oriented areas where air pollution control has been equated to public vote appeal and technological limitations have been ignored, excessively stringent regulations have created a serious impasse.

5. The abatement of pollutant emissions by the application of control equipment and/or the modification of industrial processes must be accomplished within a reasonable economic framework. The benefit-economic rationale must be applied to the assessment of all air pollution problems.

6. The development of economically effective emissions control equipment has not maintained pace with the proliferating air pollution rate. This technological gap has made it necessary to condemn and discontinue a number of industrial operations where the emission levels were intolerable and the control equipment was not economically feasible.

The foregoing statements, in essence, constitute a progress report on the current status of air pollution—what has been accomplished and some of the more important needs. As in every novel and compelling situation, the majority of industries are cooperating and meeting the emission levels prescribed by the new regulations. Others are attempting to gain time in the hope that the air pollution regulatory climate will become less demanding so that they need not consider discontinuation of their operation because of the adverse economic impact of installing control equipment.

Air pollution is really everybody's fault because everyone contributes to it. The steady and accelerated movement of people to metropolitan areas, to work for the industrial concerns located there, increases the need for more goods and services. More homes are required and more wastes are discharged into the air. More workers travel greater distances in their automobile to more plants and businesses, which results in the influx of more commercial ventures, more service facilities, and still more people and wastes, ad infinitum. Soon a situation develops where nature cannot dispose of the airborne wastes at an adequate rate and the air quality begins to degrade. The meteorological and topographical conditions might hasten or delay this condition of pollutant saturation. Both natural factors deal very harshly with Los Angeles, California.

Air pollution, then, can be seen as the result of a multitude of normal, everyday activities performed by individuals operating in a technologically advanced society [1]. Efforts undertaken to solve this problem and the various stages involved in the attempt to develop a solution are aptly described by the sociologist James Bossard [2].

1. Gradual recognition that the problem exists, accompanied by growing community discontent.
2. Discussion of the problem by individuals and groups.
3. Reforms, usually intuitively reached and ill-advised, often attempted by individuals and groups in a disorganized fashion.
4. Failure of the first efforts, followed by requests for careful studies and more information.
5. Identification of the basic factors that created the problem and development of policies.
6. Application and interpretation of the policies by administrators.

It would be interesting to conjecture which stage defines the present air pollution control activities under the Air Quality Act of 1967.

9.2. COSTS OF AIR POLLUTION

General Considerations

No serious attempts have been made to catalog the economic losses that can be charged to air pollution. The reason is simply because there do not exist any marketing forces presently being exerted against a major contributor to air pollution that would compel him to make economic restitution for airborne damages to other persons or properties. There have been individual court actions to resolve disputes between a private party who has been aggrieved and a polluter, but these are random and isolated cases in which the litigant seeks compensation for his damages or wishes to have the offending source abated. In such actions, the costs and damages are assessed for each specific case without any attempt at correlation with a centralized body of information. This deficiency of specific economic loss values is being realized, and the various national estimated values ranging from $4 to $20 billion are rapidly assuming a mythical quality. In 1969 the State of Pennsylvania carried out a statewide survey to provide estimates of air pollution damage to food, fiber, and ornamental crops within the state [3]. The primary object of this survey was to determine the nature and extent of losses so as to provide a sound basis and system of loss assessment.

The losses due to air pollution are difficult to determine. In the evaluation of these losses and the costs and benefits of control measures, it becomes necessary to obtain an adequate representation of the physical effects. The type of information developed by the State of Pennsylvania for crops and vegetation must be extended to an assessment of air pollution damage or destruction to animals, property, and the health of individuals. Such secondary effects as accelerated erosion, increased sedimentation in streams and reservoirs, and the loss of surface water must also be considered. The loss of human productivity is one of the greatest economic burdens associated with air pollution. The degree of incapacity to be charged off against the automobile commuter because of the carbon monoxide content of his blood stream is indeed difficult to evaluate.

These effects should be translated into monetary terms so that cost-benefit relationships can be developed. Thus, the polluter who maintains a fluoride emissions source capable of inflicting cattle and vegetation losses of $500,000 would certainly be receptive to the installation of control equipment at a capital cost requirement of $80,000. Again, and this fact bears reiteration, the determination of and agreement to the $500,000 figure is no small task. Deciding a monetary figure based on current values involves the use of reasonable assumptions as to the level of future prices and the adoption of acceptable rates of capitalization and discount. The basic reason for evaluating air pollution costs is to determine the effects of pollution, with or without a control program or with programs providing various performance levels. The $500,000 assessment value that must be considered by the fluorides emitter might represent the difference between the crop yield that might reasonably have been expected and the actual yields, with due allowance for any lasting injurious effects on the soil, not to mention the discomforts of prolonged negotiations that might arise during litigation proceedings [4].

The economic loss due to air pollution is real. People have died from polluted air and others have become ill. The damage caused by particulate and gaseous air pollutants was described in Chapter 2. Everyone acknowledges the harm and depredations caused by air pollution, but only in a qualitative sense. In the following section, attempts will be made to develop measurement systems that can be used to estimate the costs of pollution. Diseases, materials soiling and damage, and plant and vegetation losses will be identified as they relate to air pollution; some methods for cost measurement will be discussed.

Representation of Air Pollution Costs

If a reliable system for the measurement of air pollution costs were developed, then the representation of these costs on a benefit-cost curve would appear as shown in Fig. 9-1(a). It is assumed that all the items contributing to air pollution costs could be determined as a function of the pollution level, as depicted by curve AB and the costs of controls similarly plotted as curve CD. The relative positions of both curves represent a typical maximum-minimum composite. Therefore, it has been proposed [5] that the optimum level at which to set ambient air quality standards would be at that point where the net control costs are at a minimum equivalent to maximum net benefits. The composite curve shown in Fig. 9-1(b) indicates this point of maximum benefits, represented by the value m. This composite curve indicates that nothing is to be gained by a departure from point m; the consideration of either an increase or decrease of the pollution level would involve increasing costs and decreasing benefits.

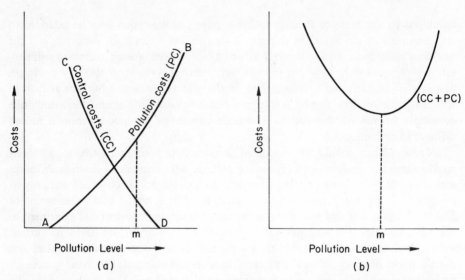

Figure 9-1. Representation of air pollution and control costs. (Adapted from Ridker, Ref. 5.)

A cost-benefit determination technique has been demonstrated by Wilson and Minnotte [6] to evaluate the optimum level of particulate emissions control. This evaluation was applied to the area of Washington, D.C., where air pollution losses were considered to be materials-soiling costs and the major emissions-contributing sources had been identified. Data for soiling costs were provided by a prior study of residential soiling costs in Washington, D.C., made by Michelson [7].

The frequency of performing certain cleaning and maintenance activities in the affected area was determined and the values were quantitatively correlated with air pollution levels. The frequency values were then converted into per-capita pollution costs, and these costs were plotted against pollution levels, as shown in Fig. 9-2. The relationship was assumed to be a straight-line function, with a lower ambient air particulates concentration of 50 $\mu g/m^3$, considered to be the background level. The slope of this curve is $Y = 1.85X - 42$. The pollution cost in each sector comprising the total area was determined by multiplying the population density of that sector by the per-capita costs. In the downtown Washington sector, the total 1968 population was determined as 442,000. Therefore, the slope of the total pollution cost (CP) curve for this sector was calculated to be $CP = 442,000(1.85X - 42)$, or $817,700X - 18,564,000$. The marginal cost of pollution is the slope of the total pollution cost curve, 817,700. Thus, a cost of $817,700 results from each change of 1 $\mu g/m^3$ in pollution level in this particular sector.

With the damage function defined for each sector, the benefit function for control of emissions by the various emitters in the total area was determined. The benefit to be derived from a reduction in emissions from a particular emitter was found by obtaining the difference in damages between a higher level of control and the existing control level. Thus, the net benefit values were determined as a function of pollution-

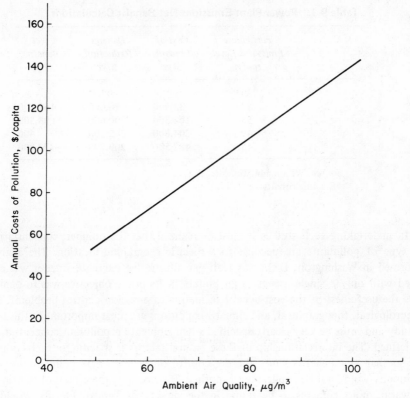

Figure 9-2. Residential per capita soiling costs. (Adapted from Wilson and Minnotte, Ref. 6.)

damage savings at each control stage and expressed as a difference between pollution costs and control costs. To calculate benefit values, it was necessary to define the damages attributable to each emitter at each control stage. Control costs for the 131 emitters in the area were calculated as annual values derived from an optimum selection of control equipment and fuel substitution alternatives.

The results of sample benefit calculations for a power-plant emitter are summarized in Table 9-1. Net benefits were all positive except for the last stage, where the costs for maximum emissions reduction would have been greater than the benefits derived from the reduction of pollution-damage costs. The point of maximum net benefit by the control of emissions from this power plant is at the second stage of abatement. Thus, the reduction of particulate emissions to a value of 58 tons/yr would result in a net benefit of $162,100. Similar analyses were accomplished for all emitters in the area, and it was determined that for a total benefits value of $22,073,900, the control costs would be $2,833,400. Thus, the net benefits for the entire area would be $19,240,500 per year. It was concluded that, for the chosen conditions of this problem, complete control of stationary sources was economically justified.

Table 9-1. Power Plant Emissions Net Benefit Calculations

Particulate Emissions Level, ton/yr	Cost of Control, $/yr	Damage Reduction, $/yr	Net Benefits, $/yr
230*	0*	0*	–
58	33,900	196,000	162,100
53	102,300	200,600	98,300
49	204,800	205,200	400
45	482,500	209,700	–272,800

Source: Wilson and Minnotte, Ref. 6.
*Existing condition.

In undertaking such studies it must be realized that the number, concentration, and type of pollutants are peculiar to a specific geographic location. The analysis performed in Washington, D.C., was really an illustrative exercise, since it was concerned with only a single aspect of air pollution. Its prime objective was to demonstrate the usefulness of the cost-benefit technique to emission control problems. The meteorological, topographical, and climatic conditions are most important to this type of study and must be very clearly specified when ambient air pollutant concentrations are defined. The interrelationship of these natural factors is specific for each community so that any benefits study for the control of even a single pollutant for any community would not be applicable to another. For example, the effects of fluorine in the warm, moist climates to be found in the vicinity of Tampa, Florida, would be much more damaging than for similar concentrations of this gaseous pollutant in northern Canada.

Measurement of Costs

To develop the costs of air pollution, a description of the damage for each object or person must first be determined as a function of the pollution level, with all other factors that might cause such damages being maintained constant. A monetary factor must then be applied as a measurement of the particular effect, and then the number of objects affected are multiplied by the unit assessed-damage cost to obtain the total cost. This, in essence, was the general procedure adopted in the illustrative Washington, D.C., example. There follows a description of a system for the application of such cost determinants for a number of deleterious air pollution effects.

Disease and Death. Studies of the effects of airborne pollutants on human health are not too numerous. Except for episodic investigations undertaken for such well-known incidents as Donora, London, and the Meuse Valley, the evidence associating various pollution levels with such diseases as lung cancer, chronic bronchitis, emphysema, asthma, and pneumonia has not been expressed quantitatively. Such parallel and more influential factors as weather conditions, work practices, economic level, and smoking habits have interfered with the various attempts to establish significant correlations.

There does seem to be general agreement that respiratory diseases are more prevalent for persons whose daily routine brings them into more constant contact with pollutants normally found in urban areas. A study of the relation of pollution to mortality was made in New York City during the period of 1963 to 1968 [8]. An attempt was made to correlate daily mortality and pollution levels where the pollution ranged from mild to severe. The three variables evaluated were over one-half million death certificates with dates and cause of death, two daily recordings of SO_2, and smoke opacity and daily weather variates, including the mean temperature. The linear model developed allowed an estimate of the excess deaths caused by pollution. Average and weighted values indicated that the 10,000 deaths occurring each year would not have occurred at the time they did if there had been no pollution on the day of death or on immediately preceding days. This value represents 12% of the total deaths; the specific diseases responsible were tuberculosis, respiratory diseases, vascular lesion, coronary heart disease, hypertensive heart disease, other circulatory disorders, respiratory cancer, and infant diseases, as well as other less common disorders. These data were not extended to a correlation of deaths and diseases with air pollution levels.

The economic burdens associated with disease comprise those incurred for prevention, those for treatment, and those due to premature death. If a person who contracts a respiratory illness continues to work, his productivity is impaired and he may infect others. If he remains at home to recuperate, his recovery may be hastened, but his productivity is reduced to zero and his medical costs rise. If he is forced to relocate to a more hospitable environment, moving costs and possibly a lower income level are involved. If the person fully recovers, the losses are eliminated; but if death takes place, additional costs such as his output as measured by his potential earnings at the time of his death, the costs for support of his family, and burial costs must be charged off as air pollution losses.

Ridker [5] developed some approximate values for the various cost elements associated with those respiratory diseases most commonly caused by air pollution. Treatment and burial costs together with loss of productivity values were estimated. The application of these estimated losses was limited by the lack of damage function. However, a very approximate method did yield an estimated figure of 18% of all respiratory diseases occurring in Pennsylvania as being caused by air pollution.

Materials Soiling and Damage. Air pollution has been responsible for considerable losses due to soiling and corrosion of materials. Self-regulation by the offending pollutant sources has failed, so it has been necessary to enlist an outside regulating force. However, before regulatory actions can be applied, there is a need to define an acceptable level of pollution. Except for the study performed by Michelson [7], there is a lack of data source to define the costs associated with materials damage for various air pollution levels. As in the case for diseases and deaths attributable to air pollution, relevant materials-damage costs must be correlated with air pollution levels and other pertinent variables must be explained or nullified.

Attempts to evaluate pure air pollution effects are usually frustrated by other uncontrollable variables. For example, power companies have amassed considerable data related to power failures occurring in their transmission equipment. Thus, the contamination of insulators by airborne dust causes flash fires and expensive interruptions of service. However, climate variations, power loads, and variable maintenance schedules interfere with the establishment of a meaningful air pollution cause-and-effect relationship.

Ridker [5] attempted to develop materials soiling and damage studies similar to those presented by Michelson [7]. Five major categories were investigated: frequency of outside painting, and inside papering and painting; time spent in routine cleaning; hours of garment utility between cleanings; cleaning and laundering supply expenditures; and frequency of car washings. The results obtained were not too conclusive, presumably because the majority of survey interviews was conducted among female household residents, each of whom strove too hard to impress the interviewer with her industriousness. In a parallel situation involving the magnitude of industrial work-clothes cleaning costs, the author had very little difficulty in obtaining accurate and complete records, inasmuch as these accrued costs were converted into back charges levied against defective control equipment.

Corrosion studies based on industrial measurements of coated metal panels and cloth are usually conducted under too severe and unrealistic conditions for the determination of deterioration costs. Actually, most industrial coatings and fabric wear tests are of an accelerated nature, with the results being expressed in relative rather than absolute units. The conclusion drawn by Ridker from these various attempts to develop materials-damage costs directly attributable to air pollution was that there was a need for more reliable testing methods. Valid data must be derived from the observation and evaluation of the physical deterioration of material goods under conditions of actual use in a known air pollution atmosphere.

Plant and Vegetation Losses. Except for various random laboratory studies, no serious and continued efforts have been undertaken to evaluate the extent of air pollution damage to plants and other vegetation. Numerous experiments have been performed to determine the effects of individual gaseous pollutants such as sulfur oxides, nitrogen oxides, and fluorides, but the results of these tests were not related with actual ambient air pollutant levels.

In one study [9], the influence of exposure to mixtures of sulfur oxides and hydrogen fluoride on sweet orange and mandarin foliage was determined under controlled greenhouse conditions. This study was undertaken to explore the possible plant-damage potential that various combinations of those two pollutants might have. However, no attempt was made to correlate the synergistic injuries observed with prevailing ambient conditions in any particular area. Similar laboratory-based investigations have been performed on a variety of plants with other individual and combinations of pollutants, but the results have not been reduced to air quality damage values.

The aluminum industries have been under regulatory pressures for some time to control the fluorides from their electrolytic reduction process. To assess their contribution of fluorides to the ambient air, many of these companies have developed their own network of monitoring stations to provide a continuous record of the air quality in the vicinity of the production facility. Some of these companies have planted various types of trees and vegetation indigenous to their particular geographical area. Vegetation and tree damage records are maintained, often as a source of background data for any possible future legal action by the surrounding agricultural community.

This situation represents a private assessment of air pollution damage to plants and vegetation for a specific pollutant, and such data are not usually made available for public use.

In 1969, the State of Pennsylvania conducted a survey [3] of air pollution damage to vegetation, to determine the impact of the problem throughout the state. A total of 92 cases of suspected air pollution damage was investigated in the 28 counties involved in the survey, with 60 of the reported cases being confirmed. Estimates of damage were based on crop values and production costs at the time of damage and were classified as direct losses. Profit decline, losses resulting from crop substitution, and relocation costs were considered to be indirect losses. Complete coverage of the state was provided by county agricultural agents, nursery inspectors, and service foresters. These personnel were required to attend a course to acquaint them with the objectives of the survey, sources of air pollution, meteorological aspects of air pollution, diagnosis of injury, recognition of symptoms caused by major and minor pollutants, pseudo-pollution damage, and damage assessment methods.

The survey was undertaken in recognition of the need to protect the farmer from crop losses and income depreciation. It was noted that attempts to recover such losses usually involved litigation costs, which added further to the farmer's economic burden. The types of crops on which damage was reported were agronomic crops; fruit and vegetable crops; forest, shade, and ornamental trees; floricultural crops; lawns, shrubs, and cover crops. It was found that in some of the cases reported that acute fumigation damage was often mistakenly considered as being caused by air pollution. In such doubtful areas, a plant pathologist was available to assist in the investigation.

Although the primary objective of the survey was stated to be the determination of the nature and extent of air pollution losses, the reference [3] material was mostly concerned with the methods employed for the continuous surveillance by experienced personnel and the means of establishing efficient communications and evaluation of the reports.

Summary

The air reservoir in urban areas has been used as a waste disposal medium for many years. Users of this facility have treated it as a free resource without regard for the damages inflicted on the receptors of this dirty air. Although regulatory bodies are actively establishing air quality standards and are enforcing compliance by air polluting sources, there is a need to define an acceptable air pollution level by the application of economic values. With the definition of air pollution damages in terms of dollars, an economic evaluation of any specific pollution situation can be made in the form of a cost-benefit analysis. Thus, the optimum degree of emissions abatement in an area can be determined as a function of the resulting economic benefits. Such a system of evaluation would provide a more valid basis for the acceptance of air pollution control costs by major polluters, rather than the present method of demanding compliance to semi-arbitrary standards.

In the development of cost-benefit relationships, two cost elements are involved: the cost of pollution and the cost of controls. Although there is a recognized need for pollution cost data, and much has been published to explain and develop systematic approaches for obtaining these data, very little of this cost information is available. Admittedly, it is very difficult to develop because of the variety of background interferences that make it almost impossible to isolate the air pollution factor. For example, to place a "due to pollution" label on a deceased person involves the evaluation of his personal habits; bodily intakes including food, alcohol, tobacco smoke, and drugs; his psychic composure; disease susceptibility; and similar physiological data. One suggested overall approach is to determine the difference between urban and rural disease rates after adjusting for known nonpollution factors.

Regardless of the difficulties, such pollution cost information is indispensable, and even some approximate crude approaches to the true values would be beneficial at the present time.

9.3. COSTS OF AIR POLLUTION ABATEMENT

Air Pollution Control Equipment and Other Costs

The cost of pollution control is well known. This situation reflects the fact that those who must comply with air pollution abatement regulations soon develop a very complete understanding of equipment costs and cost-effectiveness characteristics.

Generally, air pollution equipment costs vary considerably for the different industries. Some approximate capital costs for individual items of equipment were presented in Chapters 5 and 6, for particulates and gaseous emissions control, respectively. The one major variable common to all classifications of control equipment is the capacity in cubic feet per minute. Approximate equipment costs are usually expressed in terms of $/cfm. In Figs. 5-3 through 5-6, the capacity of the various types of particulate collection equipment is logarithmically related to the purchase cost. The curve for each equipment type has its own characteristic average slope, the values varying from 0.60 to 0.85. These slope values can be applied in exponential form to determine the costs of different sizes for each collector classification. For example, from Fig. 5-4 a low-efficiency electrostatic precipitator with a capacity of 50,000 acfm costs about $35,000. The cost curve has an average slope of 0.70 so that the price of a similar model with a capacity of 200,000 acfm would be $35,000 × (200,000/50,000)$^{0.70}$, or $92,000. The curve value at 200,000 acm reads $87,000, which indicates sufficiently close agreement for cost-benefit evaluations. This exponential estimating method is used extensively for process equipment cost work, where it is referred to as the "six-tenths rule" [10], since that is the value of the exponent most commonly accepted for plant design cost determinations.

Besides capacity, the other control equipment price determinant is related to the difficulty of the pollution control duty. For example, as discussed in Chapter 5, the removal of submicron particulates from an effluent gas stream represents a more severe transfer duty than that required for the collection of plus 2 μ particles. If a wet scrubber is to be employed, the former duty can be achieved at the expense of increased equipment capital costs for a high-efficiency model, as indicated in Fig. 5-6. Operating costs will also be greater because of the increased operating pressure drops.

Significant quantitative relationships between control costs and pollutant reduction are necessary to assess the impact of control on product prices, profits, product improvement, production rate and by-product recovery. The control costs associated with a specific emissions problem would involve consideration of such factors as raw materials and fuels, alterations in process equipment, control equipment and auxiliary hardware, and disposal/recovery of the collected pollutants. In Fig. 9-3(a) there is shown a typical cost-effectiveness relationship that is essentially identical in function as curve *CD* in Fig. 9-1(a). Point *P* represents the existing source emissions level at the present plant-operating conditions before consideration of new or additional control facilities. As control efficiency improves, marked by a decrease in emissions, the cost of control increases. The variable slope of this curve indicates that the incremental cost of control is smaller in the area of maximum emissions and becomes steeper as the removal efficiency demands become greater. This follows the characteristic performance curve of emissions control equipment, an incremental advance in collection efficiency from 98 to 99% costing disproportionately more than an increase from 60 to 70%.

Process changes, fuel substitutions, and equipment replacements may all contribute to emissions reduction without severely increasing overall process costs. Research and development expenditures directed at process improvement or modifications may improve the economies of control. The curve shown in Fig. 9-3(b) indicates the influence of such improvements. As improved control technology is developed in both the equipment and process areas, the cost of attaining a specific emission level will be reduced from C_a to C_b.

The various and alternative control methods to be considered for a particular emissions problem is well illustrated by the cost-effective study of air pollution abatement in the Washington D.C., area [11]. This study was used as the basis for the cost-benefit evaluation, previously discussed for this same city [6]. The major particulate emission sources in this study were those caused by fossil-fuel combustion for

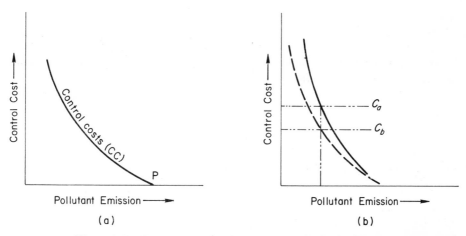

Figure 9-3. Pollution control cost effectiveness relationships.

space heating and power generation, and refuse incineration. The alternatives considered for the control of the combustion processes were the substitution of low ash-content coal or natural gas for the existing fuels and the installation of cyclonic separator, electrostatic precipitator, and wet scrubber particulate collection equipment. Although not included in the problem scope, consideration was also given to sulfur oxide emissions control with low-sulfur fuels, with alkaline scrubbing being evaluated as a viable alternative solution.

For the smaller combustion sources such as space heating, only the costs required for substitution of low-ash fuel were evaluated. Costs and the resulting emission levels were determined for the many possible combinations of alternative controls. As an example, a power plant that utilizes 9% ash coal might switch to a 7% ash content and add control equipment, either a cyclonic separator or electrostatic precipitator, depending on the severity of the collection duty. Thus, the 9% ash coal might require a high collection efficiency device such as the electrostatic precipitator, whereas the use of the lower-ash content coal might prescribe the need for only a cyclonic separator. The costs to be considered in the assessment of these various alternatives are purchase costs, installation, maintenance and operating costs for the control equipment, and fuel and boiler conversion costs for fuel substitutions.

The annual operating costs for the power-plant emissions control systems used in Table 9-1 were determined as the result of upgrading the existing electrostatic precipitator and substituting lower-ash coal as the fuel. At existing conditions, this power plant emitted 230 tons of particulates per year, when burning a coal containing 2% sulfur and 8.7% ash. The existing electrostatic precipitator operated at a collection efficiency of 98%, and at these base level conditions the control costs were taken at zero. The first control step was to upgrade the precipitator from 98 to 99.5% collection efficiency, thereby decreasing the emissions level from 230 to 58 tons/yr. The cost of this upgrading effect was $33,900/yr. The remaining three control steps involved the use of lower-ash content coals. Thus, the reduction of emissions from 58 to 53 tons/hr was accomplished by the substitution of an 8.1% ash coal for the 8.7% fuel while utilizing the upgraded precipitator. The last two steps consisted of using 7.5% and 6.9% ash coals with corresponding reduction in flyash emissions of 49 and 45 tons/yr. The incremental costs for the last three reductions in emissions increased rapidly, thereby reflecting the steeply increasing costs for this premium fuel.

Cost effectiveness relationships vary from industry to industry and are most significantly dependent on the complexity of the process. The particulate emission problems associated with power plants are readily solvable by the application of available and familiar technology and/or the substitution of fuels. Other industries have more complex problems due to the nature of their pollutants and their emissions rate, and consequently the corresponding economic impact is much greater.

Pollution Control Costs for Various Industries

Control Equipment Markets. In a marketing survey made in 1968 [12], it was estimated that electric power plants were responsible for more than one-fourth of all spending on air pollution control, or $78 million. Practically all money spent up to this time has been for particulate control. Future expenditures for sulfur oxides control, which are just beginning to be felt, are expected to be substantial, so that the total volume for both particulate and gaseous emissions has been estimated to reach about $200 million by 1977. Of this amount, it is expected that almost 40% will be for gaseous control.

The primary metals and rock-product industries together account for one-third of total air pollution control. By 1977, spending is expected to reach $122 million for the primary metals industry and $93 million for the rock-products industry. In both cases, expenditures will be largely for particulates control.

Table 9-2. Estimated Expenditures, Emissions Control Equipment (millions of dollars for U.S. industries)

Application	1958	1963	1967	1972	1977
Particulate Control					
Electric power	17	28	74	94	117
Primary metals	15	26	53	70	97
Rock products	17	30	40	57	85
Chemicals and allied products	6	9	16	22	33
Petroleum refining	2	3	5	8	13
Paper and allied products	3	6	15	22	34
Coal	2	3	6	10	13
Other	11	22	34	50	74
Total	73	127	243	333	466
Gaseous Control					
Electric power	1	1	4	37	80
Primary metals	2	3	7	15	25
Rock products	1	1	2	4	8
Chemicals and allied products	6	10	18	32	50
Petroleum refining	16	14	19	35	50
Paper and allied products	1	1	2	9	26
Coal	—	—	1	4	10
Other	2	3	7	29	71
Total	29	33	60	165	320
Total Air Pollution Control					
Electric power	18	29	78	131	197
Primary metals	17	29	60	85	122
Rock products	18	31	42	61	93
Chemicals and allied products	12	19	34	54	83
Petroleum refining	18	17	24	43	63
Paper and allied products	4	7	17	31	60
Coal	2	3	7	14	23
Other	13	25	41	79	145
Total	102	160	303	498	786

Source: Predicasts Markets Study, Ref. 12.

The chemical industry spends relatively little in proportion to its air pollution potential because of process modifications. However, spending on gaseous control is expected to increase significantly, while particulate equipment will continue to be used increasingly for material recovery as well as for abatement purposes.

The greatest amount of spending by the petroleum industry has been for gaseous control in the various refining operations. This may continue to increase, although process innovations may cut into this market. The largest gains are expected to be for equipment to reduce the sulfur content of residual oils. Similarly, the coal industry will increase spending on equipment to reduce the sulfur content of coal. Increases in paper mill expenditures will be largely for gaseous control in the kraft mills. Considerable process development efforts are being expended by this industry to develop modifications and/or alternatives for the sulfate pulping process.

Estimated capital expenditures for particulate and gaseous emissions control equipment for the various major industries is shown in Table 9-2. These values have been projected up through 1977. Because of the rapidly changing complexion of the air pollution control profile, with its present emphasis on the abatement of gaseous emissions and the enactment of increasingly severe regulations, these estimates are probably low.

There follows a description of the specific emission problems faced by the electric power and foundry industries and the costs required for their control.

Electric Power Industry. Fly ash particulates, sulfur oxides, and nitrogen oxides are the major pollutants produced in power-generating stations fired by fossil fuel. In addition, polycyclic hydrocarbons are discharged in sufficient quantities to be categorized somewhere between a nuisance and a health hazard. All current regulations have established air quality standards for the particulate and sulfur oxide emissions, while nitrogen oxide levels have been established by the majority of air pollution control agencies in the United States.

A projection of the electric power energy sources in the United States for the years 1970 through 2000 is shown in Table 9-3. The power industry presently utilizes

Table 9-3. Projected United States Electric Power Energy Sources

	Power Generation-Billions Kwhr.			
Fuel	*1970*	*1975*	*1985*	*2000*
Coal	695	940	1,390	2,340
Oil	205	242	308	110
Natural Gas	358	393	470	513
Hydropower	246	282	363	632
Nuclear	19	462	1,982	5,441
Totals	1,523	2,319	4,513	9,036

Source: Adapted from U.S. Energy, A Summary Review, Ref. 13.

control equipment for particulates abatement, fuel substitution, and/or control equipment for sulfur oxides removal and furnace-operating modifications for the control of the nitrogen oxides. Flyash collection for both coal- and oil-fired power boilers has been practiced for over 50 yr. Typical loadings from coal-fired operations and the earlier regulated emission levels made it acceptable to use a single cyclonic separator, performing at an efficiency in the range of 65 to 75%. As the regulations became increasingly strict, it was necessary to install electrostatic precipitators, either singly or in series with a cyclonic separator. The collection efficiencies achieved by the early precipitators were in the range of 95 to 98%. As the regulatory demands continued to advance, efficiencies in the range of 99 to 99.8% were found necessary to meet the revised air quality standards. When operated in series with a cyclonic separator, overall collection efficiencies up to 99.95% were obtained.

Because oil-fired combustion is usually marked by much lower particulate emissions, about one-tenth of those from coal-fired operations, the use of cyclonic separators achieved sufficiently adequate emissions control for some time. However, in recent years, some technological advances have been made in precipitator design and these have resulted in improved performance for the collection of the more difficult light, sticky, oil-fired flyash. Many power plants were therefore able to comply with the demands for reduced particulate emissions by the installation of electrostatic precipitators.

Sulfur oxides abatement for both coal- and oil-fired power boilers was initially accomplished by the substitution of low-sulfur fuels. Thus, New York City, in May 1966, demanded that the sulfur content of fuels be restricted from 2.8 wt % to 2.2% by January 1967, to 2.0% by May 1969, and to 1.0% by May 1971. The bill defining these restrictions also discouraged the use of high-ash-content bituminous coal, to limit particulate emissions.

Consolidated Edison Company, the City's supplier of power, considered many alternatives to comply with these emission goals. Prior to the new ruling, Consolidated Edison's fuel oil averaged 2.35% sulfur and coal averaged 1.58% sulfur. Since the refineries had not advanced too far in the desulfurization of oil, the direct purchase of oil with 1.0% sulfur was ruled out. However, blending Venezuelan crude oil with low sulfur-content Nigerian oil was considered a possibility. The use of natural gas was eliminated because of the current short supply. Removing sulfur oxides from the flue gases was considered the most impractical approach of all because of the commercial unavailability of a process system at that time. Low-sulfur coal was available at a premium price of $2.00/ton more than standard utility coal. Plant relocation was also considered because the reduced capital and operating costs of mine-mouth operations would permit expenditures for the "latest" air pollution control equipment. The transport of electricity from the mine to the city's distribution center was an important cost factor in the cost-benefit evaluation of this alternative. The solution most favored by the Consolidated Edison was the purchase of low-sulfur fuels for their existing generating plants and the construction of nuclear plants outside the city for expanded power requirements. One interesting and troublesome side effect immediately apparent from the use of low-sulfur fuel was the performance deterioration of the electrostatic precipitator. The reduction of SO_3 values in the effluent flue gas

Table 9-4. Commercial Sulfur Oxides Control Systems for Electric Power Generating Plants in the United States

Utility Company and Plant	Unit Size, Megawatts	Scheduled Startup	Fuel Type
A. Limestone scrubbing			
1. Union Electric Co., St. Louis Meramec No. 2	140	Sept. '68	3.0% S coal
2. Union Electric Co., St. Louis Meramec No. 1	125	Spring '73	3.0% S coal
3. Kansas Power and Light Lawrence Station No. 4	125	Dec. '68	3.5% S coal
4. Kansas Power and Light Lawrence Station No. 5	430	Nov. '71	3.5% S coal
5. Kansas City Power and Light Hawthorne Station No. 3	130	Mid-'72	3.5% S coal
6. Kansas City Power and Light Hawthorne Station No. 4	140	Mid-'72	3.5% S coal
7. Kansas City Power and Light La Cygne Station	820	Late '72	5.2% S coal
8. Detroit Edison Co. St. Clair Station No. 3	180	Nov. '72	2.5-4.5% S coal
9. Detroit Edison Co. River Rouge Station No. 1	270	Dec. '72	3-4% S coal
10. Commonwealth Edison, Chicago Will County Station No. 1	175	Feb. '72	3.5% S coal
11. Northern States Power Co., Minn. Suburban County Stations No. 1 & 2	680 680	May '76 (First unit)	0.8% S coal
12. Arizona Public Service Co. Cholla Station	115	Jan. '73	0.4-1% S coal
13. Tennessee Valley Authority Widow's Creek Station No. 8	550	April '75	3.7% S coal
14. Duquesne Light Co., Pittsburgh Phillips Station	100	Feb. '73	2.3% S coal
15. Louisville Gas and Electric Co. Paddy's Run Station No. 6	70	Late '72	3.0% S coal
16. City of Key West Stock Island Station	37	June '72	2.75% S oil
B. Sodium-based scrubbing			
17. Nevada Power Co. Reid Gardner Station	250	Mid-'73	1.0% S coal
C. Magnesium oxide scrubbing			
18. Boston Edison Mystic Station No. 6	150	March '72	2.5% S oil
19. Potomac Electric and Power, Md. Dickerson Station No. 3	195	Early '74	3.0% S coal
D. Catalytic oxidation			
20. Illinois Power Wood River	100	June '72	3.5% S coal

Source: Journal APCA, Ref. 14.

caused an appreciable increase in the flyash resistivity, thereby increasing the difficulty of its removal in the precipitator.

Of the innumerable processes currently being investigated and demonstrated for the reduction of sulfur oxides, the wet scrubbing system is the most advanced. Table 9-4 lists the various commercial sulfur oxides control systems and their scheduled startup date. A total of almost 5500 MW of power-generating facilities is being equipped with SO_x control equipment, with 87.3% of this potential capacity being controlled by limestone scrubbing, 4.6% by sodium hydroxide scrubbing, 6.3% by magnesium oxide scrubbing, and 1.8% by catalytic oxidation. All of these systems will provide a sulfur removal efficiency of 70 to 90%, as well as a high-efficiency particulate collection. The magnesium oxide scrubbing and catalytic oxidation systems have promise of sulfur value recoveries. Approximate investment values for wet scrubbing systems have been estimated in the range of $10 to $15 per installed kilowatt of plant capacity. Operating costs have been quoted for nonrecovery scrubbing processes at $0.80 to $1.00/ton of coal [15].

The need for nitrogen oxide emission controls was first realized in Los Angeles, where the original studies relating this gaseous pollutant with the photochemical formation of smog were undertaken. Up to that time, combustion of natural gas was considered the ideal solution to all pollutant emission situations. Although this fuel is still the most favored because of the absence of particulate and sulfur oxide emissions, its combustion causes maximum nitrogen oxide emission levels. Control equipment for the removal of NO_x from flue gas streams is currently unavailable, and the prescribed methods for its control are stoichiometric air-fuel variations and flue gas recirculation [16]. The design of burners that establish conditions which discourage the formation of the nitrogen oxides is receiving some attention, but commercial realization of these combustion control devices has not been accomplished as yet.

Polycyclic hydrocarbons are formed by the partial combustion of both oil and coal fuels. Although the complete identification of these pollutants has not been determined, they are suspected of having carcinogenic potential. These compounds are emitted from power-plant stacks largely in the vapor phase, but little is known of the effectiveness of wet scrubbing systems for their removal. It is believed that the highly efficient combustion characteristics of large power-plant operations minimize the formation and emission of these pollutants.

Fossil-fuel power plant expenditures for particulate controls have risen at an increased slope in recent years because of the enactment of increasingly severe emission regulations. Since many plants have now completed the installation of their flyash control systems under regulatory pressures, future capital investment in this area is expected to level off. However, as evidenced by the SO_x control systems installation list in Table 9-4, gaseous pollutant control expenditures are expected to rise under the continuing identification and restriction of their emissions. A projected air pollution control expenditure schedule, up through 1977, for the abatement of flyash and sulfur oxides, emitted by the electric power industry, is shown in Table 9-5.

Gray Iron Foundry Industry. One of the most economically hard-hit industries which has suffered under the impact of emissions control equipment expenditures is the gray iron foundry. In the melting process employed by this industry, pig iron,

Table 9-5. Electric Power Industry Expenditures for Pollution Control in the United States

(values expressed in millions of dollars)

Item	1958	1963	1967	1972	1977
Electric power capital expenditures	4,906	4,357	7,830	10,800	13,000
Generating equip., %	52.6	39.5	44.8	42.	42.
Generating equipment	2,582	1,721	3,490	4,570	5,460
Fossil-fuel fired type, %	76.6	72.0	64.7	53.0	43.0
Fossil-fuel fired generating equip.	1,979	1,239	2,257	2,420	2,350
Pollution control, %	1.0	2.7	4.3	7.1	11.2
Pollution control expenditures	19.	34	98	171	263
Air pollution control, %	95.	86	80	77	75
Air pollution control expenditures					
Particulate control	17	28	74	94	117
Gaseous control	1	1	4	37	80
Total	18	29	78	131	197

Source: Predicasts Markets Study, Ref. 12.

scrap metal, discarded casting parts, rejected castings, and small amounts of alloying metal are charged into the furnace. There are two basic types of furnaces used to melt the charge: the cupola and the electric furnace. The cupola is the predominant furnace type and is the major source of particulate air pollution emissions associated with foundry operations.

The cupola is heated by lighting a bed of coke or wood. During melting operations, air is forced into the cupola near the bottom while a mixture of metal, coke, and limestone is charged into it from an upper level. The contact between the ascending hot gases and the descending charge provides a quick and efficient melting process. The turbulence created in the cupola by forced air, hot combustion gases, and the metal charge generates particulate emissions in the range of 20 to 21 lb/ton of metal charged. The hot dust-laden combustion gases are discharged to the atmosphere at temperatures varying from 1500 to 2000° F, thus dispersing the particulates over a wide area. The greater portion of the particulates are submicron in size and therefore the collection duty is difficult.

Because of the finely dispersed submicron particles and the moderate loadings, the regulations have been particularly severe on the foundry cupola operation, demanding collection efficiencies in the range of 98 to 99[+]%. Process modifications are few and costly. The most common are conversions to electric arc or induction type furnaces. The former accomplishes melting by passing an electric current between two electrodes contained in a closed charging vessel. Emissions are much less than for the cupola, being on the order of 5 to 10 lb/ton of melt. The induction furnace provides

416

heat for melting by introducing an electromagnetic field through an enclosed charge of the metal. Emissions amount to about 2 lb of particulate per ton of melt. If clean scrap free of oils or other organic contaminants is used, the emissions are reduced to a point where control equipment is unnecessary. One recent process modification, still being evaluated, involves the use of a coke-fired cupola with natural gas being injected as the fuel just above the air inlet ports. Preliminary results on a malleable iron cupola have shown a sufficient reduction in particulates to achieve acceptable atmospheric emissions with a wet-cap control device.

Aside from process modifications, three types of control equipment are commonly used for cupola emissions control. These are the wet-cap collector, fabric filter, and high-energy wet scrubber. The wet-cap collector consists of a conical gas/water contactor supported on the cupola stack. Water pours over the conical section to produce a water curtain through which the hot gases must pass. This type of collector was considered adequate prior to the enactment of serious air pollution control regulations, but is no longer used because of unacceptable performance. The fabric filter and high-energy wet scrubber were described in Chapter 5. Performance characteristics, power requirements, and capital and operating costs for four collector types as applied to two 10 ton/hr cupolas are shown in Table 9-6. The cupolas are operated on alternate days so that the control equipment handles the effluent from a single cupola at any one time. The more strict process-weight regulations for the control of emissions from a cupola of this capacity is about 16 lb/hr. At uncontrolled emissions of 21 lb/ton capacity, equivalent to 210 lb/hr, the required collection efficiency would be 92.4%. However, to comply with emission concentration and opacity regulations, efficiencies on the order of 98 to 99%, equivalent to discharge concentrations of 0.02 to 0.01 grain/scf, are usually required.

Table 9-6. Performance Characteristics and Costs for Various Collector Types on Cupola Service

Equipment Type	Press. Drop, in. w.g.	Drive, hp	Collect. Eff., %	Discharge Emissions, grain/scf	Installed Cost, $1000	Operating Cost, $1000/yr
Wet cap	1.0	20	50	0.5	50	3
Low-energy scrubber	6.0	95	80	0.2	120	10
High-energy scrubber, venturi model	70	500	98–99	0.01	225	25
Fabric filter	6	100	99+	0.005	250	25

Source: Weber, Ref. 17.
Notes: (1) Single collector system handles effluent from two 10 ton/hr cupolas alternately. (2) Wet collectors furnished complete with accessory equipment, including sludge-handling system. (3) Fabric filter furnished complete with accessory equipment, including cooling quench section and water pump. (4) Operating costs include maintenance labor, utilities, and depreciation charges.

In the evaluation of cost-benefit relationships, the gray iron foundry operator does not have too many options available. In a survey [18] made of 1232 foundries, about 90% of these operated cupolas, 3% had electric arc furnaces, and 7% used induction or other types of furnaces. Of those foundries operating cupolas, only 180, or 15%, were found to employ some form of air pollution control. The wet-cap system was used on 95 of the 180 controlled cupolas. This hesitancy to adopt the more sophisticated control technology necessary to meet the current regulations is directly due to financial hardship. The alternative of converting to electric melting equipment has been assumed by a limited number of foundries. The major reasons for such conversions have been given by the foundrymen as compliance with air pollution regulations, economy of operation, and improved metallurgical quality control.

A study of melting operating costs for the cupola, arc furnace, and induction furnace was performed by the American Foundrymen's Society [17] to underscore the economic impact caused by the need for air pollution control equipment. Cost data were developed for three size groups of gray iron foundries: Group I, rated at production capacities of 1,500,000 lb/month and greater of hot metal; Group II, at 500,000 to 1,500,000 lb/month; and Group III, at 100,000 to 500,000 lb/month. A summary of the gross melt operating costs, including materials, labor, maintenance, and overhead is given in Table 9-7. Costs for air pollution controls are included in these values and would indicate that the cost to produce the melted product, expressed as $/Cwt, for the cupola and its emissions control equipment is still less than that for the arc and electric melt operations in all three production rate categories.

A portion of this study was devoted to the economic impact that the recently adopted Rhode Island regulations would have on five small foundries located in that

Table 9-7. Operating Costs for Various Cast Iron Melting Processes

Production Capacity, lb/month, thousands	*Gross Melt Costs, $/cwt*		
	Cupola	Electric Arc	Electric Induction
Group I, 1500 and greater	2.56	2.85	2.93
Group II, 500–1500	2.64	2.84	3.13
Group III, 100–500	2.75	2.92	2.97

Source: Weber, Ref. 17.

Notes: (1) Costs include air pollution control systems for cupola. (2) Gross melt cost values include metals, supervision and labor, utilities, supplies, maintenance, and depreciation charges.

state, whose gross sales did not exceed $500,000/yr. The net worth of each of the five industries lies in the range of $30,000 to $175,000. The installed cost of the air pollution control equipment to meet the revised 99$^+$% performance requirement imposed by the regulations was $135,000. At a 7-yr depreciation period and an interest rate of 7% on the unpaid annual balance, the total annual financial burden would amount to $24,686. The annual maintenance and operating costs for this equipment is estimated to be 10% of the installed cost, or $13,500, thereby bringing the total annual costs to $38,186. Assuming an annual net return of 5% of gross sales for the largest of these foundries, the net profit would amount to $25,000. This foundry would therefore show a cash loss of $13,186 each year, and the smaller foundries would be proportionately worse off.

Some Control Benefits

In earlier discussions, some of the benefits derived from the adoption of air pollution emission controls have been enumerated: for example, product recovery, regain of by-product values, and product quality improvement. In the foundry evaluation study, mention was made of the improved product quality accruing to electric melting methods, and it is assumed that this factor could be converted to a cost benefit. Of course the other very real benefit that should not be overlooked is that of being permitted to continue operations. Regulation enforcement is more often taking the form of pollution suits wherein the polluter must comply or be shut down.

Another recently recognized benefit of air pollution control is in the area of employer-employee relations. The in-plant working conditions of a noncomplying operation is not usually conducive to the health and welfare of the workers. Ambient in-plant air standards have been established by health authorities in all concerned communities, to ensure the well-being of the plant labor force. Labor unions have become more aware of the adverse health effects of plant emissions, both inside and outside, and are becoming more vocal in their demands that management provide "clean" working conditions. In taking steps to meet external air quality standards, employers have often improved both the plant atmosphere and their relations with the employees.

The federal government, through the U.S. Tax Reform Act of 1969, has conceded some financial benefits to cooperative industries by providing a 5 yr depreciation period for certified air pollution control facilities. The equipment must control pollution from plants in operation before Jan. 1, 1969. To qualify for this benefit, the state must certify to the Department of Health, Education, and Welfare that the equipment conforms to their programs and regulations. Thus, any control system that diffuses or dilutes pollutants, such as tall stacks, is excluded from this provision. Comparable economic aid is being offered through federal and state legislation [19] to purchasers of air pollution control facilities.

Recently proposed Internal Revenue Service regulations, published on June 5, 1971, provide additional methods for financing pollution control equipment through the use of tax-exempt industrial development revenue bonds. Interest income on these bonds may be exempt from federal income tax, in accordance with prevailing statutes and court decisions. Several pollution control projects have already been successfully financed by tax-exempt bonds. Paper industries in Montana, Michigan, and Georgia have financed emissions control facilities through this financing procedure by issuing bonds totaling $37,650,000.

The State of Florida legislature enacted a bill in 1970 to allow financing of pollution control facilities through bond issuance on a long-term basis. A section of this bill enables each county commission in Florida to create a five member Industrial Development Authority with the power to issue industrial development bonds, thus encouraging and facilitating the procurement of pollution control systems. Considerable benefits accrue to industries making use of these bond-issue plans. It is estimated that for a 30 yr serial bond issue, an average savings of $440,000 can be realized for each million dollars financed if industrial development bonds rather than corporate debt security financing are utilized [19]. Most of the states have devised some form of industrial development bond-financing procedure for the benefit of those industries making conscientious efforts to comply with the pollution control regulations.

9.4. LEGAL CONSIDERATIONS

Legal processes concerned with problems of air pollution proceed in two areas: The first settles disputes between persons regarding a specific issue and the second guides policy-making activities. The latter involve exercises in the interpretation and clarification of air pollution control legislation. The major thesis of emissions control regulations is that the polluter must prove that there is no harm in the pollutants he discharges into the atmosphere. This is treacherous ground, and its legality is often under attack as being an infringement of personal liberties.

The most all-encompassing legislation concerning the environment is the National Environmental Policy Act of 1970 (NEPA). This Act, in establishing the requirement of environmental impact statements, discussed in Chapter 3, has made environmental considerations the sole criteria for judging the value of any proposed industrial activity. An example is cited by Clingan [20] to illustrate the force of NEPA legislation. It concerns the efforts of a developer to land-fill a portion of a public body of water for the creation of a trailer park. The U.S. Corps of Engineers refused a permit to construct on the basis of adverse environmental influences. The developer sued, claiming that the only concern of the Corps should be navigation restraints, which at one time it was. The circuit court decided that NEPA had essentially modified the Corps responsibilities and therefore upheld the refusal to construct. This expansion of authority directed at the "prevention of adverse environmental impact" has influenced other governmental agencies such as Fish and Wildlife, Federal Power, Transportation, and the Atomic Energy Commission.

Most courts have concluded that the environmental impact statement procedure is enforceable by any interested citizen. This interpretation of the intent of NEPA legislation has been responsible for a host of private citizen suits directed against development groups who cannot satisfactorily prove that their proposed project does not adversely affect the environment. Whether or not a citizen has sufficient "standing" to bring such a lawsuit is presently being debated in the courts. The argument revolves around the issue of an individual's legal entity as a surrogate for all affected parties.

The first area of legal activity, that of settling disputes between persons over specific environmental issues, is more easily understood. Many decisions have favored the petitioner on the basis of a nuisance committed. This reflects the difficulties involved in establishing a satisfactory cause-effect relationship for the majority of pollutants. In 1963 the United States District Court in Oregon directed an aluminum company to install equipment for more efficient removal of fluoride gases. The case was filed by a number of orchard owners, and the decision in their favor was based on the fact that the emission of gases crossing property lines constitutes a trespass and that continuing trespass was considered a nuisance. Innumerable cases have been tried, with many decisions being given in favor of the aggrieved pollutant recipient. In very few of these cases, however, have damages been awarded solely to compensate for harmful effects derived from a specific air pollution condition.

Under the sympathetic climate that presently favors those who consider themselves distressed by industrial polluters, some rather ambitious court actions have been attempted. An example described by Clingan [20] concerns a private citizen who recently filed suit in the Los Angeles courts on behalf of all the residents of Los Angeles against a long list of companies, including GM, Ford, Chrysler, glass manufacturers, chemical plants, and others. The plaintiff claimed that the Los Angeles smog had become a health hazard and requested the court to force the responsible parties to solve the problem. The court declined to hear the case, probably sobered by the thought of the monumental task involved in the resolution of a dispute between the 7,119,184 plaintiffs, representing the entire citizenry of Los Angeles, and more than 200 defendants. Actually, the attempted suit was most likely initiated by the claimant, who was appalled by the city's mounting air pollution problem and was determined "to do something about it."

REFERENCES

1. J. P. Dixon, "Air Conservation—The Report of the Air Conservation Commission of the American Association for the Advancement of Science," Publication No. 80, American Association for the Advancement of Science, Washington, D.C. 1965. Copyright, March 28, 1974.
2. J. H. S. Bossard, "Comment on the Natural History of a Social Problem," *Amer. Sociological Rev.,* Vol. 6, No. 3 (June 1941), p. 329.
3. N. L. Lacasse and T. C. Weidensaul, "A Cooperative Extension-Based System of Detecting and Evaluating Air Pollution Damage to Vegetation," *J-APCA,* Vol. 22, No. 2 (February 1972), p. 112.
4. V. W. Johnson, "Air Pollution in Relation to Economics," in L. C. McCabe, *Proceedings of the United States Technical Conference on Air Pollution.* McGraw-Hill Book Co., Inc., New York, 1952.
5. R. G. Ridker, *Economic Costs of Air Pollution—Studies in Measurement.* Frederick A. Praeger, Publishers, New York, 1967.
6. R. D. Wilson and D. W. Minnotte, "A Cost-Benefit Approach to Air Pollution Control," *J-APCA,* Vol. 19, No. 5 (May 1969), p. 303.

7. I. Michelson, "The Household Cost of Living in Polluted Air in the Washington, D.C., Metropolitan Area," a report to the U.S. Public Health Service.

8. H. Schimmel and L. Greenburg, "A Study of the Relation of Pollution to Mortality, New York City, 1963–1968," *J-APCA*, Vol. 28, No. 2 (August 1972), p. 607.

9. J. Matsushima and R. F. Brewer, "Influence of Sulfur Dioxide and Hydrogen Fluoride as a Mix or Reciprocal Exposure on Citrus Growth and Development," *J-APCA*, No. 22, No. 9 (September 1972), p. 710.

10. R. Williams, Jr., *"Six-Tenths Factor" Aids in Approximating Costs,* Chemical Engineering Process Equipment Cost Estimation Series, May 1947 to July 1950. McGraw-Hill Book Co., Inc., New York.

11. Ernst and Ernst, "A Cost Effective Study of Air Pollution Abatement in the Washington, D.C., Area," Contract No. PH 86-68-51, U.S. Public Health Service, 1968.

12. "Industrial Markets for Air Pollution Equipment," Report No. 54, Predicasts, Inc., Cleveland, Ohio, October 1969.

13. "United States Energy—A Summary Review," Dept. of the Interior, Jan. 1972.

14. "How They're Controlling Sulfur Oxide Emissions from Electric Power Generating Plants," *J-APCA*, Vol. 26, No. 6 (June 1972), p. 474.

15. W. A. Pollock, James P. Tomany, and Garry Frieling, "Removal of Sulfur Dioxide and Flyash from Coal Burning Power Plant Flue Gases," paper 66-WA/CD-4 presented to Annual ASME Meeting, New York, December 1966.

16. J. P. Tomany, R. R. Koppang, and H. L. Burge, "A Survey of Nitrogen-Oxides Control Technology and the Development of a Low NO_x Emissions Combustor," paper 70-WA/Pwr-2 presented to the Annual ASME Meeting, New York, December 1970.

17. H. J. Weber, "The Impact of Air Pollution Laws on the Small Foundry," *J-APCA*, Vol. 20, No. 2, (February 1970), p. 67.

18. "Economic Impact of Air Pollution Controls on Gray Iron Foundry Industry," U.S. Dept. of Health, Education, and Welfare, Public Health Service, National Air Pollution Control Administration, Raleigh, N.C. November 1970.

19. B. H. Dickens, "Tax-free Financing of Pollution Control Facilities," *J-APCA*, Vol. 22, No. 1 (January 1971), p. 13.

20. T. H. Clingan, Jr., "Analysis of Important Recent Environmental Cases," *J-APCA*, Vol. 22, No. 1 (January 1972), p. 9.

COMBUSTION
GAS FLOW CALCULATIONS

A typical problem involving a flue gas stream from a coal-fired combustion process is defined as follows:

> Gas flow: 620,000 lb/hr
> Gas temperature: 460°F
> Gas analysis, vol %:
> CO_2,   12.2
> O_2,   7.7
> N_2,   74.5
> H_2O,   <u>5.6</u>
>   100.0

The problems to be considered are:

    1. Calculate the flue gas volumetric flow rates in scfm and acfm.

    2. Fuel substitution feasibility studies require the approximate fuel requirements for coal-, oil- and gas-fired operations. Determine these values and recalculate the flue gas composition and flow rate for oil and gas firing.

PART 1

Table A-1. Flue Gas Mol and Weight Analyses

Compound	Mol fraction	Molecular weight	Weight fraction, lb
CO_2	0.122	44	5.37
O_2	0.077	32	2.46
N_2	0.745	28	20.86
H_2O	0.056	18	1.01
Total	1.000	29.7	29.70

Gas flow at standard conditions 60°F, and 1 atm pressure:

$$\text{Flow} = 620{,}000 \text{ lb/hr} \times \frac{1}{29.70} \times 359 \times \left(\frac{460+60}{460+32}\right) \times \frac{1}{60} = 131{,}830 \text{ scfm}$$

423

Gas flow at $460°F$:

$$Flow = 131,830 \times \left(\frac{460+460}{460+60}\right) = 233,340 \text{ acfm}$$

PART 2

For coal combustion:

Gross calorific value of coal = 13,600 Btu/lb
Coal analysis, wt %: C, 75.1; H_2, 5.2; O_2, 13,8; ash, 13.1

Assuming 100% combustion of coal as fired,

Coal requirements = 131,830 scfm $\times$ 0.122/0.751 $\times$ 12/379.5 $\times$ 60
          = 40,600 lb/hr

For oil combustion:

Gross calorific value of fuel oil = 135,000 Btu/gal
Oil density = 7.3 lb/gal
Oil analysis, wt %: C = 86.0; H_2 = 9.0; ash = 0.08
Combustion heat input, = 40,600 $\times$ 13,600
          = 552 $\times$ 10^6 Btu/hr
Oil requirements = (552 $\times$ 10^6) $\times$ (7.3/135,000)
          = 29,800 lb/hr

For 100 lb fuel oil:

86.0 lb carbon requires 86.0 $\times$ 32/12 = 230 lb O_2
9.0 lb hydrogen requires 9.0 $\times$ 16/2 = 72
          Total theoretical   = 302 lb O_2

Assuming 10% excess combustion air, requirements are

O_2 = 332 lb
N_2 332 $\times$ 76.5/23.5 = 1080 lb
Air = 332 + 1080 = 1412 lb

Material Balance for Combustion of Oil

Bases for this calculation are: 100 lb oil and 10% excess air.

Table A-2. Oil Combustion Conversion Calculations

In, lb		Out				
C	86.0	CO_2	316 lb	44 MW	7.18 mol	14.0 vol %
H_2	9.0	H_2O	81	18	4.50	8.8
O_2	332.	O_2	30	32	0.94	1.8
N_2	1080.	N_2	1080	28	38.57	75.4
Total	1507		1507	29.4	51.19	100.0

Flue gas flow = 1507 × 29,800/100 = 450,000 lb/hr

Flue gas flow at standard conditions, 60°F, 1 atm pressure:

Flow = 450,000 × 1/29.4 × 359 × (460+60/460+32) × 1/60 = 96,700 scfm
Gas flow at 460°F = 96,700 × (460 + 460/460 + 60) = 171,000 acfm

Flue gas composition, vol %: CO_2 14.0; O_2, 1.8; N_2, 75.4; H_2O, 8.8.

For gas combustion:

Gross calorific value of natural gas = 970 Btu/scf
Gas analysis, vol %: CH_4 = 96.0; N_2 = 3.2.
Combustion system heat input = 552 × 10^6 Btu/hr
Natural gas requirements = 552 × 10^6/970 = 570,000 scfh

For 100 mols of gas:
96.0 mols CH_4 requires 96.0 × 2 = 192 mols O_2
Assuming 5% excess combustion air, requirements are

O_2 = 202 mols
N_2 = 202 × 4 = 808 mols

Material Balance for Combustion of Natural Gas

Bases for this calculation are: 100 mols gas and 5% excess air.

Table A-3. Natural Gas Combustion Calculations

In			Out				
CH_4	96.0 mols	1,536 lb	CO_2	96 mol	44 MW	4,224 lb	8.7 vol %
N_2	3.2	89.6	H_2O	192	18	3,456	17.3
O_2	202.	6,464.	O_2	10	32	320	0.9
N_2	808.	22,624.	N_2	811.2	28	22,713.6	73.1
	Total 1109.2	30,713.6		1109.2	27.8	30,713.6	100.0

Flue gas flow = 30,713. 6/100 × 570,000/379.5 = 461,000 lb/hr

Gas flow at standard conditions, 60°F and 1 atm pressure:

Flow = 461,000 × 1/27.8 × 359 × (460+60/460+32) × 1/60 = 104,800 scfm
Gas flow at 460°F = 104,800 × (460+460/460+60) = 185,000 acfm

Flue gas composition, vol %: CO_2 8.7; O_2, 0.9; N_2, 73.1; H_2O, 17.3.

DUST CONCENTRATION AND
COLLECTION EFFICIENCY DETERMINATIONS

In the illustrated problem in Appendix A, the ash contents of the coal and oil were specified. The problems to be considered are:

1. Calculate the dust loadings from the combustion source for both coal- and oil-firing in the following terms: lb/hr, grain/acf, grain/scf, and grain/dscf.

2. Assuming the process weight and dust concentration regulations for coal firing allow maximum emissions of 48 lb/hr and 0.03 grain/scf, indicate the limiting emission factor and compute the control equipment collection efficiency necessary for compliance.

PART 1

For coal combustion:
Coal firing rate = 40,600 lb/hr
At an ash content of 13.1%, ash production rate is

Rate = 40,600 $\times$ 0.131 = 5,330 lb/hr

Assuming 80% of ash is discharged as flyash,

Dust loading = 5,330 $\times$ 0.80 = 4,260 lb/hr

For dust concentration at gas conditions

Conc. = 4,260 lb/hr $\times$ 7000 grain/lb $\times$ 1 hr/60 min $\times$ 1 min/233,340 ft^3
     = 2.14 grains/acf

For dust concentration at standard conditions,

Conc. = 4260 $\times$ 7000 $\times$ 1/60 $\times$ 1/131,830 = 3.78 grains/scf

For dust concentration at dry standard conditions,

Conc. = 3.78 × 1/1.000 − 0.056 = 4.00 grains/dscf

For oil combustion:

Oil firing rate = 29,800 lb/hr

At an ash content of 0.08% ash production rate is

Rate = 29,800 × 0.008 = 23.8 lb/hr

Assuming 90% of ash is discharged as flyash,

Dust loading = 23.8 × 0.90 = 21.5 lb/hr

For dust concentration at gas conditions,

Conc. = 21.5 × 7000 × 1/60 × 1/171,000 = 0.015 grain/acf

For dust concentration at standard conditions,

Conc. = 21.5 × 7000 × 1/60 × 1/96,700 = 0.026 grain/scf

For dust concentration at dry standard conditions,

Conc. = 0.026 × 1/1.000 − 0.088 = 0.029 grain/dscf

PART 2

For coal combustion:

Allowable process weight emission = 48 lb/hr
Required collection eff = (1.000 − 48/4260)100 - 98.9%
Allowable dust conc emission = 0.03 grain/scf
Required collection eff = (1.000 − 0.03/3.78)100 - 99.2%

Judging from the collection efficiency levels required to comply with this regulation, the limiting emission factor (i.e., the most difficult to achieve) is the concentration value of 0.03 grain/scf.

Appendix C

PSYCHROMETRY AND
GAS MOISTURE CALCULATIONS

EXPLANATION

Referring to the problem solution in Appendix A, the moisture content of the flue gas for coal-firing was given as 5.6 vol %. In actual source test procedures, this volumetric value could be determined directly in an Orsat apparatus or with a sampling train. Wet-bulb/dry-bulb temperature measurement is the most simple method for determination of the moisture content of any gas stream at moderate temperatures. The dewpoint temperature is often necessary for the design of control equipment where it is desired to avoid moisture condensation.

Material balance calculations are simplified considerably by expressing the moisture content of the process gas stream as lb H_2O/lb dry gas and the specific volume of a humid gas as ft^3/lb dry gas. Gas moisture values expressed in these units can be determined from a psychrometric chart and gas densities or specific volumes at different temperature, and moisture levels can be similarly obtained.

There follows an illustrated problem relating the most commonly used of these gas-moisture factors.

PROBLEM STATEMENT

The gas stream characteristics defined by the problem stated in Appendix A are as follows:

Gas flow: 620,000 lb/hr; 233,340 acfm
Gas temperature: 460°F
Gas analysis, vol %:

CO_2 = 12.2
O_2 = 7.7
N_2 = 74.5
H_2O = 5.6
100.0

428

1. The specified moisture content was verified by wet-dry bulb temperature readings and values equivalent to 132°F/460°F were obtained.

(a) Determine the accuracy of the specified values, assuming the thermometry values to be correct.

(b) Determine the gas dewpoint, humidity, and relative humidity of the gas stream, based on the wet-bulb/dry-bulb readings.

2. Assuming that this gas stream is to be processed in a wet scrubber, it becomes necessary to determine the diminished, saturated, volumetric gas flow for the scrubber design basis. Calculate: (a) saturation temperature; (b) saturation humidity; (c) evaporation water requirements; (d) saturated volumetric flow; and (e) percentage gas flow reduction as it passes through the scrubber.

PROBLEM SOLUTION

Part 1

From the psychrometric curve, Fig. C-1, a wet-bulb/dry-bulb reading of 132/460°F is equivalent to a humidity of 0.034 lb H_2O/lb dry gas. Converting to volume percent.

0.034 lb H_2O/lb dry gas $\times$ 29/18 = 0.055 mol H_2O/mol dry gas
$(0.055 \times 1/1.055)100$ = 5.2 vol %
Accuracy = $(5.2/5.6)100$ = 92.6%

From Fig. C-1:

Dewpoint = 92°F
Humidity = 0.034 lb H_2O/lb dry gas
Relative humidity = <1%

Part 2

As the gases enter the wet scrubber, evaporative cooling is achieved, with the amount of water evaporated being equivalent to the gas cooling duty.

1. Referring to Fig. C-1, at an entering temperature of 460°F and humidity of 0.034 lb H_2O/lb dry gas, the gases become saturated and cooled to the wet bulb or saturation temperature of 132°F.

2. The humidity of the saturated gas, from Fig. C-1, is 0.119 lb H_2O/lb dry gas.

3. Data given:

Original wet gas entering scrubber = 620,000 lb/hr
Dry gas flow = 620,000 $\times$ 1.000/1.034 = 600,000 lb/hr

For moisture in gas entering scrubber,

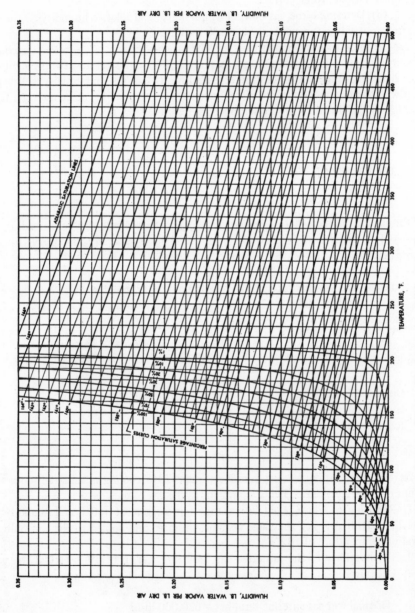

Figure C-1. Psychrometric curves for air. (From Zimmerman and Lavine, *Psychrometric Tables and Charts*, 2d. ed. Industrial Research Service, Inc.)

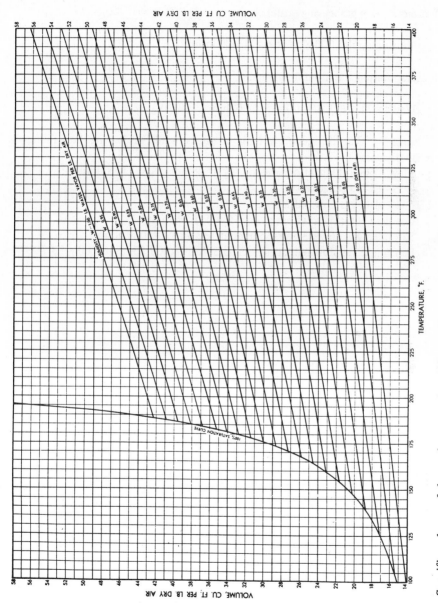

Figure C-2. Specific volume of dry and saturated air. (From Zimmerman and Lavine, *Psychrometric Tables and Charts*, 2d. ed. Industrial Research Service, Inc.)

Moisture = 600,000 lb/hr dry gas $\times$ 0.034 lb H_2O/lb dry gas
= 20,400 lb/hr

For moisture in gas leaving scrubber,

Moisture = 600,000 $\times$ 0.119 = 71,400 lb/hr

For evaporation water requirements,

Water = 71,400 − 20,400 = 51,000 lb/hr

4. From Fig. C-2, the specific volume of the saturated gas leaving the scrubber is 17.8 ft^3/lb dry gas.
Flow = 600,000 lb/hr dry gas $\times$ 17.8 ft^3/lb dry gas $\times$ 1 hr/60 min
= 178,000 acfm
5. To calculate gas flow reduction:
Gas flow entering scrubber at 460°F = 233,340 acfm
Saturated gas flow leaving scrubber at 132°F = 178,000 acfm
Percentage gas flow reduction = (233,340−178,000/233,340)100 = 23.7%

Appendix D

AIR POLLUTANT
EMISSION FACTORS

The information provided here is taken from the compilation made by the U.S. Environmental Protection Agency, published in February 1972.

Table D-1. Emission Factors for Petroleum Refineries without Controls

Type of Process	Particulates	SO$_x$	CO	HC	NO$_x$	Aldehydes	NH$_3$
Boilers & process heaters							
Lb/10^3 bbl oil burned	840	NA*	Neg	140	2,900	25	–
Lb/10^3 ft^3 gas burned	0.02	NA	Neg	0.03	0.23	0.003	–
Fluid catalytic cracking units							
Lb/10^3 bbl fresh feed	61	525	13,700	220	63	19	54
Moving-bed catalytic cracking							
Lb/10^3 bbl fresh feed	17	60	3,800	87	5	12	6
Compressor internal combustion engines							
Lb/10^3 ft^3 gas burned	–	–	Neg	1.2	0.9	0.1	0.2
Blowdown systems, lb/10^3 bbl refinery capacity							
With control	–	–	–	5	–	–	–
Without control	–	–	–	300	–	–	–
Miscellaneous losses, lb/10^3 bbl refinery capacity							
Pipeline valves & flanges	–	–	–	28	–	–	–
Vessel relief valves	–	–	–	11	–	–	–
Pump seals	–	–	–	17	–	–	–
Compressor seals	–	–	–	5	–	–	–
Others: sampling, blowdown, etc.	–	–	–	10	–	–	–

*Not available

433

Table D-2. Emission Factors for Steel Mills without Controls

Type of Operation	Particulates, lb/ton	CO, lb/ton
Iron production		
Blast furnace		
Ore charge	110	1,400–2,100
Agglomerates charge	40	–
Sintering		
Windbox*	20	–
Discharge	22	44
Steel production		
Open hearth furnace		
Oxygen lance	22	–
Without lance	12	–
Basic oxygen furnace	46	120–150 (before ignition)
Electric arc furnace		
Oxygen lance	11	18
Without lance	7	18
Scarfing	20	–

*Approximately 3 lb SO_2/ton of sinter is produced at the windbox.

Table D-3. Emission Factors for Sulfate Pulping

(Values in lb/ton of air-dried unbleached pulp)

Source	Controls	Particulates	SO_x	CO	H_2S	Mercaptans and Sulfides
Blow tank accumulator	None	–	–	–	0.1	3.0
Washers and screens	None	–	–	–	0.02	0.2
Evaporators	None	–	–	–	0.5	0.4
Recovery boilers and direct-contact	None	151	5.0	60	12	0.9
evaporators	Precipitator	15	5.0	60	12	0.9
	Wet scrubber	47	5.0	60	12	0.9
Smelt dissolving tanks	None	2	–	–	0.03	0.04
Lime kilns	None	45	–	10	1.0	0.6
	Wet scrubber	4	–	10	1.0	0.6
Turpentine condenser	None	–	–	–	0.01	0.5

434

Appendix E

TYPICAL CONTROL EQUIPMENT
PERFORMANCE CHARACTERISTICS

Table E-1. Collection Efficiencies for Flyash Control Equipment

Furnace Type	Electrostatic Precipitator	High-Efficiency Cyclone	Low-Resistance Cyclone	Settling Chamber
Cyclone furnace	65–99*	30–40	20–30	–
Pulverized-coal unit	80–99.9*	65–75	40–60	–
Spreader stoker	–	85–90	70–80	20–30
Other stoker types	–	90–95	75–85	25–50

*High values attained with high-efficiency cyclone in series with electrostatic precipitator.

Table E-2. Collection Efficiencies for Particulate Control Systems in the Steel Industry

Operation	High-Efficiency Cyclone	Electrostatic Precipitator	Wet Scrubber	Fabric Filter
Blast furnace	60	–	80–90*	–
Sintering windbox	90	–	–	–
Sintering discharge	93	–	–	–
Open hearth furnace	–	98	85–98	99
Basic oxygen furnace	–	99	99	–
Electric arc furnace	–	92–97	90–98	98–99

*High value is for wet scrubber-cyclone series system

435

Table E-3. Estimated Emission Levels from Electrostatic Precipitators and Fabric Filters Yielding Clear or Near-Clear Stacks (Equivalent to Ringelmann Numbers in Range 0-1).

		Gas Conditions	
Emission Source	*Stack Emissions Grains/acf*	*Max. Temp. °F*	*Humidity vol % H_2O*
Fuel-fired boilers			
Coal, pulverized	0.020	320	5
Coal, cyclone	0.010	320	5
Coal, stoker	0.050	450	5
Oil	0.003	400	8
Wood and bark	0.050	400	10
Bagasse	0.040	400	10
Coke	0.015	450	Dry
Pulp and paper			
Kraft recovery boiler	0.020	350	30
Soda recovery boiler	0.020	350	30
Lime sludge kiln	0.020	400	30
Rock product kilns			
Cement, dry	0.015	600	3
Cement, wet	0.015	600	25
Gypsum	0.020	500	25
Alumina	0.020	400	20
Lime	0.020	600	8
Bauxite	0.020	450	8
Magnesium oxide	0.010	550	20
Steel			
Basic oxygen furnace	0.010	450	20
Open hearth	0.010–0.015	600	8
Electric furnace	0.015	600	5
Sintering	0.025	300	10
Ore roasters	0.020	500	10
Cupola	0.015–0.020	400	20
Pyrites roaster	0.02	500	10
Hot scarfing	0.01–0.015	250	25–40

Source: From IGCI report, J-APCA, Vol. 23, No. 7, (July 1973), p. 608.

SO_2 ABSORPTION EFFICIENCIES

The absorption efficiencies shown in Fig. E-1 were graphed for various alkaline liquors in a mobile packing scrubber.

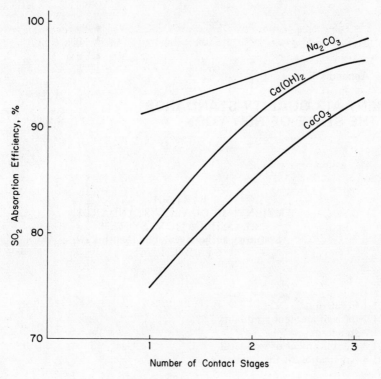

Figure E-1. SO_2 absorption efficiencies with various alkaline liquors in a mobile packing scrubber. (Adapted from Pollock et al., "Removal of Sulfur Dioxide and Flyash from Coal-Burning Power Plant Flue Gases," paper presented at ASME meeting, November 1966.)

AMBIENT AIR QUALITY STANDARDS
FOR THE STATE OF NEW YORK

PART 501
AMBIENT AIR QUALITY STANDARDS –
CLASSIFICATIONS SYSTEM
(Statutory authority: Public Health Law,
§§ 1271, 1276)

Sec.

501.1 Foreword
501.2 Basis of ambient air quality
standards
501.3 Establishment of standards
501.4 Application of the standards

Pursuant to the authority vested in the Air Pollution Control Board by §§ 1271 and 1276 of the Public Health Law, the title of Subchapter B, Chapter IV of Title 10 (Health) of the Official Compilation of Codes, Rules and Regulations of the State of New York, is hereby amended and new Part 501, entitled "Ambient Air Quality Standards—Classifications System", is hereby added to said Subchapter B, to become effective on the 3rd day of November, 1968, as follows:

Section 501.1 Foreword. (a) Ambient air quality standards describe a level of air quality designed to protect people from the adverse effects of air pollution; and they are intended further to promote maximum comfort and enjoyment and use of property consistent with the economic and social well-being of the community.

(b) The basic use of our air resource is to sustain life. Our environment is an ocean of air of which the average adult male must inhale approximately 400 cubic feet daily to obtain the necessary oxygen. Air entering the respiratory tract must not menace health. Therefore, all other uses must yield to the necessity of air which will not degrade, either acutely or chronically, the health and well-being of the populace.

(c) The only areas of compromise are the economic and the esthetic. In these areas the cost to society of providing a given level of air quality must be balanced against the benefits attained. As a consequence, in many communities, achievement of a desired air quality will be realized through a number of transition periods, during which improvements will be made as advancing technology reduces the cost of cleaner air.

(d) Ambient air quality standards, as used here, are a tool in achieving cleaner air, not as a license to permit unnecessary degradation of air quality which would

438

thwart attainment of the long-range goal to maintain a reasonable degree of air purity. They have been designed, and should serve, as a sound basis in the development of contaminant emission limitations, in the preparation of long and short-range objectives for acceptable total contaminant limitations and for the selection of air pollution control measures for planned and existing establishments which may create pollution.

(e) These standards are not intended to represent the ultimate in air quality achievement. It is anticipated that research and development will gradually make possible cleaner air at lower cost.

(f) It has not been possible at this time to establish acceptable limits for many contaminants because of the dearth of information regarding the threshold values which might be harmful to the receptors. As evidence accumulates on the deleterious effect of a contaminant, present standards will be revised or additional standards established.

501.2 Basis of ambient air quality standards. (a) The degree of air purity required may depend on the effect on any or all of the following receptors: man, animals, vegetation, and property. This is especially so, for example, with some of the pollutants, such as fluorides, which not only damage vegetation, but also may build up in forage crops concentrations toxic to grazing ruminants. When protection of human health is of concern, the standards must be set so as to assure no adverse effects. Otherwise, standards in whole or part must be based on the potential effects on susceptible receptors and accepted or potential land uses.

(a) Air pollution potentials range from the almost pristine pure air found in mountain areas to the heavily contaminated air in the highly populous and industrialized, large metropolitan complexes (the Niagara Frontier and Metropolitan New York). Within this range, the degree of social and economic development accounts for the kind and quality of contaminants emitted to the atmosphere and hence the potential effect on the receptors. These factors in combination with topographical and meterological variations would tend either to accentuate or reduce the effect of emissions.

(c) Thus, it is illogical to attempt one over-all set of air quality standards to apply to the entire State. It cannot be expected that air in a clean area—such as the Adirondack Mountains—would be degraded to a level that is reasonably attainable in an extensive area of heavy industry. Nor would it be reasonable to expect a highly industrialized area to attain economically the level of air quality prevailing in the presently clean areas.

(d) Investigations by the New York State Department of Health, Division of Air Resources, show that five general levels of air contamination exist in the State. These levels appear to be related generally to the macro-population density and related economic development, which vary with land use. The land uses associated with the five general levels of air contamination are described below:

Level I—predominantly used for timber and agricultural crops, dairy farming, and recreation. Habitation is sparse and confined to small incorporated and unincorporated communities. Industry is not developed to any degree and basically uses the natural resources of the area—e.g., pulping of wood or food processing.

Level II—single and two family residences, small farms, and limited commercial services and industrial development or sparsely inhabited land near a large metropolitan complex.

Level III—densely populated, primarily commercial, office buildings, department stores, and light industries in small and medium metropolitan complexes or suburban areas of limited commercial and industrial development near large metropolitan complexes.

Level IV—densely populated, primarily commercial, office buildings, department stores, and light industries in large metropolitan complexes, or limited areas of heavy industry such as chemical and allied products, primary ferrous and nonferrous metals, stone, clay, glass, petroleum, and coal, and their adjacent commercial and densely populated areas.

Level V—extensive areas of heavy industry, such as found in the Niagara Frontier.

(e) Extreme toxicity to humans of some contaminants (such as beryllium compounds) or to ruminants (such as soluble fluorides in rural areas) requires the establishment of one maximum allowable concentration.

(f) The public welfare and the protection of physical property and other resources require the establishment—based on what can be reasonably achieved in keeping with the social and economic development of the area—of air quality standards for other contaminants.

(g) The ambient air quality standards for the five general levels of air contamination are tabulated in Tables I and II.

(h) The laboratory procedures used in the analytical treatment of samples to determine contaminant concentrations are constantly being developed and revised. The sampling and analytical procedures to be used in evaluating the standards shall be acceptable to the commissioner.

501.3 Establishment of the standards. Based on comprehensive investigations and following public hearings, as required by law, a level of air quality will be assigned to areas of the State in accordance with existing and potential land use.

501.4 Application of the standards. (a) Evaluation of the standards will be based on avoiding adverse effects on receptors. In this respect, emissions in a classified area should be controlled to the extent so as not to contravene the standards in another classified area.

(b) It is necessary to establish attainment of the ambient air quality standards by employing any reasonable method or combination of methods, such as area sampling, source sampling, emission evaluation, and assessment of source contribution and effect. When these standards are not met appropriate action may be taken by the commissioner against the person responsible for any contributing source or sources.

(c) This Part supersedes Part 500 and Part 500 is hereby repealed except that Part 500 will continue to apply to any area of the State that has been classified pursuant thereto, until such time that such area is classified pursuant to this Part.

Appendix F

Table I. New York State Ambient Air Quality Standards (24 hour averages to be less than given values, except as noted)

	Level I	Level II	Level III	Level IV	Level V
Particulates					
Suspended Particulates (μg/m^3)					
50% of values less than	45	55	65	80	100
84% of values less than	70	85	100	120	150
Settleable Particulates (mg/cm^2/mo) Sampling period—30 days					
50% of values less than	0.30	0.30	0.40	0.60	0.80
84% of values less than	0.45	0.45	0.60	0.90	1.20
Beryllium (μg/m^3)	0.01	0.01	0.01	0.01	0.01
Total Fluorides (ppm)					
dry weight basis (as F) in and on forage for consumption by grazing ruminants					
Average concentration over 6 consecutive months	40	40	–	–	–
Average concentration over 2 consecutive months	60	60	–	–	–
Average concentration for any month	80	80	–	–	–
Sulfuric Acid Mist (mg/m^3)	0.10	0.10	0.10	0.10	0.10
Gases and Vapors					
Sulfur Dioxide (ppm)	0.10	0.10	0.10	0.10	0.15
(To be less than given values	or	or	or	or	or
99% of the time on an annual	0.25	0.25	0.25	0.25	0.40
basis)	for 1 hr.	for 1 hr.	for 1 hr.	for 1 hr.	for 1 hr.
Hydrogen Sulfide (ppm)	\multicolumn{5}{c}{0.10 for 1 hr. for all levels}				
Carbon Monoxide (ppm)					
(To be less than given value 85% of the time on an annual basis)	\multicolumn{5}{c}{15 for 8 hrs. for all levels}				
(To be less than given value 100% of the time)	\multicolumn{5}{c}{30 for 8 hrs. for all levels}				
(To be less than given value 99% of the time on an annual basis)	\multicolumn{5}{c}{60 for 1 hr. for all levels}				
Gases and Vapors					
Oxidants (ppm)	0.05	0.05	0.05	0.10	0.10
(Incl. ozone, photochemical aerosols, and other oxidant contaminants not listed	\multicolumn{5}{c}{0.15 for 1 hr. for all levels}				
separately)	\multicolumn{5}{c}{0.10 for 4 hrs. for levels I and II}				
Fluorides (ppb)					
(As HF) in air	1	1	2	3	4

Table II

Odorous Substances

Consistent with the economic and social well-being of the community, the ambient air shall not contain odorous substances in such concentrations or of such duration as will prevent enjoyment and use of property.

Other Toxic or Deleterious Substances

The ambient air shall not contain toxic or deleterious substances, in addition to those specifically listed in these objectives, in concentrations that affect human health or well-being, or unreasonably interfere with the enjoyment of property, or unreasonably and adversely affect plant or animal life. Guides for specific substances will be considered on an individual basis by the board.

CONTAMINANT EMISSIONS RULING
FOR THE STATE OF NEW YORK

PART 187

CONTAMINANT EMISSIONS
FROM
PROCESSES, AND EXHAUST AND VENTILATION SYSTEMS

(Statutory authority: Public Health Law, §§ 1271, 1276)

Sec.	Sec.
187.1 Applicability	187.3 Prohibitions
187.2 Definitions	187.4 Abatement

Historical Note

Part added, filed Jan. 12, 1968 to be eff.
Feb. 6, 1968.

Section 187.1 **Applicability.** This Part shall apply throughout the State of New York to contaminant emissions from processes, and exhaust and ventilation systems, except that when another Part applies to a specific air contaminant or a specific air contamination source, that Part shall take precedence and shall be applied in place of this Part.

Historical Note

Sec. added, filed Jan. 12, 1968 to be eff.
Feb. 6, 1968.

187.2 **Definitions.** (a) *Environmental rating.* A rating indicated by the letter A, B, C or D, considers the environmental effects of an air contamination source. A rating takes into account properties and quantities of contaminants emitted; effects on human, plant, or animal life, or property; meteorological parameters, stack heights, characteristics of the community; and ambient air quality classification of the area in which the source is located or which it affects.

(b) *Emission rate potential.* The rate in pounds per hour at which air contaminants would be emitted to the outer air in the absence of air pollution control facilities or other control measures. The emission rate potential for cyclic operations shall be determined by considering both the instantaneous emission potential and the total emission potential over the time period of the cycle.

(c) *Emission source.* Any point at which air contaminants enter the outer air from processes, and exhaust and ventilation systems.

(d) *Exhaust and ventilation system.* Any system which removes and transports any gaseous or gas borne products from their point of generation to the outer air.

(e) *Permissible emission rate.* The maximum rate in pounds per hour at which air contaminants are allowed to be emitted to the outer air.

(f) *Process weight.* The total weight of all materials introduced into any specific process which may cause any discharge into the atmosphere. Solid fuels used in the process will be considered as part of the process weight, but liquid and gaseous fuels, uncombined water and combustion air will not.

(g) *Process weight per hour.* The total process weight divided by the number of hours in one complete operation from the beginning of a cycle to the completion thereof. For continuous processes, process weight should be determined on a daily basis.

Historical Note

Sec. added, filed Jan. 12, 1968 to be eff.
Feb. 6, 1968.

187.3 Prohibitions. (a) No person shall cause, permit or allow the emission of air contaminants from an emission source resulting from an operation begun or modified, after the effective date of this Part, which exceeds the permissible emission rates specified in tables 2 and 3‡, for the environmental rating as determined in accordance with table 1‡.

(b) On January 1, 1971, or such later date as established by an order of the commissioner, the permissible emission rates specified in subdivision (a) shall become applicable to emission sources in existence on or prior to the effective date of this Part.

(c) The provisions of this section shall not be construed to allow or permit any person to emit air contaminants in quantities which alone or in combination with other sources would contravene any established air quality standards.

Historical Note

Sec. added, filed Jan. 12, 1968 to be eff.
Feb. 6, 1968.

187.4 Abatement. (a) The commissioner may require the person operating or maintaining emission sources to provide pertinent data concerning emissions so as to show compliance with the requirements of section 187.3.

(b) When required by the commissioner, the person operating or maintaining emission sources in operation before the effective date of this Part shall submit a detailed report including emission data, pertinent environmental factors and a proposed environmental rating so as to show conformity with this Part of proposed corrective measures and schedule for compliance. If this report is acceptable, the commissioner will so notify the person operating or maintaining the emission source. If the report is not acceptable, the commissioner will notify the person operating or maintaining the emission source as to the reasons together with an environmental rating that is acceptable and a time schedule for compliance. Upon petition to the commissioner within 30 days of such notice, the commissioner shall grant a hearing to the petitioner.

(c) Persons beginning or modifying operations after the effective date of this Part are required to submit to the commissioner or his representative, either prior to* or concurrently with submission of plans and/or specifications, an appraisal of the items mentioned in table 1‡ in the form of a report including the proposed rating to be used for design purposes.

(d) The commissioner may seal any process equipment or prohibit any operation in accordance with a determination made under the provisions of section 1282 of article 12-A of the Public Health Law. The seal may be removed from the equipment only upon receipt of written notice from the commissioner.

Historical Note

Sec. added, filed Jan. 12, 1968 to be eff.
Feb. 6, 1968.

‡ See Appendix 4.
* It is recommended that for large installations the report be submitted prior to submission of plans. Following approval of the preliminary report, final detailed plans and/or specifications will be completed and submitted to the commissioner or his representative for approval.

886.4b   H   1-31-68

APPENDIX 4

TABLE 1

Environmental Rating

Rating   *Criteria*

A   Includes processes, and exhaust and ventilation systems where the discharge of a contaminant or contaminants results, or would reasonably be expected to result, in serious adverse effects on receptors or the environment. These effects may be of a health, economic or aesthetic nature or any combination of these.

B   Includes processes, and exhaust and ventilation systems where the discharge of a contaminant or contaminants results, or would reasonably be expected to result, in only moderate and essentially localized effects; or where the multiplicity of sources of the contaminant or contaminants in any given area is such as to require an overall reduction of the atmospheric burden of that contaminant or contaminants.

C   Includes processes, and exhaust and ventilation systems where the discharge of a contaminant or contaminants would reasonably be expected to result in localized adverse effects of an aesthetic or nuisance nature.

D   Includes processes, and exhaust and ventilation systems where, in view of properties and concentrations of the emissions, isolated conditions, stack height, and other factors, it can be clearly demonstrated that discharge of the contaminant or contaminants will not result in measurable or observable effects on receptors, nor add to an existing or predictable atmospheric burden of that contaminant or contaminants which would reasonably be expected to cause adverse effects.

The following items will be considered in making a determination of the environmental rating to be applied to a particular source:

a)   properties, quantities and rates of the emission

b)   physical surroundings of emission source

c)   population density of surrounding area, including anticipated future growth

d)   dispersion characteristics at or near source

e)   location of emission source relative to ground level and surrounding buildings, mountains, hills, etc.

f)   current or anticipated ambient air quality in vicinity of source

g)   latest findings relating to effects of ground-level concentrations of the emission on receptors

h)   possible hazardous side effects of contaminant in question mixing with contaminants already in ambient air

i)   engineering guides which are acceptable to the commissioner

911   H   1-31-68

TABLE 2

Usual Degree of Air Cleaning Required (1)

from

Processes, and Exhaust and Ventilation Systems

for

Gases and Liquid Particulate Emissions (Environmental Rating A, B, C and D)

and

Solid Particulate Emissions (Environmental Rating A and D)

Environment Rating	EMISSION RATE POTENTIAL (LB/HR)									
	Less than 1.0	1 to 10	10 to 20	20 to 100	100 to 500	500 to 1,000	1,000 to 1,500	1,500 to 4,000	4,000 to 10,000	10,000 and Greater
A	See Note (2)	99% OR GREATER								
B	*		90-91%	91-94%	94-96%		96-97%	97-98%	98-99%	99% or Greater
C	*		70-75%	75-85%	85-90%		90-93%	93-95%	95-98%	98% or Greater
D										

* Degree of air cleaning may be specified by the commissioner providing satisfactory dispersion is achieved.

(1) Where multiple emission sources are connected to a common air cleaning device, the degree of air cleaning required will be that which would be required if each individual emission source were considered separately.

(2) For an average Emission Rate Potential less than 1.0 lb/hr, the desired air cleaning efficiency shall be determined by the expected concentration of the air contaminant in the emission stream. Where it is uneconomical to employ air cleaning devices, other methods of control should be considered.

APPENDIX 4

TABLE 3

Allowable Emissions

from

Processes, and Exhaust and Ventilation Systems

for

Solid Particulates (Environmental Rating B & C)

Process Weight (lb/hr)	Maximum Weight Discharge*** (lb/hr)
100	.50
500	1.46
1,000	2.30
5,000	6.70
10,000	10.80
25,000	20.00
50,000	31.80
75,000	43.00
100,000**	50.00
250,000**	58.20
500,000**	64.30
750,000**	68.40
1,000,000**	71.10
2,000,000**	78.30
5,000,000**	88.10

* In cases where process weight is not applicable (such as grinding and woodworking) the concentration of solid particulates in the effluent gas stream shall not exceed 0.3 lb/1000 lb of undiluted exhaust gas at actual conditions.

** For process weights in excess of 100,000 lb/hr, the permissible maximum weight discharge may exceed tabular value if the concentration of particulate matter in the effluent gas stream is less than 0.1 lb/1000 lb of undiluted exhaust gas at actual conditions.

*** To determine intermediate values of maximum weight discharge:
for process weights up to 100,000 lb/hr use $E = 0.024P^{0.665}$
for process weights in excess of 100,000 lb/hr use $E = 39P^{0.062}-50$
where E = maximum weight discharge in lb/hr; P = process weight in lb/hr

913   H   1-31-68

447

Appendix H

EMISSION SOURCE EVALUATION—

STATE OF NEW YORK
DEPARTMENT OF HEALTH

GUIDELINES FOR THE PREPARATION OF THE ENVIRONMENTAL ANALYSIS REPORT FOR COMPLIANCE WITH 10 NYCRR 187

The Environmental Analysis Report should contain four sections. Prepare each in triplicate. Cover each report with completed form AIR 111. Consult the field representative of the New York State Department of Health for assistance in preparing the report. A list of references is included on page 451 of these guidelines.

The sections should consist of the following:

Section 1. Plot Plan

Prepare a plot plan to scale which includes:

a. north orientation

b. plant property line(s)

c. direction and, if known, velocity of prevailing wind

d. land use. Describe surrounding property and use; e.g., commercial, residential or industrial.

e. location of all existing and proposed emission sources (including those associated with the generation of steam, hot water, heat and electricity). Distinguish between existing and proposed emission points and assign a reference number to each point.

f. horizontal distance and bearing to nearest off-property receptor for each emission point.

AIR 100.2 (6-68)

g. location of refuse disposal sites. Describe nature of refuse, quantity (tons/yr), disposal method and method of transport to disposal site.

Section 2. Explanation of Plant Process

Prepare an explanation of each plant process, including flow diagrams and indicating emission sources by reference numbers assigned on plot plan.

Section 3. Industrial Emission Source Inventory

Prepare form AIR 302A for each emission source identified on plot plan.

Some definitions and calculations which may be helpful are:

Process weight—the total weight of all materials introduced into any specific process which may cause any discharge into the outer air. Liquid fuels, gaseous fuels, uncombined water, combustion air and internally recirculated materials are not to be considered in the calculation of process weight.

Emission rate potential—the rate in pounds per hour at which air contaminants would be emitted to the outer air in the absence of air pollution control facilities or other control measures. Two examples of the calculation of emission rate potential for the use of solvents are listed below:

A. Spray Coating. A spray coating operation uses one gallon of paint per hour. The paint is composed of 30% solids and 70% solvents. The solvent is a mixture of toluene, xylene and other aromatics. Most of the solvent mixture is toluene and the specific gravity of 0.87 for toluene is assumed to be applicable to the solvent mixture. It is also assumed that 90% of the solvent is released to the outer air through the spray booth exhaust from the operation.

Use emission rate potential - $Qsg\rho_{H_2O}$

where Q = gallons of solvent used in one hour,

sg = specific gravity of solvent

and ρ_{H_2O} = density of water (equals 8.3 lb/gal at normal conditions)

therefore, emission rate potential = 1 × 0.7 × 0.9 × 0.87 × 8.3

= 4.5 lb/hr

B. Vapor Degreaser. A gas heated vapor degreaser has a vapor condenser and a slot type local exhaust hood. Twenty gallons of perchlorethylene solvent with a density of 13.5 lb/gal is added as make-up each week. The degreaser is operated 30 hours each week. It is assumed that 95% of the make-up solvent is exhausted to the outer air by the degreaser exhaust system.

Use emission rate potential = $Q\rho_{solvent}$

where Q = gallons of solvent used in one hour

and $\rho_{solvent}$ = $sg_{solvent}$ × ρ_{H_2O}

therefore, emission rate potential = 20 × 1/30 × 13.5 × 0.95

= 8.6 lb/hr

Section 4. Environmental Rating Proposal and Permissible Emissions

Propose environmental rating (s) in accordance with Table 1, 10 NYCRR 187; see Appendix G. Then calculate permissible emissions using Tables 2 and 3.

For sources emitting gases and liquid particulates and for sources emitting solid particulates with an environmental rating of "A" or "D", use Table 2 to determine the degree of cleaning or treatment required.

For sources emitting solid particulates with an environmental rating of "B" or "C", refer to Table 3 to calculate the permissible emission rate (maximum weight discharge). Two equations are provided. One is to be used if the process weight exceeds 100,000 lb/hr and the other if it does not.

AIR 100.2 (6-68)

Sample calculation:

A. Process weight does not exceed 100,000 lb/hr

Calculate the maximum weight discharge allowed from a production cupola with an environmental rating of "B" and a process weight of 28,500 lb/hr.

Use $E = 0.024\ P^{0.665}$

where E = maximum weight discharge (lb/hr)

and   P = process weight (lb/hr)

$$E = 0.024\ (28,500)^{0.665}$$

$$E = 22.2\ \text{lb/hr}$$

B. Process weight exceeds 100,000 lb/hr

Calculate the maximum weight discharge allowed from an asphalt plant with an environmental rating of "B" and a process weight of 325 ton/hr.

Use $E = 39\ P^{0.082} - 50$

$$E = 39\ (325 \times 2,000)^{0.082} - 50$$

$$E = 117 - 50$$

$$E = 67\ \text{lb/hr}$$

Linear interpolation should not be used for intermediate values in Table 3.

REFERENCES

1. *Air Pollution,* Stern, A., Vol. 1 and 2, Academic Press 1962.

2. *Air Pollution Field Operations Manual,* 1962, U.S.P.H.S. Publication No. 937.

3. *Air Pollution Handbook,* Magill, P. L., Holden, C. Ackley, C. McGraw Hill.

4. *Chemical Engineer's Handbook,* Perry, R. H., McGraw Hill, 1967.

5. *Chemical Safety Data Sheets,* Manufacturing Chemists Association.

AIR 100.2 (6-68)

6. *Dangerous Properties of Industrial Materials,* Sax, N. I., Reinhold, (1963).

7. *Foundry Air Pollution Control Manual,* American Foundryman's Society, Second Edition, 1967;

8. *Handbook of Air Pollution,* U.S. Public Health Service.

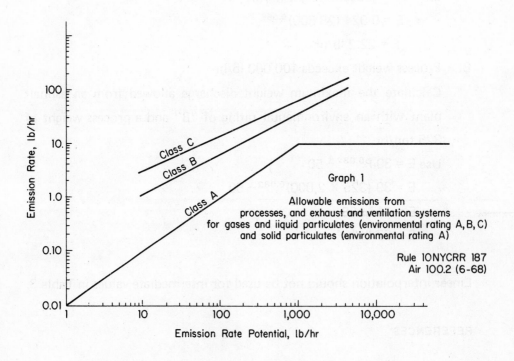

Graph 1

Allowable emissions from
processes, and exhaust and ventilation systems
for gases and iiquid particulates (environmental rating A, B, C)
and solid particulates (environmental rating A)

Rule 10NYCRR 187
Air 100.2 (6-68)

452

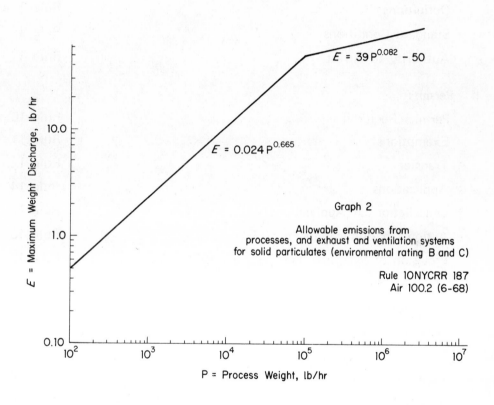

E = 39 P$^{0.082}$ − 50

E = 0.024 P$^{0.665}$

Graph 2

Allowable emissions from
processes, and exhaust and ventilation systems
for solid particulates (environmental rating B and C)

Rule 10NYCRR 187
Air 100.2 (6-68)

E = Maximum Weight Discharge, lb/hr

P = Process Weight, lb/hr

453

RULES AND REGULATIONS—
COUNTY OF LOS ANGELES
AIR POLLUTION CONTROL DISTRICT

TABLE OF CONTENTS

TABLE OF CONTENTS (Cont.)

TABLE OF CONTENTS (Cont.)

IV - Prohibitions

Rule 50   Ringelmann Chart.

(Effective until January 1, 1973 for all sources completed and put into service before January 6, 1972.   See amended Rule below)

A person shall not discharge into the atmosphere from any single source of emission whatsoever any air contaminants for a period or periods aggregating more than three minutes in any one hour which is:

a.   As dark or darker in shade as that designated as No. 2 on the Ringelmann Chart, as published by the United States Bureau of Mines, or

b.   Of such opacity as to obscure an observer's view to a degree equal to or greater than does smoke described in subsection (a) of this Rule.

Rule 50.   Ringelmann Chart.

(Effective January 6, 1972 for any source not completed and put into service. Effective for all sources on January 1, 1973.)

A person shall not discharge into the atmosphere from any single source of emission whatsoever any air contaminant for a period or periods aggregating more than three minutes in any one hour which is:

a.   As dark or darker in shade as that designated No. 1 on the Ringelmann Chart, as published by the United States Bureau of Mines, or

b.   Of such opacity as to obscure an observer's view to a degree equal to or greater than does smoke described in subsection (a) of this Rule.

This amendment shall be effective on the date of its adoption for any source of emission not then completed and put into service. As to all other sources of emission this amendment shall be effective on January 1, 1973.

Rule 51. Nuisance.

A person shall not discharge from any source whatsoever such quantities of air contaminants or other material which cause injury, detriment, nuisance or annoyance to any considerable number of persons or to the public or which endanger the comfort, repose, health or safety of any such persons or the public or which cause or have a natural tendency to cause injury or damage to business or property.

Rule 52. Particulate Matter.

(Effective until January 1, 1973 for all equipment completed and put into service before January 6, 1972. See amended Rule below)

Except as otherwise provided in Rules 53 and 54, a person shall not discharge into the atmosphere from any source particulate matter in excess of 0.3 grain per cubic foot of gas at standard conditions.

Rule 52. Particulate Matter - Concentration.

(Effective January 6, 1972 for any equipment not completed and put into service. Effective for all equipment on January 1, 1973.)

A person shall not discharge into the atmosphere from any source particulate matter in excess of the concentration shown in the following table: (See Rule 52 Table)

Where the volume discharged falls between figures listed in the table, the exact concentration permitted to be discharged shall be determined by linear interpolation.

The provisions of this rule shall not apply to emissions resulting from the combustion of liquid or gaseous fuels in steam generators or gas turbines.

For the purposes of this rule "particulate matter" includes any material which would become particulate matter if cooled to standard conditions.

This amendment shall be effective on the date of its adoption for any equipment not then completed and put into service. As to all other equipment this amendment shall be effective on January 1, 1973.

Table For Rule 52

Volume Discharged-- Cubic Feet Per Minute Calculated as Dry Gas at Standard Conditions	Maximum Concentration of Particulate Matter Allowed in Discharged Gas-Grains Per Cubic Foot of Dry Gas at Standard Conditions	Volume Discharged-- Cubic Feet Per Minute Calculated as Dry Gas at Standard Conditions	Maximum Concentration of Particulate Matter Allowed in Discharged Gas-Grains Per Cubic Foot of Dry Gas at Standard Conditions
1000 or less	0.200	20000	0.0635
1200	.187	30000	.0544
1400	.176	40000	.0487
1600	.167	50000	.0447
1800	.160	60000	.0417
2000	.153	70000	.0393
2500	.141	80000	.0374
3000	.131	100000	.0343
3500	.124	200000	.0263
4000	.118	400000	.0202
5000	.108	600000	.0173
6000	.101	800000	.0155
7000	.0949	1000000	.0142
8000	.0902	1500000	.0122
10000	.0828	2000000	.0109
15000	.0709	2500000 or more	.0100

Rule 54. Dust and Fumes.

(Effective until January 1, 1973 for all equipment completed and put into service before January 6, 1972. See amended Rule below)

A person shall not discharge in any one hour from any source whatsoever dust or fumes in total quantities in excess of the amount shown in the following table: (see next page)

To use the following table, take the process weight per hour as such is defined in Rule 2(j). Then find this figure on the table, opposite which is the maximum number of pounds of contaminants which may be discharged into the atmosphere in any one hour. As an example, if A has a process which emits contaminants into the atmosphere and which process takes 3 hours to complete, he will divide the weight of all materials in the specific process, in this example, 1,500 lbs. by 3 giving a process weight per hour of 500 lbs. The table shows that A may not discharge more than 1.77 lbs. in any one hour during the process. Where the process weight per hour falls between figures in the left hand column, the exact weight of permitted discharge may be interpolated.

* (You will find Table for Rule 54 with amended Rule following)

Rule 54. Solid Particulate Matter - Weight.

(Effective January 6, 1972 for any equipment not completed and put into service. Effective for all equipment on January 1, 1973.)

A person shall not discharge into the atmosphere from any source solid particulate matter, including lead and lead compounds, in excess of the rate shown in the following table: (See Rule 54 Table)

Where the process weight per hour falls between figures listed in the table, the exact weight of permitted discharge shall be determined by linear interpolation.

For the purposes of this rule "solid particulate matter" includes any material which would become solid particulate matter if cooled to standard conditions.

This amendment shall be effective on the date of its adoption for any equipment not then completed and put into service. As to all other equipment this amendment shall be effective on January 1, 1973.

TABLE FOR RULE 54

*Process Wt/hr(lbs)	Maximum Weight Disch/hr(lbs)	*Process Wt/hr(lbs)	Maximum Weight Disch/hr(lbs)
50	.24	3400	5.44
100	.46	3500	5.52
150	.66	3600	5.61
200	.85	3700	5.69
250	1.03	3800	5.77
300	1.20	3900	5.85
350	1.35	4000	5.93
400	1.50	4100	6.01
450	1.63	4200	6.08
500	1.77	4300	6.15
550	1.89	4400	6.22
600	2.01	4500	6.30
650	2.12	4600	6.37
700	2.24	4700	6.45
750	2.34	4800	6.52
800	2.43	4900	6.60
850	2.53	5000	6.67
900	2.62	5500	7.03
950	2.72	6000	7.37
1000	2.80	6500	7.71
1100	2.97	7000	8.05
1200	3.12	7500	8.39
1300	3.26	8000	8.71
1400	3.40	8500	9.03
1500	3.54	9000	9.36
1600	3.66	9500	9.67
1700	3.79	10000	10.0
1800	3.91	11000	10.63
1900	4.03	12000	11.28
2000	4.14	13000	11.89
2100	4.24	14000	12.50
2200	4.34	15000	13.13
2300	4.44	16000	13.74
2400	4.55	17000	14.36
2500	4.64	18000	14.97
2600	4.74	19000	15.58
2700	4.84	20000	16.19
2800	4.92	30000	22.22
2900	5.02	40000	28.3
3000	5.10	50000	34.3
3100	5.18	60000	40.0
3200	5.27	or	
3300	5.36	more	

*See Definition in Rule 2(j).

TABLE FOR RULE 54
(Amended January 6, 1972)

Process Weight Per Hour-- Pounds Per Hour	Maximum Discharge Rate Allowed for Solid Particulate Matter (Aggregate Discharged From All Points of Process)--Pounds Per Hour	Process Weight Per Hour-- Pounds Per Hour	Maximum Discharge Rate Allowed for Solid Particulate Matter (Aggregate Discharged From All Points of Process)--Pounds Per Hour
250 or less	1.00	12000	10.4
300	1.12	14000	10.8
350	1.23	16000	11.2
400	1.34	18000	11.5
450	1.44	20000	11.8
500	1.54	25000	12.4
600	1.73	30000	13.0
700	1.90	35000	13.5
800	2.07	40000	13.9
900	2.22	45000	14.3
1000	2.38	50000	14.7
1200	2.66	60000	15.3
1400	2.93	70000	15.9
1600	3.19	80000	16.4
1800	3.43	90000	16.9
2000	3.66	100000	17.3
2500	4.21	120000	18.1
3000	4.72	140000	18.8
3500	5.19	160000	19.4
4000	5.64	180000	19.9
4500	6.07	200000	20.4
5000	6.49	250000	21.6
5500	6.89	300000	22.5
6000	7.27	350000	23.4
6500	7.64	400000	24.1
7000	8.00	450000	24.8
7500	8.36	500000	25.4
8000	8.70	600000	26.6
8500	9.04	700000	27.6
9000	9.36	800000	28.4
9500	9.68	900000	29.3
10000	10.00	1000000 or more	30.0

Rule 67. Fuel Burning Equipment.

A person shall not build, erect, install or expand any non-mobile fuel burning equipment unit unless the discharge into the atmosphere of contaminants   will not and does not exceed any one or more of the following rates:

1. 200 pounds per hour of sulfur compounds, calculated as sulfur dioxide (SO_2);

2. 140 pounds per hour of nitrogen oxides, calculated as nitrogen dioxide (NO_2);

3. 10 pounds per hour of combustion contaminants as defined in Rule 2m and derived from the fuel.

For the purpose of this rule, a fuel burning equipment unit shall be comprised of the minimum number of boilers, furnaces, jet engines or other fuel burning equipment, the simultaneous operations of which are required for the production of useful heat or power.

Fuel burning equipment serving primarily as air pollution control equipment by using a combustion process to destroy air contaminants shall be exempt from the provisions of this rule.

Nothing in this rule shall be construed as preventing the maintenance or preventing the alteration or modification of an existing fuel burning equipment unit which will reduce its mass rate of air contaminant emissions.

Rule 68. Fuel Burning Equipment -- Oxides of Nitrogen.

A person shall not discharge into the atmosphere from any non-mobile fuel burning article, machine, equipment or other contrivance, having a maximum heat input rate of more than 1775 million British Thermal Units (BTU) per hour (gross), flue gas having a concentration of nitrogen oxides, calculated as nitrogen dioxide (NO_2) at 3 per cent oxygen, in excess of that shown in the following table:

NITROGEN OXIDES - PARTS PER MILLION PARTS OF FLUE GAS		
	EFFECTIVE DATE	
FUEL	DECEMBER 31, 1971	DECEMBER 31, 1974
Gas	225	125
Liquid or Solid	325	225

Rule 71. Carbon Monoxide.

A person shall not, after December 31, 1971, discharge into the atmosphere carbon monoxide (CO) in concentrations exceeding 0.2 per cent by volume measured on a dry basis.

The provisions of this rule shall not apply to emissions from internal combustion engines.

INDEX

466